VOLUME NINETY SIX

ADVANCES IN
VIRUS RESEARCH
Coronaviruses

VOLUME NINETY SIX

ADVANCES IN
VIRUS RESEARCH
Coronaviruses

Edited by

JOHN ZIEBUHR
Institute of Medical Virology,
Justus Liebig University Giessen,
Giessen, Germany

AMSTERDAM • BOSTON • HEIDELBERG • LONDON
NEW YORK • OXFORD • PARIS • SAN DIEGO
SAN FRANCISCO • SINGAPORE • SYDNEY • TOKYO
Academic Press is an imprint of Elsevier

ISBN: 978-0-12-804736-1
ISSN: 0065-3527

For information on all Academic Press publications
visit our website at https://www.elsevier.com/

  Working together
to grow libraries in
developing countries

www.elsevier.com • www.bookaid.org

Publisher: Zoe Kruze
Acquisition Editor: Alex White
Editorial Project Manager: Helene Kabes
Production Project Manager: Surya Narayanan Jayachandran
Designer: Greg Harris

Typeset by SPi Global, India

IN MEMORIAM

K. Maramorosch

Jan 16, 1915–May 9, 2016

Karl Maramorosch, eminent virologist, entomologist and plant patholo-gist/virologist, and longtime serial editor of *Advances in Virus Research* sadly passed away aged 101 on May 9, 2016. Karl was born on January 16, 1915 in Vienna in the then Austro-Hungarian Empire in the middle of World War I. His father originated from Poland and his polyglot mother from Croatia. He grew up in Poland and received his university degree from Warsaw Agricultural University in Warsaw in 1938. In 1939 he fled from the Nazi invasion of Poland with his Polish wife Irene to Romania where both were interred. Most of his family perished in the Holocaust. In 1947 Karl and Irene emigrated to the United States where Karl received his Ph.D. at Columbia University in 1949. He started his scientific career at Rockefeller University in New York working on insect-transmitted plant viruses, a topic he continued throughout his life. From 1961 to 1973 he was a Program Director of Virology at the Boyce Thompson Institute in Ithaca, NY, and then moved to the Waksman Institute of Microbiology at Rutgers University, New Jersey in 1974. In 1984 he joined the Entomology Department at Rutgers which remained his scientific home. He pioneered work on insect-transmitted plant viruses including microinjection into vectors as well as insect tissue culture to make major advances in our understanding of the replication of plant pathogens in insect vectors and the interaction between insects, viruses, and plants. Among numerous other distinctions, in 1980 Karl received the Wolf Prize in Agriculture, considered the "Agriculture Nobel Prize," for his work on insect vectors and plant pathogens. Karl has authored or coauthored an impressive number of scientific papers (>800) and books (>100). Since 1973 Karl has served as a serial editor for *Advances in Virus Research*, the oldest review series in virology. Karl lived to see his autobiographical book **"The Thorny Road to Success: A Memoir"** published recently, a lasting testimony of his life in the context of the 20th and early 21st centuries.

Karl was a role model in his perseverance and his dedication to and enthusiasm for science. Together with Fred Murphy and Aaron Shatkin, Karl significantly influenced *Advances in Virus Research*, which retains its place among the premier virology reviews. In 2007 Karl's engaging account of his life and career, entitled "Viruses, Vectors, and Vegetation: An autobiography," was published in Volume 70. We mourn his death and honor his memory!

MARGARET KIELIAN
THOMAS C. METTENLEITER

CONTENTS

CONTRIBUTORS

B.-J. Bosch
Faculty of Veterinary Medicine, Utrecht University, Utrecht, The Netherlands

M.J. Buchmeier
University of California, Irvine, Irvine, CA, United States

J. Canton
National Center of Biotechnology (CNB-CSIC), Campus Universidad Autónoma de Madrid, Madrid, Spain

C. Castaño-Rodriguez
National Center of Biotechnology (CNB-CSIC), Campus Universidad Autónoma de Madrid, Madrid, Spain

C.A.M. de Haan
Faculty of Veterinary Medicine, Utrecht University, Utrecht, The Netherlands

E. Decroly
Aix-Marseille Université; CNRS, AFMB UMR 7257, Marseille, France

L. Enjuanes
National Center of Biotechnology (CNB-CSIC), Campus Universidad Autónoma de Madrid, Madrid, Spain

M. Fricke
Faculty of Mathematics and Computer Science, Friedrich Schiller University Jena, Jena, Germany

J. Gutierrez-Alvarez
National Center of Biotechnology (CNB-CSIC), Campus Universidad Autónoma de Madrid, Madrid, Spain

R.J.G. Hulswit
Faculty of Veterinary Medicine, Utrecht University, Utrecht, The Netherlands

E. Kindler
University of Bern, Bern; Institute of Virology and Immunology, Bern and Mittelhäusern, Switzerland

K.G. Lokugamage
The University of Texas Medical Branch, Galveston, TX, United States

R. Madhugiri
Institute of Medical Virology, Justus Liebig University Giessen, Giessen, Germany

S. Makino
The University of Texas Medical Branch; Center for Biodefense and Emerging Infectious Diseases; UTMB Center for Tropical Diseases; Sealy Center for Vaccine Development;

Institute for Human Infections and Immunity, The University of Texas Medical Branch, Galveston, TX, United States

M. Marz
Faculty of Mathematics and Computer Science, Friedrich Schiller University Jena; FLI Leibniz Institute for Age Research, Jena, Germany

K. Nakagawa
The University of Texas Medical Branch, Galveston, TX, United States

B.W. Neuman
School of Biological Sciences, University of Reading, Reading, United Kingdom; College of STEM, Texas A&M University, Texarkana, Texarkana, TX, United States

E.J. Snijder
Leiden University Medical Center, Leiden, The Netherlands

I. Sola
National Center of Biotechnology (CNB-CSIC), Campus Universidad Autónoma de Madrid, Madrid, Spain

G. Tekes
Institute of Virology, Faculty of Veterinary Medicine, Justus Liebig University Giessen, Giessen, Germany

H.-J. Thiel
Institute of Virology, Faculty of Veterinary Medicine, Justus Liebig University Giessen, Giessen, Germany

V. Thiel
University of Bern, Bern; Institute of Virology and Immunology, Bern and Mittelhäusern, Switzerland

F. Weber
Institute of Virology, Faculty of Veterinary Medicine, Justus Liebig University Giessen, Giessen, Germany

J. Ziebuhr
Institute of Medical Virology, Justus Liebig University Giessen, Giessen, Germany

S. Zuñiga
National Center of Biotechnology (CNB-CSIC), Campus Universidad Autónoma de Madrid, Madrid, Spain

PREFACE

Coronaviruses are important human and animal pathogens, with severe acute respiratory syndrome (SARS) and Middle East respiratory syndrome (MERS) coronavirus being prominent examples. Although animal coronavirus-associated diseases, such as feline infectious peritonitis, have been reported more than a century ago, the causative agents were only recognized in the late 1960s as a group of related enveloped RNA viruses. The International Committee on Taxonomy of Viruses approved a separate virus family *Coronaviridae* in 1975. More recently, an additional taxonomic rank was introduced in this family, with coronaviruses being classified as members of the subfamily *Coronavirinae* that, currently, is comprised of four genera. The first complete genome sequence of a coronavirus (avian infectious bronchitis virus) was published in 1987. The study revealed that, compared to other RNA viruses, coronaviruses have extremely large genomes of approximately 30 kb. In 1989, Alexander Gorbalenya and colleagues published the first comprehensive sequence analysis of a coronavirus replicase gene. Functional predictions arising from this seminal work provided a framework for many subsequent biochemical, structural, and phylogenetic studies of corona- and related viruses. The studies identified several phylogenetically related lineages of plus-strand RNA viruses that, despite profound divergence at the sequence level, were revealed to encode a conserved array of functional domains in their replicase genes and to use similar strategies to express and replicate their large polycistronic genomes. These viruses are now recognized as members of the order *Nidovirales* and belong to one of the four established families *Arteriviridae*, *Mesoniviridae*, *Coronaviridae*, and *Roniviridae*. Nidovirus RNA synthesis and processing was revealed to involve an unusually large number of proteins, and also the enormous complexity of interactions between nidoviral and host cell functions is exceptional in the RNA virus world. Coronaviruses and other nidoviruses with genome sizes of more than 20 kb also stick out from all other RNA viruses by encoding a $3'$-to-$5'$ exoribonuclease that was shown to increase the fidelity of viral RNA synthesis and is thought to be a key factor in the evolution of RNA genomes of this large size. Additional interest in studying coronaviruses was sparked by two newly emerging zoonotic coronaviruses, SARS-CoV and MERS-CoV, that are able to cause severe or even fatal respiratory disease in humans. Over the past few years, a large number of previously unknown corona- and related nidoviruses were discovered in mammals, birds, insects, fish, and reptiles, while other studies provided a wealth of new information on the biology and pathogenesis of human and animal coronaviruses. In several cases, the studies also revealed unique properties not reported previously for other RNA viruses outside the *Nidovirales*.

The chapters included in this volume review our current understanding of important aspects of coronavirus biology and pathogenesis, including virus–host interactions, and they give an overview of new strategies in the development of vaccines and antivirals suitable to combat coronavirus infections. I hope that this book will be useful to academic researchers and their students as well as clinicians with an interest in coronavirus-related diseases and the biology of these viruses. I thank all colleagues who contributed to this book and would like to express my gratitude to Prof. Thomas Mettenleiter for his encouragement and Ms. Helene Kabes and her staff for guidance and help in the preparation of this volume.

JOHN ZIEBUHR
Giessen, Germany
August 2016

CHAPTER ONE

Supramolecular Architecture of the Coronavirus Particle

B.W. Neuman*,†,1, M.J. Buchmeier‡
*School of Biological Sciences, University of Reading, Reading, United Kingdom
†College of STEM, Texas A&M University, Texarkana, Texarkana, TX, United States
‡University of California, Irvine, Irvine, CA, United States
1Corresponding author: e-mail address: bneuman@tamut.edu

Contents

Abstract

Coronavirus particles serve three fundamentally important functions in infection. The virion provides the means to deliver the viral genome across the plasma membrane of a host cell. The virion is also a means of escape for newly synthesized genomes. Lastly, the virion is a durable vessel that protects the genome on its journey between cells. This review summarizes the available X-ray crystallography, NMR, and cryoelectron microscopy structural data for coronavirus structural proteins, and looks at the role of each of the major structural proteins in virus entry and assembly. The potential wider conservation of the nucleoprotein fold identified in the *Arteriviridae* and *Coronaviridae* families and a speculative model for the evolution of corona-like virus architecture are discussed.

1. INTRODUCTION

A virus particle is essentially a ruggedized viral genome that contains at least the minimal set of components necessary to propagate a virus infection. While virions from most viruses are considered to be metabolically inactive, some undergo internal structural changes after release, including

Advances in Virus Research, Volume 96
ISSN 0065-3527
http://dx.doi.org/10.1016/bs.aivir.2016.08.005

protease-dependent retrovirus maturation (Konvalinka et al., 2015) and bicaudovirus elongation (Haring et al., 2005; Scheele et al., 2011). In addition to the viral genome and four conserved virally encoded structural proteins, coronavirus particles have been shown to contain a variety of packaged host-encoded proteins (Dent et al., 2015; Kong et al., 2010; Neuman et al., 2008; Nogales et al., 2012) including enzymes that may play important roles in promoting or preventing infection such as protein kinases (Neuman et al., 2008; Siddell et al., 1981), cyclophilin A (Neuman et al., 2008; Pfefferle et al., 2011), and APOBEC3G (Wang and Wang, 2009). Purified virions can also contain low levels of some virus-encoded replicase proteins (Neuman et al., 2008; Nogales et al., 2012), though packaged replicase proteins have not been shown to enhance infectivity. While the positive sense ssRNA genome is infectious for members of the genera *Alphacoronavirus* (Almazan et al., 2000; Donaldson et al., 2008b; Jengarn et al., 2015; Tekes et al., 2008; Thiel et al., 2001; Yount et al., 2000), *Betacoronavirus* (Donaldson et al., 2008a; Scobey et al., 2013; Yount et al., 2003), and *Gammacoronavirus* (Casais et al., 2001), it has been shown that expression of the viral nucleoprotein and nsp3 can together or separately promote infection (Hurst et al., 2010, 2013; Pan et al., 2008; Schelle et al., 2005; Thiel et al., 2001). Taken together, this data demonstrate that packaged virion proteins are not essential for infection, but suggest that packaged proteins may confer a small replication advantage.

Coronaviruses encode three conserved membrane-associated proteins that are incorporated in virions: spike (S), envelope (E), membrane (M), and nucleoprotein (N; Fig. 1). These four proteins occur in the order S–E–M–N in every known coronavirus lineage (Woo et al., 2014). In between the S–E–M–N genes, coronaviruses encode species-specific accessory proteins, many of which appear to be incorporated in virions at low levels, ranging from one accessory in alphacoronaviruses including human coronavirus NL63 (Pyrc et al., 2004) to a predicted nine accessories in the gammacoronavirus HKU22 (Woo et al., 2014). The genomic position of these accessory genes varies, with accessories encoded before S in some betacoronaviruses, between S and E in most lineages, between M and N in most lineages, and after N rarely in alphacoronaviruses and gammacoronaviruses and commonly in deltacoronaviruses. Interestingly, the M gene appears to directly follow the E gene throughout the *Coronaviridae*, though there is not an obvious transcriptional or translational reason why this should necessarily be the case. Interestingly, in one study, deletion of E resulted in the evolution of spontaneously joined gene

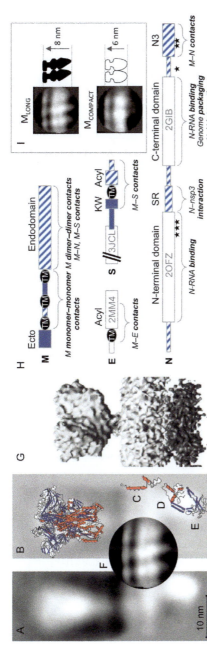

Fig. 1 Structure and organization of proteins in the virion. Cryo-EM reconstruction of the SARS-CoV structural proteins from virions (A; Neuman et al., 2006) superimposed with solved structures of the MHV S ectodomain (B; 3JCL), SARS-CoV E (C; 2MM4), SARS-CoV C-terminal domain (D; 2GIB), SARS-CoV N-terminal domain (E; 2OFZ), and a cryo-EM reconstruction of the M proteins from MHV VLPs (F; Neuman et al., 2011). An alternative shorter, wider cryo-EM reconstruction of SARS-CoV virion proteins is shown for comparison (G; EMD-1423). Schematics are based on MHV proteins, following the annotation of Kuo et al. (2016). The orientation of N protein domains shown here is hypothetical, and is intended for illustrative purposes only. (H) Domain structure and annotation of MHV S, E, M, and N proteins showing domains outside the virion (*solid blue*), inside the virion (*striped blue*), and the position of solved structures. Transmembrane regions (TM), sites of palmitoylation (Acyl), a conserved sequence preceding the transmembrane of S (KW), a serine–arginine-rich unstructured region (SR), phosphorylation sites (*stars*), and the C-terminal M-interacting domain of N (N3) are marked. Comparison of the appearance of dimeric M$_{LONG}$ and M$_{COMPACT}$ (I).

fragments of M encoded in the position normally occupied by E (Kuo and Masters, 2010).

The coronavirus particle provides three important things for the genome: a durable transport vessel, a means of escape for newly synthesized genomes, and a means of entry into a cell. The following sections will examine each of the key functions of the coronavirus particle in durability, budding, and entry.

2. VIRION STRUCTURE AND DURABILITY

The genome likely comprises a relatively small part of the internal volume of a virion. MHV produces virions that are approximately 80 nm in diameter, which is typical for hydrated coronavirus particles in vitreous ice (Barcena et al., 2009; Neuman et al., 2006, 2011). The average radius of MHV virions is about 42 nm (Neuman et al., 2011); subtracting 8 nm occupied by the viral membrane and M protein at each side (Neuman et al., 2011), and assuming a spherical virion, gives a predicted interior volume of 1.6×10^5 nm^3 for a coronavirus particle. The partial specific volume of a coronavirus genome, calculated using a density of 0.57 cm^3/g for ssRNA (Voss and Gerstein, 2005) and the molecular mass of the smallest coronavirus genome (porcine deltacoronavirus HKU15, 25.4 kb) and the largest (bottlenose dolphin coronavirus HKU22, 31.8 kb), gives a volume of $0.8–1.0 \times 10^4$ nm^3 per genome, equivalent to 5–6% of the virion. From the estimated $0.7–2.2 \times 10^3$ N proteins per virion (Neuman et al., 2011) and the volume of about 60 nm^3 per N protein (Neuman et al., 2006), we can calculate that N proteins should occupy $0.4–1.3 \times 10^5$ nm^3 or 25–80% of the virion interior, and that each N is associated with 14–40 nt of genomic RNA (Neuman et al., 2011).

In their native state, virions are filled with water, and preimaging procedures that remove the water from the virion have led to the mistaken but strangely persistent ideas that both arteriviruses (Horzinek et al., 1971) and coronaviruses (Risco et al., 1996) contain an icosahedrally organized ribonucleoprotein core. The shape of hydrated coronavirus particles in water ice as observed by cryoelectron microscopy is roughly spherical, and shows no sign of icosahedral organization (Barcena et al., 2009; Beniac et al., 2006, 2007; Neuman et al., 2006, 2011). The same is true of *Arteriviridae* (Spilman et al., 2009), and probably also of *Mesoniviridae* (Nga et al., 2011). Other members of the *Nidovirales* produce bacillus-shaped particles, including *Bafinivirus* (Schutze et al., 2006) and *Okavirus* (Spann et al., 1997).

In cell culture toroviruses produce a mixture of spherical and bacilliform particles (B. Neuman, unpublished data). Careful measurement of coronavirus particles in cryoelectron micrographs and tomograms reveals that most coronavirus particles are slightly prolate spheroids that differ from the shape of more spherical exosomal vesicles that appear in the same images (Neuman et al., 2011). While the rigidity of coronavirus particles remains to be investigated, the shape difference, along with the observation that coronavirus particles remain roughly spherical despite repeated prodding with a carbon nanotube during atomic force microscopy (Liu et al., 2012; Ng et al., 2004), suggests that coronavirus particles are relatively resistant to deformation, as recently reported for influenza A virus (Li et al., 2011).

Just because a virion is enveloped, does not mean it is necessarily fragile or quickly inactivated, as demonstrated by the enveloped pithovirus that was successfully recovered from 30,000-year-old permafrost (Legendre et al., 2014). Coronavirus particles are relatively robust compared to HIV-1, with SARS-CoV virions remaining infectious for 1–4 days on the relatively harsh environment of hard surfaces (reviewed in Sobsey and Meschke, 2003). MERS-CoV virions are somewhat less robust than SARS-CoV, with half lives on the order of an hour on hard surfaces and a maximum survival time of 2–3 days, but are considerably more durable than pandemic influenza A virus under the same conditions (van Doremalen et al., 2013). The persistent infectivity of coronaviruses outside the body has been used as evidence to suggest that direct contact with contaminated surfaces as well as respiratory droplets is a potential route of MERS-CoV spread (Assiri et al., 2013; Goh et al., 2013).

3. VIRAL PROTEINS IN ASSEMBLY AND FUSION

3.1 Membrane Protein

The M protein facilitates viral assembly by interacting with other M (Arndt et al., 2010; de Haan et al., 2000), E (Boscarino et al., 2008; Corse and Machamer, 2003; Lim and Liu, 2001), S (de Haan et al., 1999; Godeke et al., 2000), and N proteins (Escors et al., 2001; Hurst et al., 2005; Narayanan et al., 2000; Sturman et al., 1980). The MHV M protein may also interact with the RNA packaging signal that mediates incorporation of the viral genome into viral particles (Narayanan et al., 2003), but direct M-genome interactions are not efficient enough to rescue viruses in which the C-terminal region of N that interacts with M has been perturbed,

suggesting that M-genome interactions are less important than M–N interactions for recovery of infectious virus (Kuo et al., 2016). Communication
between the carboxyl termini of the M and N proteins has also been
observed from mutagenesis and second-site reversion studies (Hurst et al.,
2005; Kuo and Masters, 2002).

M proteins in SARS-CoV, FCoV, and MHV virions and virus-like particles (VLPs) form homodimers (Neuman et al., 2011), which appear to be
functionally analogous to the M-GP5 heterodimers of *Arteriviridae* (de Vries
et al., 1995; Faaberg et al., 1995; Snijder et al., 2003). Coronavirus M dimers
resemble the shape of a Greek amphora, with the ectodomain forming the
lip, transmembrane region forming narrow neck, and the endodomain forming the lower chamber (Fig. 1I). In virions it appears that the transmembrane region does not make contact between adjacent M dimers, suggesting
that reported M–M interaction domains in the transmembrane region
(de Haan et al., 2000) are between the two monomers that make up an
M dimer. M protein deletion mutants were not assembly competent in
the absence of wild-type M (de Haan et al., 2000). Thus M–M interactions
are necessary, but not always sufficient for VLP assembly (de Haan
et al., 2000).

In contrast, the endodomains of M dimers appeared to make close contact in the cryo-EM reconstructions (Neuman et al., 2011), suggesting that
interactions between M dimers are likely to occur in the endodomain. The
endodomain of MHV M is important for M–M interactions that could
involve monomers or dimers. Purified SARS-CoV M endodomains can
dimerize and can form M dimer–dimer interactions (Neuman et al.,
2011), suggesting that the endodomain is the primary site of M dimer–dimer
interaction.

Cryo-EM and cryoelectron tomography analysis also suggests that M can
exist in two forms (Neuman et al., 2011). The main form found on virions
and VLPs is an elongated conformation that makes contact with the ribonucleoprotein and imparts a spherical membrane curvature of about 5–6
degrees per M dimer. The minor form is more compact, has indistinct
boundaries suggestive of a disordered aggregate, and does not appear to
impart membrane curvature. The long conformation could be partially
converted to the shorter conformation by transient acidification, weakening
M–RNP interactions. This suggests a model in which formation of
M_{LONG}–RNP interactions drives the budding process, and membrane
fusion is preceded by release of M_{LONG}–RNP interactions. However, further work is needed to test the accuracy of this model.

The structure of the M protein is not known, but may be partially inferred from sequence comparison and secondary structure prediction algorithms. M proteins possess a short glycosylated ectodomain of variable sequence (Oostra et al., 2006), followed by three closely spaced hydrophobic transmembrane helix signatures, and a relatively long cytoplasmic tail region that may fold into a compact beta-dominated structure (Masters et al., 2006).

The weight of evidence suggests that coronavirus M and E proteins are the critical components required for assembly of coronavirus virions and VLPs (Bos et al., 1996; Corse and Machamer, 2000; Vennema et al., 1996b). However the SARS-CoV M protein appears to readily form VLPs in the absence of E (Tseng et al., 2010), as does the M and S protein of IBV (Liu et al., 2013). M protein sequences from different coronaviruses are highly conserved, approaching the level of conservation of some of the viral enzymes and replicase accessory proteins from pp1a (Stadler et al., 2003). During the assembly process, N protein contributes to the formation of M_{LONG}, narrowing the size range of resulting VLPs (Fig. 2; Neuman et al., 2011) and increasing the efficiency of VLP production (Boscarino et al., 2008; Siu et al., 2008). A study of VLP size and organization showed

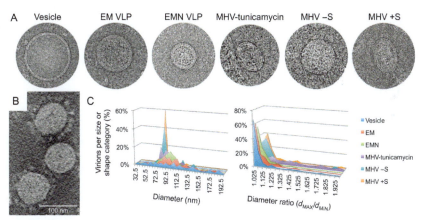

Fig. 2 Appearance and characteristics of MHV virions, MHV VLPs, and copurified exosomal vesicles. Cryo-EM images show a vesicle, VLPs produced after coexpression of E and M (EM VLP) or E, M and N (EMN VLP), a virion from tunicamycin-treated cells, and virions from the same preparation that spontaneously formed lacking (−S) or displaying (+S) visible spikes. A cryoelectron tomography image of MHV is shown to highlight the variation in spike incorporation in virions from the same preparation (B). Sizes (diameter) and shapes (d_{MAX}/d_{MIN}) of virions, VLPs, and vesicles were categorized to illustrate differences in the shape of particles that incorporate different combinations of structural proteins (C).

that MHV VLPs and virions with different components formed particles with a constrained minimum size, and that the size range of particles became narrower and approached the minimum size of VLPs as more structural components were incorporated (Fig. 2; Neuman et al., 2011). This suggests that incorporation of S and the genome, which are both essential for virion infectivity but dispensable for VLP production, affected the efficiency of budding in the sense that larger particles incorporate more M proteins.

3.2 Nucleoprotein

Coronavirus nucleoproteins are phosphoproteins, and are encoded near the 3′ end of the genome. MHV N is phosphorylated at six sites (S162, S170, T177, S389, S424, and T428; White et al., 2007) by host kinases like cyclin-dependent kinase, glycogen synthase kinase, mitogen-activated protein kinase, and casein kinase II (Surjit et al., 2005). The protein is also sumoylated (at Lys62 in the SARS-CoV N protein), a posttranslational modification that enhances the protein's tendency to homooligomerize and affects typical N protein-mediated interference in host cell division (Li et al., 2005b). Several groups have successfully expressed soluble protein (both full-length and partial domains), but its unusually high positive charge, tendency to oligomerize, structural flexibility, and extremely low stability have hampered structural studies of the whole protein (Chang et al., 2006). N possesses two RNA-binding domains: an N-terminal domain with adjacent S/R-rich motif (He et al., 2004) and the C-terminal 209 amino acids (Ma et al., 2010; Surjit et al., 2004; Yu et al., 2005). The N protein also binds with nanomolar affinity to human cyclophilin A, though the physiological significance of this finding is still unknown (Luo et al., 2004; Pfefferle et al., 2011).

N protein supports coronavirus infection in several ways: the C-terminal domain (CTD) of N is important for binding the genomic RNA packaging signal leading to selective genome incorporation (Kuo et al., 2014; Molenkamp and Spaan, 1997), the N3 domain interacts with the endodomain of M to form virions (Kuo et al., 2016), and the serine–arginine repeat region of N (SR) interacts with the first ubiquitin-like domain of nsp3 in a critical early replication step (Hurst et al., 2010, 2013). It has also been demonstrated that N can oligomerize through interactions in the CTD (Chang et al., 2013), bind viral RNA through the N-terminal domain (Fan et al., 2005), unwind double-stranded nucleic acid in the manner of an RNA chaperone (Neuman et al., 2008; Zuniga et al., 2007), and pack in a helix through the N-terminal domain (Saikatendu et al., 2007), though none of these other functions has yet been demonstrated to be important for infection. N protein is dynamically associated with sites of viral RNA replication, suggesting that N may also function to protect the genome or possibly mediate genome transport to the budding site (Verheije et al., 2010).

The nucleoprotein of coronavirus is not a good match for nucleoproteins of other members of the *Nidovirales* at the level of amino acid sequence. However, the small N protein of arteriviruses EAV (Deshpande et al., 2007) and PRRSV (Doan and Dokland, 2003) adopts a similar fold to the CTD of coronavirus N (Yu et al., 2006). Alignment of predicted protein secondary structures from N proteins from other nidoviruses with the structures of coronavirus and arenavirus N suggests that a helix–strand–strand–helix motif may form a conserved functional domain near the C-terminus of all nidovirus N proteins (Fig. 3).

3.3 Envelope Protein

E proteins are encoded by all known coronavirus genomes and are found at low levels in the virion (Godet et al., 1992; Liu and Inglis, 1991). As pointed out by Kuo and Masters (Kuo et al., 2016), E appears to have three distinct functions that contribute to infection: regulating aggregation-prone M–M interactions (Boscarino et al., 2008), disrupting Golgi organization in a way that produces larger vesicles capable of transporting virions (Machamer and Youn, 2006; Ruch and Machamer, 2011, 2012), and interacting with host factors in a way that affects pathogenesis (DeDiego et al., 2007, 2011; Dediego et al., 2008; Nieto-Torres et al., 2015; Regla-Nava et al., 2015; Teoh et al., 2010). E proteins of several coronaviruses have been

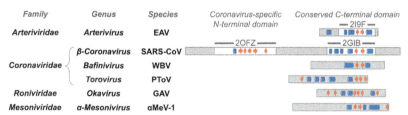

Fig. 3 Evidence for conservation of the N protein C-terminal domain fold across the *Nidovirales*. Proteins are depicted as *rectangles* with resolved regions of solved protein structures (*white* regions) and unsolved or unresolved regions (*gray* regions) indicated. Alpha helix (*blue cylinders*) and beta strand (*red arrows*) regions of solved protein secondary structures or JPRED secondary structure predictions are shown manually aligned by secondary structure to facilitate comparison.

reported to have ion channel activity (Liao et al., 2004; Madan et al., 2005; Wilson et al., 2004), which appears to enhance viral growth (Wilson et al., 2006; Ye and Hogue, 2007). In addition to these three roles, E proteins have been speculated to be involved in scission of the membrane to free newly budded virions based on evidence from MHV VLP formation in the presence of E (Vennema et al., 1996a) and aberrant virion formation in E mutant viruses (Fischer et al., 1998), analogous to the function of M2 in influenza A virus (Rossman et al., 2010). However, it is not clear whether the role of E in virion release is distinct from its role in limiting M aggregation (Boscarino et al., 2008) and SARS-CoV M does not appear to require the presence of E to generate VLPs (Tseng et al., 2010), suggesting that the role of E differs, or differs in importance, in some coronaviruses.

E is probably best viewed as a multifunctional accessory gene that contributes to both virus growth and pathogenesis. Expression of SARS-CoV E protein is dispensable for coronavirus growth, but deletion results in severe defects in virus growth, presumably related to inefficient assembly (DeDiego et al., 2007; Kuo and Masters, 2003). Interestingly, deletion of the SARS-CoV E gene has less effect on replication than the corresponding E gene deletion in MHV, suggesting that the function of E may be duplicated elsewhere in the genome. A candidate for the compensating factor could be the SARS-CoV 3a protein, which similarly displays ion channel activity (Lu et al., 2006) and serves to increase the viral growth rate when present (Yount et al., 2005), or possibly the SARS-CoV ORF6-encoded protein, which is predicted to adopt a similar membrane topology to E and induces intracellular membrane rearrangement (Zhou et al., 2010). In MHV, duplication of the amino-terminal part of M can partially compensate for the deletion of E, suggesting that the functions of MHV M and E proteins may overlap (Kuo and Masters, 2010). The 229E-4a accessory protein also has ion channel activity, but it is not known whether this can compensate for deletion of E (Zhang et al., 2014). In contrast, deletion of E in TGEV prevented the formation of infectious virions (Ortego et al., 2007).

A number of general structural features can be identified in all coronavirus E proteins, including a short hydrophilic N-terminal region, followed by a hydrophobic putative transmembrane region, and a relatively long hydrophilic C-terminal tail (Liu et al., 2007). SARS-CoV E protein forms pentamers (Pervushin et al., 2009) that conduct cations (Pervushin et al., 2009), which is likely a conserved feature of E proteins. There is evidence that IBV E protein has different functions that correlate to different oligomerization states in mutant E proteins: monomeric E is sufficient to disrupt the Golgi, while oligomerization-competent E supports VLP release (Westerbeck and Machamer, 2015). The target of ion channel activity may be disruption of intracellular processes by releasing calcium ions stored in the endoplasmic reticulum, ultimately driving an inflammatory immune response (Nieto-Torres et al., 2015). The E protein is palmitoylated in at least some coronaviruses (Corse and Machamer, 2002; Liao et al., 2006), and palmitylation is important for the role of MHV E protein in limiting M–M aggregation (Boscarino et al., 2008). The transmembrane and palmitoylated domains of E are each sufficient to colocalize with M protein (Nieto-Torres et al., 2011).

3.4 Spike Protein

The first open reading frame downstream of the SARS-CoV replicase encodes the S glycoprotein which is conserved in all coronaviruses. The ~180-kDa spike protein plays a central role in the host cell attachment and entry processes. The spike is organized into an amino-terminal S1 domain that contains receptor-binding determinants, and a carboxyl-terminal S2 domain that contains the membrane anchor and fusion motor domains (Belouzard et al., 2012; Heald-Sargent and Gallagher, 2012). Coronavirus S proteins contain short amino-terminal hydrophobic signal sequence motifs (von Heijne, 1984). Although some coronavirus spike proteins are cleaved between the S1 and S2 regions as part of the activation process, the SARS-CoV spike is not appreciably cleaved before it is internalized in a host cell (Xiao et al., 2003). The more variable amino-terminal region of the spike protein (S1) has been demonstrated to contain the receptor-binding activity (Wong et al., 2004). The more conserved S2 region contains the transmembrane anchor, palmitic acid acylation site (Thorp et al., 2006) that is important for membrane fusion (McBride and Machamer, 2010; Shulla and Gallagher, 2009), and the coiled-coil fusion motor domain (Bosch et al., 2003; Duquerroy et al., 2005; Liu et al., 2004; Tripet et al., 2004; Xu et al., 2004a,b). One or more protease cleavage events are necessary to prime S for membrane fusion. An apparent fusion peptide of coronaviruses resides in the S2 region (Bosch et al., 2003; Sainz et al., 2005), where it is exposed by cleavage that takes place on tetraspanin-enriched membranes where host proteases including TMPRSS2 and HAT localize for some coronaviruses including MERS-CoV, while other spikes can be primed for entry by cathepsins (Bertram et al., 2013; Earnest et al., 2015; Glowacka et al., 2011; Heurich et al., 2014; Huang et al., 2006; Shulla et al., 2011; Simmons et al., 2005). This is consistent with host gene knockdown experiments that show that coronavirus entry is dependent on several elements that are important in the endosomal and lysosomal trafficking (Burkard et al., 2014; Wong et al., 2015). Expression of IFITM proteins inhibits entry driven by several coronavirus spike proteins (Huang et al., 2011; Wrensch et al., 2014), but paradoxically appears to promote infection by HCoV-OC43 (Zhao et al., 2014).

Near-atomic resolution cryo-EM structures have been published for the ectodomains of trimeric MHV (Walls et al., 2016) and HKU1 (Kirchdoerfer et al., 2016) up to the second heptad repeat region. The high-resolution spike ectodomain structures are similar to the profile to the upper part of SARS-CoV S in one study (Neuman et al., 2006) and taller and less square than the spikes reconstructed by another group (Beniac et al., 2006, 2007). High-resolution X-ray crystallography structures have also been obtained for two domains of the coronavirus spike protein. The structure of the minimal receptor-binding domain from SARS-CoV was solved first in conjunction with angiotensin-I converting enzyme 2 (ACE2; Towler et al., 2004), the primary cellular receptor for SARS-CoV (Li et al., 2003) and human coronavirus NL63 (Hofmann et al., 2005). More recently, S1 receptor structures have been solved for betacoronaviruses MHV (Peng et al., 2011) and MERS-CoV (Wang et al., 2013) and alphacoronaviruses NL63 (Wu et al., 2009), TGEV (Reguera et al., 2012), and PRCoV (Reguera et al., 2012). The receptor-binding domain structure of SARS-CoV consists of a core subdomain containing a five-stranded antiparallel β sheet with three short connecting α helices, and an extended loop subdomain that contacts ACE2 (Li et al., 2005a), and the domains that mediate S1 receptor contact in other betacoronaviruses also seem to involve curved β sheets at the center with stabilizing loop interactions at the side. In contrast, the receptor-binding domain of alphacoronaviruses makes contact via a series of loops positioned at the edge of a β sheet. Image analysis of cryoelectron micrographs of SARS-CoV (Beniac et al., 2006; Neuman et al., 2006) and other coronaviruses (Neuman et al., 2006) confirms that spikes exist as a homotrimer in the native prefusion state on virions. However, some biochemical characterizations have revealed that S1 interacts with the receptor protein as a dimer, even within the context of the trimeric spike (Lewicki and Gallagher, 2002; Xiao et al., 2004).

In the model of coronavirus spike protein-mediated fusion, receptor binding triggers conformational changes including disulfide reshuffling (Gallagher, 1996; Lavillette et al., 2006) that release HR1 and HR2 to form the coiled-coil fusion motor structure, thereby driving fusion of the viral and host cell membranes and release of the viral ribonucleoprotein into the cytosol (Bosch et al., 2003). The fusion motor complex of S2 consisting of two hydrophobic amino acid 4-3 heptad repeat regions (HR1 and HR2) which form amphipathic helices of a coiled-coil structure has also been solved in several forms (Bosch et al., 2003; Duquerroy et al., 2005; Liu et al., 2004; Tripet et al., 2004; Xu et al., 2004a,b). In the structure representing the largest region of S2 structure, HR1 forms a trimeric 120 Å coiled coil (Duquerroy et al., 2005). Coordinated chloride ions in the structure are instrumental in the formation of hydrogen bonds that stretch both ends of the HR2 region into an extended conformation surrounding the central alpha helix (Duquerroy et al., 2005). After insertion of the fusion peptide into the target membrane, a single-particle fusion study revealed that it takes about 15 s to go from membrane insertion to hemifusion, 15–60 more seconds until a pore is formed, then another 30 s for complete lipid mixing between the virion and cell (Costello et al., 2013).

Spike protein is incorporated into virions through interactions with the membrane protein M (Godeke et al., 2000). It can be generally accepted that all the determinants for virion incorporation reside in the transmembrane and carboxyl-terminal regions of the spike protein (Godeke et al., 2000; Kuo et al., 2000). However, a recent study found that a chimeric MHV with SARS-CoV M and S transmembrane and endodomain was severely deficient in incorporating S into virions, suggesting that cellular localization signals or more complex interactions among the structural proteins may help support S incorporation (Kuo et al., 2016).

4. EVOLUTION OF THE STRUCTURAL PROTEINS

Working on nidoviruses is a mixed blessing—on one hand, there is sufficient evolutionary divergence and complexity with a few common threads to make it seem possible to reconstruct an evolutionary path from a simple coronavirus-like progenitor to the present array of nidoviruses. However, the extreme divergence between homologous proteins means that common ancestry is sometimes only evident at the level of protein fold and conserved function, meaning that homologs cannot necessarily be recognized by amino acid alignment.

Two structural proteins stand out as being conserved: coronavirus M and N. One or more M-like three-pass transmembrane proteins with endodomains rich in predicted β structure are found in coronaviruses, toroviruses, bafiniviruses, arteriviruses, the arteri-like possum nidovirus (Dunowska et al., 2012), and the newly discovered toro-like ball python nidovirus (Stenglein et al., 2014). *Roniviridae* also encode a structural poly-protein of GP116 and GP64 that includes a superficially similar three-pass transmembrane protein known as 3N at its amino terminus, though 3N has not yet been detected in infected cells or virions to date. *Mesoniviridae* lack a three-pass M protein, but two single-pass transmembrane proteins with predicted β-rich endodomains known as M and 3b may serve the same function as coronavirus M in the virion. Every member of the *Nidovirales* also encodes a positively charged N-like protein. Although the proteins differ in size, common predicted structure elements near the C-terminus suggest that nidovirus N proteins may in fact be homologous, as shown in Fig. 3. In comparison, proposed S-like proteins are highly divergent, and E-like proteins are absent in several nidovirus lineages, suggesting S and E are later refinements to nidovirus virion architecture.

We can therefore imagine two evolutionary paths that gradually built up to the current complexity of nidovirus structural proteins (Fig. 4). The first

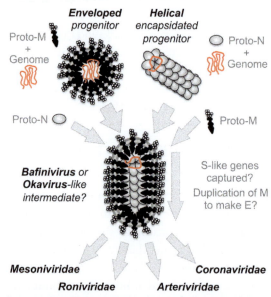

Fig. 4 Models of coronavirus virion evolution by gradual accumulation of structural proteins. Evolution of potential progenitors with enveloped pleomorphic or helical encapsidated virion architecture is shown leading to a filamentous enveloped intermediate stage, superficially resembling virions of the genus *Bafinivirus* or the family *Roniviridae*. Structural diversification by capture of attachment and fusion proteins from an unknown source, and partial duplication of M to make E in some lineages then leads to modern nidovirus lineages.

model involves an ancestral enveloped virus that encodes a hypothetical protein similar to coronavirus M (but capable of RNA packaging and attachment to host cells) served as the original structural protein in a progenitor of the last common ancestor of nidoviruses. In this model, first proto-N, then S-like and E-like proteins would be added. M protein equivalents of coronavirus, arterivirus, and torovirus are similar in appearance, and are consistent with the size of two protein chains (B. Neuman, unpublished data) suggesting that networks formed of dimers may be an ancestral trait that originated in the hypothetical proto-M. M has many of the characteristics that would be expected of a primitive movement factor. M is generally the most abundant protein in virions, is essential for VLP formation, and for S and N incorporation into VLPs. MHV M protein has been reported to mediate incorporation of RNA containing the genomic packaging signal into VLPs in the absence of N inefficiently (Narayanan et al., 2003), and the structurally similar 3a accessory protein of SARS-CoV may also bind RNA (Sharma et al., 2007), suggesting that RNA packaging could be an ancestral feature of M-like proteins. The M-like GP5 protein of arteriviruses may also be involved in attachment to host cells via its glycosylated ectodomain (Tian et al., 2012), though studies with chimeric PRRSV proteins demonstrated that M and GP5 are not solely responsible for differences in tropism (Lu et al., 2012). These observations, together with the previously mentioned instance of M protein duplication compensating for deletion of E (Kuo and Masters, 2010), suggest that M-like proteins as a group have the potential to carry out some of the essential functions of E, N, and S.

The other potential evolutionary path we will consider is an encapsidated helical progenitor virus that encoded a proto-N related to the CTD of coronavirus N. Evidence for this is the potential conservation of fold in the CTD, suggesting a common origin for nidovirus N proteins. As described earlier, N selectively binds the genomic RNA and interacts with other N proteins to form a helical ribonucleoprotein (Barcena et al., 2009) reminiscent of the encapsidated forms of helical +RNA viruses of the *Tymovirales*, *Virgaviridae*, or *Closteroviridae*. In this model, a primitive virus with a helical capsid would gain an advantage in attaching to host cells by capturing a gene encoding a membrane-spanning protein like M.

Considered individually, or as a group, viral particles formed by nidoviruses are pleomorphic. Whatever their origin, coronavirus structural proteins demonstrate a remarkable plasticity to accommodate gene deletion, gene duplication, and genetic divergence, while still facilitating the entry, egress, and protection of the genome. It seems fitting, therefore, that pleomorphic nidovirus particles should be formed from a set of structural components that could themselves collectively also be described as pleomorphic.

REFERENCES

Almazan, F., Gonzalez, J.M., Penzes, Z., Izeta, A., Calvo, E., Plana-Duran, J., Enjuanes, L., 2000. Engineering the largest RNA virus genome as an infectious bacterial artificial chromosome. Proc. Natl. Acad. Sci. U.S.A. 97 (10), 5516–5521.

Arndt, A.L., Larson, B.J., Hogue, B.G., 2010. A conserved domain in the coronavirus membrane protein tail is important for virus assembly. J. Virol. 84 (21), 11418–11428. http://dx.doi.org/10.1128/JVI.01131-10.

Assiri, A., McGeer, A., Perl, T.M., Price, C.S., Al Rabeeah, A.A., Cummings, D.A., et al., 2013. Hospital outbreak of Middle East respiratory syndrome coronavirus. N. Engl. J. Med. 369 (5), 407–416. http://dx.doi.org/10.1056/NEJMoa1306742.

Barcena, M., Oostergetel, G.T., Bartelink, W., Faas, F.G., Verkleij, A., Rottier, P.J., et al., 2009. Cryo-electron tomography of mouse hepatitis virus: insights into the structure of the coronavirion. Proc. Natl. Acad. Sci. U.S.A. 106 (2), 582–587. http://dx.doi.org/10.1073/pnas.0805270106.

Belouzard, S., Millet, J.K., Licitra, B.N., Whittaker, G.R., 2012. Mechanisms of coronavirus cell entry mediated by the viral spike protein. Viruses 4 (6), 1011–1033. http://dx.doi.org/10.3390/v4061011.

Beniac, D.R., Andonov, A., Grudeski, E., Booth, T.F., 2006. Architecture of the SARS coronavirus prefusion spike. Nat. Struct. Mol. Biol. 13 (8), 751–752. http://dx.doi.org/10.1038/nsmb1123.

Beniac, D.R., deVarennes, S.L., Andonov, A., He, R., Booth, T.F., 2007. Conformational reorganization of the SARS coronavirus spike following receptor binding: implications for membrane fusion. PLoS One 2 (10), e1082. http://dx.doi.org/10.1371/journal.pone.0001082.

Bertram, S., Dijkman, R., Habjan, M., Heurich, A., Gierer, S., Glowacka, I., et al., 2013. TMPRSS2 activates the human coronavirus 229E for cathepsin-independent host cell entry and is expressed in viral target cells in the respiratory epithelium. J. Virol. 87 (11), 6150–6160. http://dx.doi.org/10.1128/JVI.03372-12.

Bos, E.C., Luytjes, W., van der Meulen, H.V., Koerten, H.K., Spaan, W.J., 1996. The production of recombinant infectious DI-particles of a murine coronavirus in the absence of helper virus. Virology 218 (1), 52–60.

Boscarino, J.A., Logan, H.L., Lacny, J.J., Gallagher, T.M., 2008. Envelope protein palmitoylations are crucial for murine coronavirus assembly. J. Virol. 82 (6), 2989–2999. http://dx.doi.org/10.1128/JVI.01906-07.

Bosch, B.J., van der Zee, R., de Haan, C.A., Rottier, P.J., 2003. The coronavirus spike protein is a class I virus fusion protein: structural and functional characterization of the fusion core complex. J. Virol. 77 (16), 8801–8811.

Burkard, C., Verheije, M.H., Wicht, O., van Kasteren, S.I., van Kuppeveld, F.J., Haagmans, B.L., et al., 2014. Coronavirus cell entry occurs through the endo-/lysosomal pathway in a proteolysis-dependent manner. PLoS Pathog. 10 (11), e1004502. http://dx.doi.org/10.1371/journal.ppat.1004502.

Casais, R., Thiel, V., Siddell, S.G., Cavanagh, D., Britton, P., 2001. Reverse genetics system for the avian coronavirus infectious bronchitis virus. J. Virol. 75 (24), 12359–12369. http://dx.doi.org/10.1128/JVI.75.24.12359-12369.2001.

Chang, C.K., Sue, S.C., Yu, T.H., Hsieh, C.M., Tsai, C.K., Chiang, Y.C., et al., 2006. Modular organization of SARS coronavirus nucleocapsid protein. J. Biomed. Sci. 13 (1), 59–72. http://dx.doi.org/10.1007/s11373-005-9035-9.

Chang, C.K., Chen, C.M., Chiang, M.H., Hsu, Y.L., Huang, T.H., 2013. Transient oligomerization of the SARS-CoV N protein—implication for virus ribonucleoprotein packaging. PLoS One 8 (5), e65045. http://dx.doi.org/10.1371/journal.pone.0065045.

Corse, E., Machamer, C.E., 2000. Infectious bronchitis virus E protein is targeted to the Golgi complex and directs release of virus-like particles. J. Virol. 74 (9), 4319–4326.

Corse, E., Machamer, C.E., 2002. The cytoplasmic tail of infectious bronchitis virus E protein directs Golgi targeting. J. Virol. 76 (3), 1273–1284.

Corse, E., Machamer, C.E., 2003. The cytoplasmic tails of infectious bronchitis virus E and M proteins mediate their interaction. Virology 312 (1), 25–34.

Costello, D.A., Millet, J.K., Hsia, C.Y., Whittaker, G.R., Daniel, S., 2013. Single particle assay of coronavirus membrane fusion with proteinaceous receptor-embedded supported bilayers. Biomaterials 34 (32), 7895–7904. http://dx.doi.org/10.1016/j.biomaterials.2013.06.034.

de Haan, C.A., Smeets, M., Vernooij, F., Vennema, H., Rottier, P.J., 1999. Mapping of the coronavirus membrane protein domains involved in interaction with the spike protein. J. Virol. 73 (9), 7441–7452.

de Haan, C.A., Vennema, H., Rottier, P.J., 2000. Assembly of the coronavirus envelope: homotypic interactions between the M proteins. J. Virol. 74 (11), 4967–4978.

de Vries, A.A., Post, S.M., Raamsman, M.J., Horzinek, M.C., Rottier, P.J., 1995. The two major envelope proteins of equine arteritis virus associate into disulfide-linked heterodimers. J. Virol. 69 (8), 4668–4674.

DeDiego, M.L., Alvarez, E., Almazan, F., Rejas, M.T., Lamirande, E., Roberts, A., et al., 2007. A severe acute respiratory syndrome coronavirus that lacks the E gene is attenuated in vitro and in vivo. J. Virol. 81 (4), 1701–1713. http://dx.doi.org/10.1128/JVI.01467-06.

Dediego, M.L., Pewe, L., Alvarez, E., Rejas, M.T., Perlman, S., Enjuanes, L., 2008. Pathogenicity of severe acute respiratory coronavirus deletion mutants in hACE-2 transgenic mice. Virology 376 (2), 379–389. http://dx.doi.org/10.1016/j.virol.2008.03.005.

DeDiego, M.L., Nieto-Torres, J.L., Jimenez-Guardeno, J.M., Regla-Nava, J.A., Alvarez, E., Oliveros, J.C., et al., 2011. Severe acute respiratory syndrome coronavirus envelope protein regulates cell stress response and apoptosis. PLoS Pathog. 7 (10), e1002315. http://dx.doi.org/10.1371/journal.ppat.1002315.

Dent, S.D., Xia, D., Wastling, J.M., Neuman, B.W., Britton, P., Maier, H.J., 2015. The proteome of the infectious bronchitis virus Beau-R virion. J. Gen. Virol. 96, 3499–3506.

Deshpande, A., Wang, S., Walsh, M.A., Dokland, T., 2007. Structure of the equine arteritis virus nucleocapsid protein reveals a dimer-dimer arrangement. Acta Crystallogr. Sect. D 63 (Pt. 5), 581–586. http://dx.doi.org/10.1107/S0907444907008372.

Doan, D.N., Dokland, T., 2003. Structure of the nucleocapsid protein of porcine reproductive and respiratory syndrome virus. Structure 11 (11), 1445–1451.

Donaldson, E.F., Sims, A.C., Baric, R.S., 2008a. Systematic assembly and genetic manipulation of the mouse hepatitis virus A59 genome. Methods Mol. Biol. 454, 293–315. http://dx.doi.org/10.1007/978-1-59745-181-9_21.

Donaldson, E.F., Yount, B., Sims, A.C., Burkett, S., Pickles, R.J., Baric, R.S., 2008b. Systematic assembly of a full-length infectious clone of human coronavirus NL63. J. Virol. 82 (23), 11948–11957. http://dx.doi.org/10.1128/JVI.01804-08.

Dunowska, M., Biggs, P.J., Zheng, T., Perrott, M.R., 2012. Identification of a novel nidovirus associated with a neurological disease of the Australian brushtail possum (Trichosurus vulpecula). Vet. Microbiol. 156 (3–4), 418–424. http://dx.doi.org/10.1016/j.vetmic.2011.11.013.

Duquerroy, S., Vigouroux, A., Rottier, P.J., Rey, F.A., Bosch, B.J., 2005. Central ions and lateral asparagine/glutamine zippers stabilize the post-fusion hairpin conformation of the SARS coronavirus spike glycoprotein. Virology 335 (2), 276–285. http://dx.doi.org/10.1016/j.virol.2005.02.022. S0042-6822(05)00120-0 [pii].

Earnest, J.T., Hantak, M.P., Park, J.E., Gallagher, T., 2015. Coronavirus and influenza virus proteolytic priming takes place in tetraspanin-enriched membrane microdomains. J. Virol. 89 (11), 6093–6104. http://dx.doi.org/10.1128/JVI.00543-15.

Escors, D., Ortego, J., Laude, H., Enjuanes, L., 2001. The membrane M protein carboxy terminus binds to transmissible gastroenteritis coronavirus core and contributes to core stability. J. Virol. 75 (3), 1312–1324. http://dx.doi.org/10.1128/JVI.75.3.1312-1324.2001.

Faaberg, K.S., Even, C., Palmer, G.A., Plagemann, P.G., 1995. Disulfide bonds between two envelope proteins of lactate dehydrogenase-elevating virus are essential for viral infectivity. J. Virol. 69 (1), 613–617.

Fan, H., Ooi, A., Tan, Y.W., Wang, S., Fang, S., Liu, D.X., Lescar, J., 2005. The nucleocapsid protein of coronavirus infectious bronchitis virus: crystal structure of its N-terminal domain and multimerization properties. Structure 13 (12), 1859–1868. http://dx.doi.org/10.1016/j.str.2005.08.021.

Fischer, F., Stegen, C.F., Masters, P.S., Samsonoff, W.A., 1998. Analysis of constructed E gene mutants of mouse hepatitis virus confirms a pivotal role for E protein in coronavirus assembly. J. Virol. 72 (10), 7885–7894.

Gallagher, T.M., 1996. Murine coronavirus membrane fusion is blocked by modification of thiols buried within the spike protein. J. Virol. 70 (7), 4683–4690.

Glowacka, I., Bertram, S., Muller, M.A., Allen, P., Soilleux, E., Pfefferle, S., et al., 2011. Evidence that TMPRSS2 activates the severe acute respiratory syndrome coronavirus spike protein for membrane fusion and reduces viral control by the humoral immune response. J. Virol. 85 (9), 4122–4134. http://dx.doi.org/10.1128/JVI.02232-10.

Godeke, G.J., de Haan, C.A., Rossen, J.W., Vennema, H., Rottier, P.J., 2000. Assembly of spikes into coronavirus particles is mediated by the carboxy-terminal domain of the spike protein. J. Virol. 74 (3), 1566–1571.

Godet, M., L'Haridon, R., Vautherot, J.F., Laude, H., 1992. TGEV corona virus ORF4 encodes a membrane protein that is incorporated into virions. Virology 188 (2), 666–675.

Goh, G.K., Dunker, A.K., Uversky, V., 2013. Prediction of intrinsic disorder in MERS-CoV/HCoV-EMC supports a high oral-fecal transmission. PLoS Curr. 5. http://dx.doi.org/10.1371/currents.outbreaks.22254b58675cdebc256dbe3c5aa6498b.

Haring, M., Vestergaard, G., Rachel, R., Chen, L., Garrett, R.A., Prangishvili, D., 2005. Virology: independent virus development outside a host. Nature 436 (7054), 1101–1102. http://dx.doi.org/10.1038/4361101a.

He, R., Dobie, F., Ballantine, M., Leeson, A., Li, Y., Bastien, N., et al., 2004. Analysis of multimerization of the SARS coronavirus nucleocapsid protein. Biochem. Biophys. Res. Commun. 316 (2), 476–483.

Heald-Sargent, T., Gallagher, T., 2012. Ready, set, fuse! The coronavirus spike protein and acquisition of fusion competence. Viruses 4 (4), 557–580. http://dx.doi.org/10.3390/v4040557.

Heurich, A., Hofmann-Winkler, H., Gierer, S., Liepold, T., Jahn, O., Pohlmann, S., 2014. TMPRSS2 and ADAM17 cleave ACE2 differentially and only proteolysis by TMPRSS2 augments entry driven by the severe acute respiratory syndrome coronavirus spike protein. J. Virol. 88 (2), 1293–1307. http://dx.doi.org/10.1128/JVI.02202-13.

Hofmann, H., Pyrc, K., van der Hoek, L., Geier, M., Berkhout, B., Pohlmann, S., 2005. Human coronavirus NL63 employs the severe acute respiratory syndrome coronavirus receptor for cellular entry. Proc. Natl. Acad. Sci. U.S.A. 102 (22), 7988–7993. http://dx.doi.org/10.1073/pnas.0409465102. 0409465102 [pii].

Horzinek, M., Maess, J., Laufs, R., 1971. Studies on the substructure of togaviruses. II. Analysis of equine arteritis, rubella, bovine viral diarrhea, and hog cholera viruses. Arch. Gesamte Virusforschung 33 (3), 306–318.

Huang, I.C., Bosch, B.J., Li, F., Li, W., Lee, K.H., Ghiran, S., et al., 2006. SARS coronavirus, but not human coronavirus NL63, utilizes cathepsin L to infect ACE2-expressing cells. J. Biol. Chem. 281 (6), 3198–3203. http://dx.doi.org/10.1074/jbc.M508381200.

Huang, I.C., Bailey, C.C., Weyer, J.L., Radoshitzky, S.R., Becker, M.M., Chiang, J.J., et al., 2011. Distinct patterns of IFITM-mediated restriction of filoviruses, SARS coronavirus, and influenza A virus. PLoS Pathog. 7 (1), e1001258. http://dx.doi.org/10.1371/journal.ppat.1001258.

Hurst, K.R., Kuo, L., Koetzner, C.A., Ye, R., Hsue, B., Masters, P.S., 2005. A major determinant for membrane protein interaction localizes to the carboxy-terminal domain of the mouse coronavirus nucleocapsid protein. J. Virol. 79 (21), 13285–13297. http://dx.doi.org/10.1128/JVI.79.21.13285-13297.2005. 79/21/13285 [pii].

Hurst, K.R., Ye, R., Goebel, S.J., Jayaraman, P., Masters, P.S., 2010. An interaction between the nucleocapsid protein and a component of the replicase-transcriptase complex is crucial for the infectivity of coronavirus genomic RNA. J. Virol. 84 (19), 10276–10288. http://dx.doi.org/10.1128/JVI.01287-10.

Hurst, K.R., Koetzner, C.A., Masters, P.S., 2013. Characterization of a critical interaction between the coronavirus nucleocapsid protein and nonstructural protein 3 of the viral replicase-transcriptase complex. J. Virol. 87 (16), 9159–9172. http://dx.doi.org/10.1128/JVI.01275-13.

Jengarn, J., Wongthida, P., Wanasen, N., Frantz, P.N., Wanitchang, A., Jongkaewwattana, A., 2015. Genetic manipulation of porcine epidemic diarrhoea virus recovered from a full-length infectious cDNA clone. J. Gen. Virol. 96 (8), 2206–2218. http://dx.doi.org/10.1099/vir.0.000184.

Kirchdoerfer, R.N., Cottrell, C.A., Wang, N., Pallesen, J., Yassine, H.M., Turner, H.L., et al., 2016. Pre-fusion structure of a human coronavirus spike protein. Nature 531 (7592), 118–121. http://dx.doi.org/10.1038/nature17200.

Kong, Q., Xue, C., Ren, X., Zhang, C., Li, L., Shu, D., et al., 2010. Proteomic analysis of purified coronavirus infectious bronchitis virus particles. Proteome Sci. 8, 29. http://dx.doi.org/10.1186/1477-5956-8-29.

Konvalinka, J., Krausslich, H.G., Muller, B., 2015. Retroviral proteases and their roles in virion maturation. Virology 479–480, 403–417. http://dx.doi.org/10.1016/j.virol.2015.03.021.

Kuo, L., Masters, P.S., 2002. Genetic evidence for a structural interaction between the carboxy termini of the membrane and nucleocapsid proteins of mouse hepatitis virus. J. Virol. 76 (10), 4987–4999.

Kuo, L., Masters, P.S., 2003. The small envelope protein E is not essential for murine coronavirus replication. J. Virol. 77 (8), 4597–4608.

Kuo, L., Masters, P.S., 2010. Evolved variants of the membrane protein can partially replace the envelope protein in murine coronavirus assembly. J. Virol. 84 (24), 12872–12885. http://dx.doi.org/10.1128/JVI.01850-10.

Kuo, L., Godeke, G.J., Raamsman, M.J., Masters, P.S., Rottier, P.J., 2000. Retargeting of coronavirus by substitution of the spike glycoprotein ectodomain: crossing the host cell species barrier. J. Virol. 74 (3), 1393–1406.

Kuo, L., Koetzner, C.A., Hurst, K.R., Masters, P.S., 2014. Recognition of the murine coronavirus genomic RNA packaging signal depends on the second RNA-binding domain of the nucleocapsid protein. J. Virol. 88 (8), 4451–4465. http://dx.doi.org/10.1128/JVI.03866-13.

Kuo, L., Hurst-Hess, K.R., Koetzner, C.A., Masters, P.S., 2016. Analyses of coronavirus assembly interactions with interspecies membrane and nucleocapsid protein chimeras. J. Virol. 90, 4357–4368. http://dx.doi.org/10.1128/JVI.03212-15.

Lavillette, D., Barbouche, R., Yao, Y., Boson, B., Cosset, F.L., Jones, I.M., Fenouillet, E., 2006. Significant redox insensitivity of the functions of the SARS-CoV spike glycoprotein: comparison with HIV envelope. J. Biol. Chem. 281 (14), 9200–9204. http://dx.doi.org/10.1074/jbc.M512529200. M512529200 [pii].

Legendre, M., Bartoli, J., Shmakova, L., Jeudy, S., Labadie, K., Adrait, A., et al., 2014. Thirty-thousand-year-old distant relative of giant icosahedral DNA viruses with a pandoravirus morphology. Proc. Natl. Acad. Sci. U.S.A. 111 (11), 4274–4279. http://dx.doi.org/10.1073/pnas.1320670111.

Lewicki, D.N., Gallagher, T.M., 2002. Quaternary structure of coronavirus spikes in complex with carcinoembryonic antigen-related cell adhesion molecule cellular receptors. J. Biol. Chem. 277 (22), 19727–19734. http://dx.doi.org/10.1074/jbc.M201837200.

Li, W., Moore, M.J., Vasilieva, N., Sui, J., Wong, S.K., Berne, M.A., et al., 2003. Angiotensin-converting enzyme 2 is a functional receptor for the SARS coronavirus. Nature 426 (6965), 450–454. http://dx.doi.org/10.1038/nature02145.

Li, F., Li, W., Farzan, M., Harrison, S.C., 2005a. Structure of SARS coronavirus spike receptor-binding domain complexed with receptor. Science 309 (5742), 1864–1868. http://dx.doi.org/10.1126/science.1116480. 309/5742/1864 [pii].

Li, F.Q., Xiao, H., Tam, J.P., Liu, D.X., 2005b. Sumoylation of the nucleocapsid protein of severe acute respiratory syndrome coronavirus. FEBS Lett. 579 (11), 2387–2396.

Li, S., Eghiaian, F., Sieben, C., Herrmann, A., Schaap, I.A., 2011. Bending and puncturing the influenza lipid envelope. Biophys. J. 100 (3), 637–645. http://dx.doi.org/10.1016/j.bpj.2010.12.3701.

Liao, Y., Lescar, J., Tam, J.P., Liu, D.X., 2004. Expression of SARS-coronavirus envelope protein in Escherichia coli cells alters membrane permeability. Biochem. Biophys. Res. Commun. 325 (1), 374–380. http://dx.doi.org/10.1016/j.bbrc.2004.10.050. S0006-291X(04)02295-8 [pii].

Liao, Y., Yuan, Q., Torres, J., Tam, J.P., Liu, D.X., 2006. Biochemical and functional characterization of the membrane association and membrane permeabilizing activity of the severe acute respiratory syndrome coronavirus envelope protein. Virology 349 (2), 264–275. http://dx.doi.org/10.1016/j.virol.2006.01.028. S0042-6822(06)00056-0 [pii].

Lim, K.P., Liu, D.X., 2001. The missing link in coronavirus assembly. Retention of the avian coronavirus infectious bronchitis virus envelope protein in the pre-Golgi compartments and physical interaction between the envelope and membrane proteins. J. Biol. Chem. 276 (20), 17515–17523. http://dx.doi.org/10.1074/jbc.M009731200.

Liu, D.X., Inglis, S.C., 1991. Association of the infectious bronchitis virus 3c protein with the virion envelope. Virology 185 (2), 911–917.

Liu, S., Xiao, G., Chen, Y., He, Y., Niu, J., Escalante, C.R., et al., 2004. Interaction between heptad repeat 1 and 2 regions in spike protein of SARS-associated coronavirus: implications for virus fusogenic mechanism and identification of fusion inhibitors. Lancet 363 (9413), 938–947. http://dx.doi.org/10.1016/S0140-6736(04)15788-7.

Liu, D.X., Yuan, Q., Liao, Y., 2007. Coronavirus envelope protein: a small membrane protein with multiple functions. Cell. Mol. Life Sci. 64 (16), 2043–2048. http://dx.doi.org/10.1007/s00018-007-7103-1.

Liu, Y., Li, X., Liu, M., Cao, B., Tan, H., Wang, J., Li, X., 2012. Responses of three different ecotypes of reed (Phragmites communis Trin.) to their natural habitats: leaf surface micro-morphology, anatomy, chloroplast ultrastructure and physio-chemical characteristics. Plant Physiol. Biochem. 51, 159–167. http://dx.doi.org/10.1016/j.plaphy.2011.11.002.

Liu, G., Lv, L., Yin, L., Li, X., Luo, D., Liu, K., et al., 2013. Assembly and immunogenicity of coronavirus-like particles carrying infectious bronchitis virus M and S proteins. Vaccine 31 (47), 5524–5530. http://dx.doi.org/10.1016/j.vaccine.2013.09.024.

Lu, W., Zheng, B.J., Xu, K., Schwarz, W., Du, L., Wong, C.K., et al., 2006. Severe acute respiratory syndrome-associated coronavirus 3a protein forms an ion channel and modulates virus release. Proc. Natl. Acad. Sci. U.S.A. 103 (33), 12540–12545. http://dx.doi.org/10.1073/pnas.0605402103. 0605402103 [pii].

Lu, Z., Zhang, J., Huang, C.M., Go, Y.Y., Faaberg, K.S., Rowland, R.R., et al., 2012. Chimeric viruses containing the N-terminal ectodomains of GP5 and M proteins of porcine reproductive and respiratory syndrome virus do not change the cellular tropism of equine arteritis virus. Virology 432 (1), 99–109. http://dx.doi.org/10.1016/j.virol.2012.05.022.

Luo, C., Luo, H., Zheng, S., Gui, C., Yue, L., Yu, C., et al., 2004. Nucleocapsid protein of SARS coronavirus tightly binds to human cyclophilin A. Biochem. Biophys. Res. Commun. 321 (3), 557–565. http://dx.doi.org/10.1016/j.bbrc.2004.07.003.

Ma, Y., Tong, X., Xu, X., Li, X., Lou, Z., Rao, Z., 2010. Structures of the N- and C-terminal domains of MHV-A59 nucleocapsid protein corroborate a conserved RNA-protein binding mechanism in coronavirus. Protein Cell 1 (7), 688–697. http://dx.doi.org/10.1007/s13238-010-0079-x.

Machamer, C.E., Youn, S., 2006. The transmembrane domain of the infectious bronchitis virus E protein is required for efficient virus release. Adv. Exp. Med. Biol. 581, 193–198. http://dx.doi.org/10.1007/978-0-387-33012-9_33.

Madan, V., Garcia Mde, J., Sanz, M.A., Carrasco, L., 2005. Viroporin activity of murine hepatitis virus E protein. FEBS Lett. 579 (17), 3607–3612. http://dx.doi.org/10.1016/j.febslet.2005.05.046. S0014-5793(05)00654-X [pii].

Masters, P.S., Kuo, L., Ye, R., Hurst, K.R., Koetzner, C.A., Hsue, B., 2006. Genetic and molecular biological analysis of protein-protein interactions in coronavirus assembly. Adv. Exp. Med. Biol. 581, 163–173.

McBride, C.E., Machamer, C.E., 2010. Palmitoylation of SARS-CoV S protein is necessary for partitioning into detergent-resistant membranes and cell-cell fusion but not interaction with M protein. Virology 405 (1), 139–148. http://dx.doi.org/10.1016/j.virol.2010.05.031.

Molenkamp, R., Spaan, W.J., 1997. Identification of a specific interaction between the coronavirus mouse hepatitis virus A59 nucleocapsid protein and packaging signal. Virology 239 (1), 78–86. http://dx.doi.org/10.1006/viro.1997.8867.

Narayanan, K., Maeda, A., Maeda, J., Makino, S., 2000. Characterization of the coronavirus M protein and nucleocapsid interaction in infected cells. J. Virol. 74 (17), 8127–8134.

Narayanan, K., Chen, C.J., Maeda, J., Makino, S., 2003. Nucleocapsid-independent specific viral RNA packaging via viral envelope protein and viral RNA signal. J. Virol. 77 (5), 2922–2927.

Neuman, B.W., Adair, B.D., Yoshioka, C., Quispe, J.D., Orca, G., Kuhn, P., et al., 2006. Supramolecular architecture of severe acute respiratory syndrome coronavirus revealed by electron cryomicroscopy. J. Virol. 80 (16), 7918–7928. http://dx.doi.org/10.1128/JVI.00645-06.

Neuman, B.W., Joseph, J.S., Saikatendu, K.S., Serrano, P., Chatterjee, A., Johnson, M.A., et al., 2008. Proteomics analysis unravels the functional repertoire of coronavirus nonstructural protein 3. J. Virol. 82 (11), 5279–5294. http://dx.doi.org/10.1128/JVI.02631-07.

Neuman, B.W., Kiss, G., Kunding, A.H., Bhella, D., Baksh, M.F., Connelly, S., et al., 2011. A structural analysis of M protein in coronavirus assembly and morphology. J. Struct. Biol. 174 (1), 11–22. http://dx.doi.org/10.1016/j.jsb.2010.11.021.

Ng, M.L., Lee, J.W., Leong, M.L., Ling, A.E., Tan, H.C., Ooi, E.E., 2004. Topographic changes in SARS coronavirus-infected cells at late stages of infection. Emerg. Infect. Dis. 10 (11), 1907–1914. http://dx.doi.org/10.3201/eid1011.040195.

Nga, P.T., Parquet Mdel, C., Lauber, C., Parida, M., Nabeshima, T., Yu, F., et al., 2011. Discovery of the first insect nidovirus, a missing evolutionary link in the emergence of the largest RNA virus genomes. PLoS Pathog. 7 (9), e1002215. http://dx.doi.org/10.1371/journal.ppat.1002215.

Nieto-Torres, J.L., Dediego, M.L., Alvarez, E., Jimenez-Guardeno, J.M., Regla-Nava, J.A., Llorente, M., et al., 2011. Subcellular location and topology of severe acute respiratory syndrome coronavirus envelope protein. Virology 415 (2), 69–82. http://dx.doi.org/10.1016/j.virol.2011.03.029.

Nieto-Torres, J.L., Verdia-Baguena, C., Jimenez-Guardeno, J.M., Regla-Nava, J.A., Castano-Rodriguez, C., Fernandez-Delgado, R., et al., 2015. Severe acute respiratory syndrome coronavirus E protein transports calcium ions and activates the NLRP3 inflammasome. Virology 485, 330–339. http://dx.doi.org/10.1016/j.virol.2015.08.010.

Nogales, A., Marquez-Jurado, S., Galan, C., Enjuanes, L., Almazan, F., 2012. Transmissible gastroenteritis coronavirus RNA-dependent RNA polymerase and nonstructural proteins 2, 3, and 8 are incorporated into viral particles. J. Virol. 86 (2), 1261–1266. http://dx.doi.org/10.1128/JVI.06428-11.

Oostra, M., de Haan, C.A., de Groot, R.J., Rottier, P.J., 2006. Glycosylation of the severe acute respiratory syndrome coronavirus triple-spanning membrane proteins 3a and M. J. Virol. 80 (5), 2326–2336. http://dx.doi.org/10.1128/JVI.80.5.2326-2336.2006. 80/5/2326 [pii].

Ortego, J., Ceriani, J.E., Patino, C., Plana, J., Enjuanes, L., 2007. Absence of E protein arrests transmissible gastroenteritis coronavirus maturation in the secretory pathway. Virology 368 (2), 296–308. http://dx.doi.org/10.1016/j.virol.2007.05.032.

Pan, J., Peng, X., Gao, Y., Li, Z., Lu, X., Chen, Y., et al., 2008. Genome-wide analysis of protein-protein interactions and involvement of viral proteins in SARS-CoV replication. PLoS One 3 (10), e3299. http://dx.doi.org/10.1371/journal.pone.0003299.

Peng, G., Sun, D., Rajashankar, K.R., Qian, Z., Holmes, K.V., Li, F., 2011. Crystal structure of mouse coronavirus receptor-binding domain complexed with its murine receptor. Proc. Natl. Acad. Sci. U.S.A. 108 (26), 10696–10701. http://dx.doi.org/10.1073/pnas.1104306108.

Pervushin, K., Tan, E., Parthasarathy, K., Lin, X., Jiang, F.L., Yu, D., et al., 2009. Structure and inhibition of the SARS coronavirus envelope protein ion channel. PLoS Pathog. 5 (7), e1000511. http://dx.doi.org/10.1371/journal.ppat.1000511.

Pfefferle, S., Schopf, J., Kogl, M., Friedel, C.C., Muller, M.A., Carbajo-Lozoya, J., et al., 2011. The SARS-coronavirus-host interactome: identification of cyclophilins as target for pan-coronavirus inhibitors. PLoS Pathog. 7 (10), e1002331. http://dx.doi.org/10.1371/journal.ppat.1002331.

Pyrc, K., Jebbink, M.F., Berkhout, B., van der Hoek, L., 2004. Genome structure and transcriptional regulation of human coronavirus NL63. Virol. J. 1, 7. http://dx.doi.org/10.1186/1743-422X-1-7.

Regla-Nava, J.A., Nieto-Torres, J.L., Jimenez-Guardeno, J.M., Fernandez-Delgado, R., Fett, C., Castano-Rodriguez, C., et al., 2015. Severe acute respiratory syndrome coronaviruses with mutations in the E protein are attenuated and promising vaccine candidates. J. Virol. 89 (7), 3870–3887. http://dx.doi.org/10.1128/JVI.03566-14.

Reguera, J., Santiago, C., Mudgal, G., Ordono, D., Enjuanes, L., Casasnovas, J.M., 2012. Structural bases of coronavirus attachment to host aminopeptidase N and its inhibition by neutralizing antibodies. PLoS Pathog. 8 (8), e1002859. http://dx.doi.org/10.1371/journal.ppat.1002859.

Risco, C., Anton, I.M., Enjuanes, L., Carrascosa, J.L., 1996. The transmissible gastroenteritis coronavirus contains a spherical core shell consisting of M and N proteins. J. Virol. 70 (7), 4773–4777.

Rossman, J.S., Jing, X., Leser, G.P., Lamb, R.A., 2010. Influenza virus M2 protein mediates ESCRT-independent membrane scission. Cell 142 (6), 902–913. http://dx.doi.org/10.1016/j.cell.2010.08.029.

Ruch, T.R., Machamer, C.E., 2011. The hydrophobic domain of infectious bronchitis virus E protein alters the host secretory pathway and is important for release of infectious virus. J. Virol. 85 (2), 675–685. http://dx.doi.org/10.1128/JVI.01570-10.

Ruch, T.R., Machamer, C.E., 2012. A single polar residue and distinct membrane topologies impact the function of the infectious bronchitis coronavirus E protein. PLoS Pathog. 8 (5), e1002674. http://dx.doi.org/10.1371/journal.ppat.1002674.

Saikatendu, K.S., Joseph, J.S., Subramanian, V., Neuman, B.W., Buchmeier, M.J., Stevens, R.C., Kuhn, P., 2007. Ribonucleocapsid formation of severe acute respiratory syndrome coronavirus through molecular action of the N-terminal domain of N protein. J. Virol. 81 (8), 3913–3921. http://dx.doi.org/10.1128/JVI.02236-06.

Sainz Jr., B., Rausch, J.M., Gallaher, W.R., Garry, R.F., Wimley, W.C., 2005. Identification and characterization of the putative fusion peptide of the severe acute respiratory syndrome-associated coronavirus spike protein. J. Virol. 79 (11), 7195–7206. http://dx.doi.org/10.1128/JVI.79.11.7195-7206.2005. 79/11/7195 [pii].

Scheele, U., Erdmann, S., Ungewickell, E.J., Felisberto-Rodrigues, C., Ortiz-Lombardia, M., Garrett, R.A., 2011. Chaperone role for proteins p618 and p892 in the extracellular tail development of Acidianus two-tailed virus. J. Virol. 85 (10), 4812–4821. http://dx.doi.org/10.1128/JVI.00072-11.

Schelle, B., Karl, N., Ludewig, B., Siddell, S.G., Thiel, V., 2005. Selective replication of coronavirus genomes that express nucleocapsid protein. J. Virol. 79 (11), 6620–6630. http://dx.doi.org/10.1128/JVI.79.11.6620-6630.2005.

Schutze, H., Ulferts, R., Schelle, B., Bayer, S., Granzow, H., Hoffmann, B., et al., 2006. Characterization of White bream virus reveals a novel genetic cluster of nidoviruses. J. Virol. 80 (23), 11598–11609. http://dx.doi.org/10.1128/JVI.01758-06.

Scobey, T., Yount, B.L., Sims, A.C., Donaldson, E.F., Agnihothram, S.S., Menachery, V.D., et al., 2013. Reverse genetics with a full-length infectious cDNA of the Middle East respiratory syndrome coronavirus. Proc. Natl. Acad. Sci. U.S.A. 110 (40), 16157–16162. http://dx.doi.org/10.1073/pnas.1311542110.

Sharma, K., Surjit, M., Satija, N., Liu, B., Chow, V.T., Lal, S.K., 2007. The 3a accessory protein of SARS coronavirus specifically interacts with the 5'UTR of its genomic RNA, using a unique 75 amino acid interaction domain. Biochemistry 46 (22), 6488–6499. http://dx.doi.org/10.1021/bi062057p.

Shulla, A., Gallagher, T., 2009. Role of spike protein endodomains in regulating coronavirus entry. J. Biol. Chem. 284 (47), 32725–32734. http://dx.doi.org/10.1074/jbc.M109.043547.

Shulla, A., Heald-Sargent, T., Subramanya, G., Zhao, J., Perlman, S., Gallagher, T., 2011. A transmembrane serine protease is linked to the severe acute respiratory syndrome coronavirus receptor and activates virus entry. J. Virol. 85 (2), 873–882. http://dx.doi.org/10.1128/JVI.02062-10.

Siddell, S.G., Barthel, A., ter Meulen, V., 1981. Coronavirus JHM: a virion-associated protein kinase. J. Gen. Virol. 52 (Pt. 2), 235–243. http://dx.doi.org/10.1099/0022-1317-52-2-235.

Simmons, G., Gosalia, D.N., Rennekamp, A.J., Reeves, J.D., Diamond, S.L., Bates, P., 2005. Inhibitors of cathepsin L prevent severe acute respiratory syndrome coronavirus entry. Proc. Natl. Acad. Sci. U.S.A. 102 (33), 11876–11881. http://dx.doi.org/10.1073/pnas.0505577102.

Siu, Y.L., Teoh, K.T., Lo, J., Chan, C.M., Kien, F., Escriou, N., et al., 2008. The M, E, and N structural proteins of the severe acute respiratory syndrome coronavirus are required for efficient assembly, trafficking, and release of virus-like particles. J. Virol. 82 (22), 11318–11330. http://dx.doi.org/10.1128/JVI.01052-08.

Snijder, E.J., Dobbe, J.C., Spaan, W.J., 2003. Heterodimerization of the two major envelope proteins is essential for arterivirus infectivity. J. Virol. 77 (1), 97–104.

Sobsey, M.D., Meschke, J.S., 2003. WHO Virus Survival Report: Survival in the Environment with Special Attention to Survival in Sewage Droplets and Other Media of Fecal or Respiratory Origin. WHO, Geneva, Switzerland. https://http://www.unc.edu/courses/2008spring/envr/421/001/WHO_VirusSurvivalReport_21Aug2003.pdf.

Spann, K.M., Cowley, J.A., Walker, P.J., Lester, R.J.G., 1997. A yellow-head-like virus from Penaeus monodon cultured in Australia. Dis. Aquat. Org. 31, 169–179.

Spilman, M.S., Welbon, C., Nelson, E., Dokland, T., 2009. Cryo-electron tomography of porcine reproductive and respiratory syndrome virus: organization of the nucleocapsid. J. Gen. Virol. 90 (Pt. 3), 527–535. http://dx.doi.org/10.1099/vir.0.007674-0.

Stadler, K., Masignani, V., Eickmann, M., Becker, S., Abrignani, S., Klenk, H.D., Rappuoli, R., 2003. SARS—beginning to understand a new virus. Nat. Rev. Microbiol. 1 (3), 209–218.

Stenglein, M.D., Jacobson, E.R., Wozniak, E.J., Wellehan, J.F., Kincaid, A., Gordon, M., et al., 2014. Ball python nidovirus: a candidate etiologic agent for severe respiratory disease in Python regius. MBio 5 (5), e01484–e01514. http://dx.doi.org/10.1128/mBio.01484-14.

Sturman, L.S., Holmes, K.V., Behnke, J., 1980. Isolation of coronavirus envelope glycoproteins and interaction with the viral nucleocapsid. J. Virol. 33 (1), 449–462.

Surjit, M., Liu, B., Kumar, P., Chow, V.T., Lal, S.K., 2004. The nucleocapsid protein of the SARS coronavirus is capable of self-association through a C-terminal 209 amino acid interaction domain. Biochem. Biophys. Res. Commun. 317 (4), 1030–1036. http://dx.doi.org/10.1016/j.bbrc.2004.03.154.

Surjit, M., Kumar, R., Mishra, R.N., Reddy, M.K., Chow, V.T., Lal, S.K., 2005. The severe acute respiratory syndrome coronavirus nucleocapsid protein is phosphorylated and localizes in the cytoplasm by 14-3-3-mediated translocation. J. Virol. 79 (17), 11476–11486.

Tekes, G., Hofmann-Lehmann, R., Stallkamp, I., Thiel, V., Thiel, H.J., 2008. Genome organization and reverse genetic analysis of a type I feline coronavirus. J. Virol. 82 (4), 1851–1859. http://dx.doi.org/10.1128/JVI.02339-07.

Teoh, K.T., Siu, Y.L., Chan, W.L., Schluter, M.A., Liu, C.J., Peiris, J.S., et al., 2010. The SARS coronavirus E protein interacts with PALS1 and alters tight junction formation and epithelial morphogenesis. Mol. Biol. Cell 21 (22), 3838–3852. http://dx.doi.org/10.1091/mbc.E10-04-0338.

Thiel, V., Herold, J., Schelle, B., Siddell, S.G., 2001. Infectious RNA transcribed in vitro from a cDNA copy of the human coronavirus genome cloned in vaccinia virus. J. Gen. Virol. 82 (Pt. 6), 1273–1281. http://dx.doi.org/10.1099/0022-1317-82-6-1273.

Thorp, E.B., Boscarino, J.A., Logan, H.L., Goletz, J.T., Gallagher, T.M., 2006. Palmitoylations on murine coronavirus spike proteins are essential for virion assembly and infectivity. J. Virol. 80 (3), 1280–1289. http://dx.doi.org/10.1128/JVI.80.3.1280-1289.2006. 80/3/1280 [pii].

Tian, D., Wei, Z., Zevenhoven-Dobbe, J.C., Liu, R., Tong, G., Snijder, E.J., Yuan, S., 2012. Arterivirus minor envelope proteins are a major determinant of viral tropism in cell culture. J. Virol. 86 (7), 3701–3712. http://dx.doi.org/10.1128/JVI.06836-11.

Towler, P., Staker, B., Prasad, S.G., Menon, S., Tang, J., Parsons, T., et al., 2004. ACE2 X-ray structures reveal a large hinge-bending motion important for inhibitor binding and catalysis. J. Biol. Chem. 279 (17), 17996–18007. http://dx.doi.org/10.1074/jbc.M311191200. M311191200 [pii].

Tripet, B., Howard, M.W., Jobling, M., Holmes, R.K., Holmes, K.V., Hodges, R.S., 2004. Structural characterization of the SARS-coronavirus spike S fusion protein core. J. Biol. Chem. 279 (20), 20836–20849. http://dx.doi.org/10.1074/jbc.M400759200. M400759200 [pii].

Tseng, Y.T., Wang, S.M., Huang, K.J., Lee, A.I., Chiang, C.C., Wang, C.T., 2010. Self-assembly of severe acute respiratory syndrome coronavirus membrane protein. J. Biol. Chem. 285 (17), 12862–12872. http://dx.doi.org/10.1074/jbc.M109.030270.

van Doremalen, N., Bushmaker, T., Munster, V.J., 2013. Stability of Middle East respiratory syndrome coronavirus (MERS-CoV) under different environmental conditions. Euro Surveill. 18 (38), pii: 20590.

Vennema, H., Godeke, G.J., Rossen, J.W., Voorhout, W.F., Horzinek, M.C., Opstelten, D.J., Rottier, P.J., 1996. Nucleocapsid-independent assembly of coronavirus-like particles by co-expression of viral envelope protein genes. EMBO J. 15 (8), 2020–2028.

Verheije, M.H., Hagemeijer, M.C., Ulasli, M., Reggiori, F., Rottier, P.J., Masters, P.S., de Haan, C.A., 2010. The coronavirus nucleocapsid protein is dynamically associated with the replication-transcription complexes. J. Virol. 84 (21), 11575–11579. http://dx.doi.org/10.1128/JVI.00569-10.

von Heijne, G., 1984. Analysis of the distribution of charged residues in the N-terminal region of signal sequences: implications for protein export in prokaryotic and eukaryotic cells. EMBO J. 3 (10), 2315–2318.

Voss, N.R., Gerstein, M., 2005. Calculation of standard atomic volumes for RNA and comparison with proteins: RNA is packed more tightly. J. Mol. Biol. 346 (2), 477–492. http://dx.doi.org/10.1016/j.jmb.2004.11.072.

Walls, A.C., Tortorici, M.A., Bosch, B.J., Frenz, B., Rottier, P.J., DiMaio, F., et al., 2016. Cryo-electron microscopy structure of a coronavirus spike glycoprotein trimer. Nature 531 (7592), 114–117. http://dx.doi.org/10.1038/nature16988.

Wang, S.M., Wang, C.T., 2009. APOBEC3G cytidine deaminase association with coronavirus nucleocapsid protein. Virology 388 (1), 112–120. http://dx.doi.org/10.1016/j.virol.2009.03.010.

Wang, N., Shi, X., Jiang, L., Zhang, S., Wang, D., Tong, P., et al., 2013. Structure of MERS-CoV spike receptor-binding domain complexed with human receptor DPP4. Cell Res. 23 (8), 986–993. http://dx.doi.org/10.1038/cr.2013.92.

Westerbeck, J.W., Machamer, C.E., 2015. A coronavirus E protein is present in two distinct pools with different effects on assembly and the secretory pathway. J. Virol. 89 (18), 9313–9323. http://dx.doi.org/10.1128/JVI.01237-15.

White, T.C., Yi, Z., Hogue, B.G., 2007. Identification of mouse hepatitis coronavirus A59 nucleocapsid protein phosphorylation sites. Virus Res. 126 (1–2), 139–148. http://dx.doi.org/10.1016/j.virusres.2007.02.008.

Wilson, L., McKinlay, C., Gage, P., Ewart, G., 2004. SARS coronavirus E protein forms cation-selective ion channels. Virology 330 (1), 322–331. http://dx.doi.org/10.1016/j.virol.2004.09.033. S0042-6822(04)00644-0 [pii].

Wilson, L., Gage, P., Ewart, G., 2006. Hexamethylene amiloride blocks E protein ion channels and inhibits coronavirus replication. Virology 353 (2), 294–306. http://dx.doi.org/10.1016/j.virol.2006.05.028. S0042-6822(06)00359-X [pii].

Wong, S.K., Li, W., Moore, M.J., Choe, H., Farzan, M., 2004. A 193-amino acid fragment of the SARS coronavirus S protein efficiently binds angiotensin-converting enzyme 2. J. Biol. Chem. 279 (5), 3197–3201. http://dx.doi.org/10.1074/jbc.C300520200. C300520200 [pii].

Wong, H.H., Kumar, P., Tay, F.P., Moreau, D., Liu, D.X., Bard, F., 2015. Genome-wide-screen reveals valosin-containing protein requirement for coronavirus exit from endosomes. J. Virol. 89 (21), 11116–11128. http://dx.doi.org/10.1128/JVI.01360-15.

Woo, P.C., Lau, S.K., Lam, C.S., Tsang, A.K., Hui, S.W., Fan, R.Y., et al., 2014. Discovery of a novel bottlenose dolphin coronavirus reveals a distinct species of marine mammal coronavirus in Gammacoronavirus. J. Virol. 88 (2), 1318–1331. http://dx.doi.org/10.1128/JVI.02351-13.

Wrensch, F., Winkler, M., Pohlmann, S., 2014. IFITM proteins inhibit entry driven by the MERS-coronavirus spike protein: evidence for cholesterol-independent mechanisms. Viruses 6 (9), 3683–3698. http://dx.doi.org/10.3390/v6093683.

Wu, K., Li, W., Peng, G., Li, F., 2009. Crystal structure of NL63 respiratory coronavirus receptor-binding domain complexed with its human receptor. Proc. Natl. Acad. Sci. U.S.A. 106 (47), 19970–19974. http://dx.doi.org/10.1073/pnas.0908837106.

Xiao, X., Chakraborti, S., Dimitrov, A.S., Gramatikoff, K., Dimitrov, D.S., 2003. The SARS-CoV S glycoprotein: expression and functional characterization. Biochem. Biophys. Res. Commun. 312 (4), 1159–1164. S0006291X03024136 [pii].

Xiao, X., Feng, Y., Chakraborti, S., Dimitrov, D.S., 2004. Oligomerization of the SARS-CoV S glycoprotein: dimerization of the N-terminus and trimerization of the ectodomain. Biochem. Biophys. Res. Commun. 322 (1), 93–99. http://dx.doi.org/10.1016/j.bbrc.2004.07.084. S0006-291X(04)01554-2 [pii].

Xu, Y., Liu, Y., Lou, Z., Qin, L., Li, X., Bai, Z., et al., 2004a. Structural basis for coronavirus-mediated membrane fusion. Crystal structure of mouse hepatitis virus spike protein fusion core. J. Biol. Chem. 279 (29), 30514–30522. http://dx.doi.org/10.1074/jbc.M403760200. M403760200 [pii].

Xu, Y., Lou, Z., Liu, Y., Pang, H., Tien, P., Gao, G.F., Rao, Z., 2004b. Crystal structure of severe acute respiratory syndrome coronavirus spike protein fusion core. J. Biol. Chem. 279 (47), 49414–49419. http://dx.doi.org/10.1074/jbc.M408782200. M408782200 [pii].

Ye, Y., Hogue, B.G., 2007. Role of the coronavirus E viroporin protein transmembrane domain in virus assembly. J. Virol. 81 (7), 3597–3607. http://dx.doi.org/10.1128/JVI.01472-06. JVI.01472-06 [pii].

Yount, B., Curtis, K.M., Baric, R.S., 2000. Strategy for systematic assembly of large RNA and DNA genomes: transmissible gastroenteritis virus model. J. Virol. 74 (22), 10600–10611.

Yount, B., Curtis, K.M., Fritz, E.A., Hensley, L.E., Jahrling, P.B., Prentice, E., et al., 2003. Reverse genetics with a full-length infectious cDNA of severe acute respiratory syndrome coronavirus. Proc. Natl. Acad. Sci. U.S.A. 100 (22), 12995–13000. http://dx.doi.org/10.1073/pnas.1735582100.

Yount, B., Roberts, R.S., Sims, A.C., Deming, D., Frieman, M.B., Sparks, J., et al., 2005. Severe acute respiratory syndrome coronavirus group-specific open reading frames encode nonessential functions for replication in cell cultures and mice. J. Virol. 79 (23), 14909–14922. http://dx.doi.org/10.1128/JVI.79.23.14909-14922.2005. 79/23/14909 [pii].

Yu, I.M., Gustafson, C.L., Diao, J., Burgner 2nd, J.W., Li, Z., Zhang, J., Chen, J., 2005. Recombinant severe acute respiratory syndrome (SARS) coronavirus nucleocapsid protein forms a dimer through its C-terminal domain. J. Biol. Chem. 280 (24), 23280–23286. http://dx.doi.org/10.1074/jbc.M501015200.

Yu, I.M., Oldham, M.L., Zhang, J., Chen, J., 2006. Crystal structure of the severe acute respiratory syndrome (SARS) coronavirus nucleocapsid protein dimerization domain reveals evolutionary linkage between corona- and arteriviridae. J. Biol. Chem. 281 (25), 17134–17139. http://dx.doi.org/10.1074/jbc.M602107200.

Zhang, R., Wang, K., Lv, W., Yu, W., Xie, S., Xu, K., et al., 2014. The ORF4a protein of human coronavirus 229E functions as a viroporin that regulates viral production. Biochim. Biophys. Acta 1838 (4), 1088–1095. http://dx.doi.org/10.1016/j.bbamem.2013.07.025.

Zhao, X., Guo, F., Liu, F., Cuconati, A., Chang, J., Block, T.M., Guo, J.T., 2014. Interferon induction of IFITM proteins promotes infection by human coronavirus OC43. Proc. Natl. Acad. Sci. U.S.A. 111 (18), 6756–6761. http://dx.doi.org/10.1073/pnas.1320856111.

Zhou, H., Ferraro, D., Zhao, J., Hussain, S., Shao, J., Trujillo, J., et al., 2010. The N-terminal region of severe acute respiratory syndrome coronavirus protein 6 induces membrane rearrangement and enhances virus replication. J. Virol. 84 (7), 3542–3551. http://dx.doi.org/10.1128/JVI.02570-09.

Zuniga, S., Sola, I., Moreno, J.L., Sabella, P., Plana-Duran, J., Enjuanes, L., 2007. Coronavirus nucleocapsid protein is an RNA chaperone. Virology 357 (2), 215–227. http://dx.doi.org/10.1016/j.virol.2006.07.046.

CHAPTER TWO

Coronavirus Spike Protein and Tropism Changes

R.J.G. Hulswit, C.A.M. de Haan[1], B.-J. Bosch[1]
Faculty of Veterinary Medicine, Utrecht University, Utrecht, The Netherlands
[1]Corresponding authors: e-mail address: c.a.m.dehaan@uu.nl; b.j.bosch@uu.nl

Contents

Abstract

Coronaviruses (CoVs) have a remarkable potential to change tropism. This is particularly illustrated over the last 15 years by the emergence of two zoonotic CoVs, the severe acute respiratory syndrome (SARS)- and Middle East respiratory syndrome (MERS)-CoV. Due to their inherent genetic variability, it is inevitable that new cross-species transmission events of these enveloped, positive-stranded RNA viruses will occur. Research into these medical and veterinary important pathogens—sparked by the SARS and MERS outbreaks—revealed important principles of inter- and intraspecies tropism changes. The primary determinant of CoV tropism is the viral spike (S) entry protein. Trimers of the S glycoproteins on the virion surface accommodate binding to a cell surface receptor and fusion of the viral and cellular membrane. Recently, high-resolution structures of two CoV S proteins have been elucidated by single-particle cryo-electron microscopy. Using this new structural insight, we review the changes in the S protein that relate to changes in virus tropism. Different concepts underlie these tropism changes at the cellular, tissue, and host species level, including the promiscuity or

Advances in Virus Research, Volume 96
ISSN 0065-3527
http://dx.doi.org/10.1016/bs.aivir.2016.08.004

29

adaptability of S proteins to orthologous receptors, alterations in the proteolytic cleavage activation as well as changes in the S protein metastability. A thorough understanding of the key role of the S protein in CoV entry is critical to further our understanding of virus cross-species transmission and pathogenesis and for development of intervention strategies.

1. INTRODUCTION

Coronaviruses (CoVs) (order *Nidovirales*, family *Coronaviridae*, subfamily *Coronavirinae*) are enveloped, positive-sense RNA viruses that contain the largest known RNA genomes with a length of up to 32 kb. The subfamily *Coronavirinae*, which contains viruses of both medical and veterinary importance, can be divided into the four genera *alpha-*, *beta-*, *gamma-* and *deltacoronavirus* (*α-*, *β-*, *γ-* and *δ-CoV*). The coronavirus particle comprises at least the four canonical structural proteins E (envelope protein), M (membrane protein), N (nucleocapsid protein), and S (spike protein). In addition, viruses belonging to lineage A of the *betacoronaviruses* express the membrane-anchored HE (hemagglutinin–esterase) protein. The S glycoprotein contains both the receptor-binding domain (RBD) and the domains involved in fusion, rendering it the pivotal protein in the CoV entry process.

Coronaviruses primarily infect the respiratory and gastrointestinal tract of a wide range of animal species including many mammals and birds. Although individual virus species mostly appear to be restricted to a narrow host range comprising a single animal species, genome sequencing and phylogenetic analyses testify that CoVs have crossed the host species barrier frequently (Chan et al., 2013; Woo et al., 2012). In fact most if not all human coronaviruses seem to originate from bat CoVs (BtCoVs) that transmitted to humans directly or indirectly through an intermediate host. It therefore appears inevitable that similar zoonotic infections will occur in the future.

In the past 15 years, the world witnessed two such zoonotic events. In 2002–2003 cross-species transmissions from bats and civet cats were at the base of the SARS (severe acute respiratory syndrome)-CoV epidemic that found its origin in the Chinese Guangdong province (Li et al., 2006; Song et al., 2005). The SARS-CoV nearly became a pandemic and led to over 700 deaths, before it disappeared when the appropriate hygiene and quarantine precautions were taken. In 2012, the MERS (Middle East respiratory syndrome)-CoV emerged in the human population on the Arabian

Peninsula and currently continues to make a serious impact on the local but also global health system with 1800 laboratory confirmed cases and 640 deaths as of September 1, 2016 (WHO | Middle East respiratory syndrome coronavirus (MERS-CoV) – Saudi Arabia, 2016). The natural reservoir of MERS-CoV is presumed to be in dromedary camels from which zoonotic transmissions repeatedly give rise to infections of the lower respiratory tract in humans (Alagaili et al., 2014; Azhar et al., 2014; Briese et al., 2014; Reusken et al., 2013; Widagdo et al., 2016). Besides these two novel CoVs, four other CoVs were previously identified in humans which are found in either the *alphacoronavirus* (HCoV-NL63 and HCoV-229E) or the *betacoronavirus* genera (HCoV-OC43 and HCoV-HKU1). Phylogenetic analysis has shown that the bovine CoV (BCoV) has been the origin for HCoV-OC43 following a relatively recent cross-species transmission event (Vijgen et al., 2006). Moreover, HCoV-NL63, HCoV-229E, SARS-CoV, and MERS-CoV also have been predicted to originate from bats (Annan et al., 2013; Bolles et al., 2011; Corman et al., 2015; Hu et al., 2015; Huynh et al., 2012).

In general, four major criteria determine cross-species transmission of a particular virus (Racaniello et al., 2015). The cellular tropism of a virus is determined by the susceptibility of host cells (i.e., presence of the receptor needed for entry) as well as by the permissiveness of these host cells to allow the virus to replicate and to complete its life cycle. A third determinant consists of the accessibility of susceptible and permissive cells in the host. Finally, the innate immune response may restrict viral replication in a host species-specific manner. The above-mentioned criteria may play a critical role in the success of a cross-species transmission event. However, for CoVs, it seems that host tropism and changes therein are particularly determined by the susceptibility of host cells to infection. While CoV accessory genes, including the HE proteins, are thought to play a role in host tropism and adaptation to a new host, the S glycoprotein appears to be the main determinant for the success of initial cross-species infection events. In this review, we focus on the molecular changes in the S protein that underlie tropism changes at the cellular, tissue, and host species level and put these in perspective of the recently published cryo-EM structures.

2. STRUCTURE OF THE CORONAVIRUS S PROTEIN

The CoV S protein is a class I viral fusion protein (Bosch et al., 2003) similar to the fusion proteins of influenza, retro-, filo-, and paramyxoviruses

(Baker et al., 1999; Bartesaghi et al., 2013; Lee et al., 2008; Lin et al., 2014). Like other class I viral fusion proteins, the S protein folds into a metastable prefusion conformation following translation. The size of the abundantly N-glycosylated S protein varies greatly between CoV species ranging from approximately 1100 to 1600 residues in length, with an estimated molecular mass of up to 220 kDa. Trimers of the S protein form the 18–23-nm long, club-shaped spikes that decorate the membrane surface of the CoV particle. Besides being the primary determinant in CoV host tropism and pathogenesis, the S protein is also the main target for neutralizing antibodies elicited by the immune system of the infected host (Hofmann et al., 2004).

The S protein can be divided into two functionally distinct subunits: the globular S_1 subunit is involved in receptor recognition, whereas the S_2 subunit facilitates membrane fusion and anchors S into the viral membrane (Fig. 1A). The S_1 and S_2 domains may be separated by a cleavage site that is recognized by furin-like proteases during S protein biogenesis in the infected cell. X-ray crystal structures of several S domains have furthered our understanding of the S protein in the past. In addition, recent elucidation of the high-resolution structures of the spike ectodomain of two betacoronaviruses—MHV and HCoV-HKU1—by single-particle cryo-electron microscopy (Kirchdoerfer et al., 2016; Walls et al., 2016) has provided novel insights into the architecture of the S trimer in its prefusion state (Fig. 1B and C).

2.1 Structure of the S_1 Subunit

The S_1 subunit of the betacoronavirus spike proteins displays a multidomain architecture and is structurally organized in four distinct domains A–D of which domains A and B may serve as a RBD (Fig. 1C). The core structure of domain A displays a galectin-like β-sandwich fold, whereas domain B contains a structurally conserved core subdomain of antiparallel β-sheets (Kirchdoerfer et al., 2016; Li et al., 2005a; Walls et al., 2016; Wang et al., 2013). Importantly, domain B is decorated with an extended loop on the viral membrane-distal side. This loop may differ greatly in size and structure between virus species of the betacoronavirus genus and is therefore also referred to as hypervariable region (HVR). The cryo-EM structures of the MHV-A59 and HCoV-HKU1 S trimers show an intricate interlocking of the three S_1 subunits (Fig. 1B). Oligomerization of the S protomers results in a closely clustered trimer of the individual B domains close to the three-fold axis of the spike on top of the S_2 trimer, whereas the three A domains are

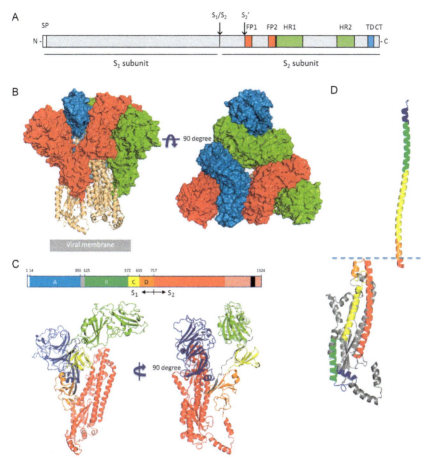

Fig. 1 Spike protein features and structure of the mouse hepatitis coronavirus spike glycoprotein trimer. (A) Schematic linear representation of the coronavirus S protein with relevant domains/sites indicated: signal peptide (SP), two proteolytic cleavage sites (S_1/S_2 and S_2'), two proposed fusion peptides (FP1 and FP2), two heptad repeat regions (HR1 and HR2), transmembrane domain (TD), and cytoplasmic tail (CT). (B) *Front* and *top view* of the trimeric mouse hepatitis coronavirus (strain A59) spike glycoprotein ectodomain obtained by cryo-electron microscopy analysis (Walls et al., 2016; PDB: 3JCL). Three S_1 protomers (surface presentation) are colored in *red*, *blue*, and *green*. The S_2 trimer (cartoon presentation) is colored in *light orange*. (C) Schematic representation of MHV spike protein sequence (drawn to scale), the S_1 domains A, B, C, and D are colored in *blue*, *green*, *yellow*, and *orange*, respectively, and the linker region connecting domains A and B in *gray*, the S_2 region is colored in *red*, and the TM region is indicated as a *black box*. *Red-shaded* region indicates spike region that was

(Continued)

ordered more distally of the center. In contrast to domains A and B, the S_1 C-terminal domains C and D are made up of discontinuous parts of the primary protein sequence and form β-sheet-rich structures directly adjacent to the S_2 stalk core, while the separate S_1 domains are interconnected by loops covering the S_2 surface. Compared to the S_2 subunit, the S_1 subunit displays low level of sequence conversation among species of different CoV genera. Moreover, S_1 subunits vary considerably in sequence length ranging from 544 (infectious bronchitis virus (IBV) S) to 944 (229-related bat coronavirus S) residues in length (Fig. 2), indicating differences in architecture of the spikes of species from different CoV genera. Structural information from the spikes of *gamma-* and *deltacoronavirus* species is currently lacking. Two independently folding domains have been assigned in the S_1 subunit of alphacoronavirus spikes, that can interact with host cell surface molecules, an N-terminal domain (in transmissible gastroenteritis virus (TGEV) S residues 1–245) and a more C-terminal domain (in TGEV S residues 506–655). Contrary to betacoronaviruses, these two receptor-interacting domains in alphacoronavirus spikes are separated in sequence by some 275 residues, which may fold into one or more separate domains. Structural information is only available for the C-terminal S_1 RBD of two α-CoV S proteins, which differs notably from that of betacoronaviruses. The RBD in the S_1 CTR of alphacoronaviruses displays a β-sandwich core structure, whereas a β-sheet core structure is seen for betacoronaviruses (Reguera et al., 2012; Wu et al., 2009).

2.2 Structure of the S_2 Subunit

The highly conserved S_2 subunit contains the key protein segments that facilitate virus–cell fusion. These include the fusion peptide, two heptad

---

Fig. 1—Cont'd not resolved in the cryo-EM structure. (*Lower panel*) Two views on the structure of the mouse hepatitis virus spike glycoprotein protomer (cartoon representation); domains are colored as depicted earlier. (D) Comparison of the S_2 HR1 region in its pre- and postfusion conformation. (*Lower left*) Structure of the MHV S_2 protomer (cartoon presentation) with four helices of the HR1 region (and consecutive linker region) and the downstream central helix colored in *blue, green, yellow, orange,* and *red,* respectively. (*Upper right*) The structure of a single SARS-CoV S HR1 helix of the postfusion six-helix bundle structure (PDB: 1WYY) is colored according to the homologous HR1 region in the MHV S_2 prefusion structure shown in the *lower left panel*. Structures are aligned based on the N-terminal segment of the central helix (in *red*). Figures were generated with PyMOL.

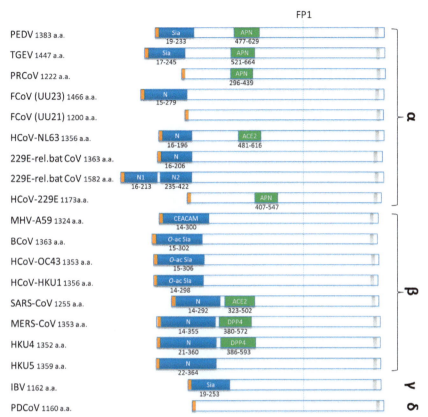

Fig. 2 Overview of currently known receptors and their binding domains within S_1. Schematic representation of coronavirus spike proteins drawn to scale. *Yellow boxes* indicate signal peptides. *Blue boxes* indicate the N-terminal regions in alpha- and betacoronavirus spike proteins, which were mapped based on sequence homology between viruses within the same genus. *Green boxes* indicate known receptor-binding domains in the C-terminal region of S_1. Known receptors are indicated in the *boxes*: *APN*, aminopeptidase N; *ACE2*, angiotensin-converting enzyme 2; *CEACAM*, carcinoembryonic antigen-related cell adhesion molecule 1; *Sia*, sialic acid; *O-ac Sia*, O-acetylated sialic acid; *DPP4*, dipeptidyl peptidase-4. *Gray boxes* indicate transmembrane domains. Spikes proteins are shown of PEDV strain CV777 (GB: AAK38656.1), TGEV strain Purdue P115 (GB: ABG89325.1), PRCoV strain ISU-1 (GB: ABG89317.1), Feline CoV strain UU23 (GB: ADC35472.1), Feline CoV strain UU21 (GB: ADL71466.1), Human CoV NL63 (GB: YP_003767.1), 229E-related bat CoV with one N domains (GB: ALK28775.1), 229E-related bat CoV with two N domains (GB: ALK28765.1), Human CoV 229E strain inf-1 (GB: NP_073551.1), MHV strain A59 (GB: ACO72893), BCoV strain KWD1 (GB: AAX38489), HCoV-OC43 strain Paris (GB: AAT84362), HCoV-HKU1 (GB: AAT98580), SARS-CoV strain Urbani (GB: AAP13441), MERS-CoV strain EMC/2012 (GB: YP_009047204), HKU4 (GB: AGP04928), HKU5 (GB: AGP04943), IBV strain Beaudette (GB: ADP06471), and PDCoV

(Continued)

repeat regions (HR1 and HR2) and the transmembrane domains which are well conserved among CoV species across different genera. In the MHV and HKU1 S prefusion structures, the S_2 domain consists of multiple α-helical segments and a three-stranded antiparallel β-sheet at the viral membrane-proximal end. A 75 Å long central helix located immediately downstream of the HR1 region stretches along the threefold axis over the entire length of the S_2 trimer. The HR1 motif itself folds as four individual α-helices along the length of the S_2 subunit, in contrast to the 120 Å long α-helix formed by this region in postfusion structures (Duquerroy et al., 2005; Gao et al., 2013; Xu et al., 2004). A 55 Å long helix upstream of the S_2' cleavage site runs parallel to and is packed against the central helix via hydrophobic interactions (Fig. 1C). The fusion peptide forms a short helix of which the strictly conserved hydrophobic residues are buried in an interface with other elements of S_2. Unlike other class I fusion proteins, this conserved fusion peptide (FP1) is not directly upstream of HR1 but located some 65 residues upstream of this region (Fig. 1A). Intriguingly, a recent published report provided experimental evidence for the existence of another fusion peptide (FP2) immediately upstream of the HR1 region (Ou et al., 2016), that had been predicted earlier based on the position, hydrophobicity profile and amino acid composition canonical for class I viral fusion peptides (Bosch and Rottier, 2008; Bosch et al., 2004; Chambers et al., 1990). The HR2 region locates closely to the C-terminal end of the S ectodomain, but it appeared to be disordered in both cryo-EM structures and therefore its prefusion conformation remains unknown.

The metastable prefusion conformation of S_2 is locked by the cap formed by the intertwined S_1 protomers. The distal tip of the S_2 trimer connects via hydrophobic interactions with domains B. This distal tip of the S_2 trimer consists of the C-terminal region of HR1 in the prefusion conformation, while the entire HR1 rearranges to form a central three-helix coiled coil in the postfusion structure (Duquerroy et al., 2005; Lu et al., 2014; Supekar et al., 2004). Interactions between this region of the S_2 trimer and domain B may therefore prevent premature conformational changes resulting in the conversion of the prefusion S protein into the very stable

Fig. 2—Cont'd strain USA/Ohio137/2014 (GB: AIB07807). PSI-BLAST analysis using the NTR of the HCoV-NL63 S protein (residues 16–196) as a query detected two homologous regions in the first 425 residues of the 229E-related bat coronavirus spike protein (GB: ALK28765.1)—designated N1 (residues 32–213) and N2 (residues 246–422) with 32% and 35% amino acid sequence identity, respectively, suggesting a duplication of the NTR. Spike proteins are drawn to scale and aligned at the position of the conserved fusion peptide (FP1).

postfusion structure. Also domains C and D of the betacoronavirus S_1 subunit and the linker region connecting domain A and B interact with the surface of the adjacent S_2 protomer and may hence play a role in stabilizing the prefusion S_2 trimer. Domain A appears to play a minor role in this respect in view of its relatively small a surface area that interacts with the S_2 trimer.

3. SPIKE–RECEPTOR INTERACTIONS

3.1 Different Domains Within S_1 May Act as RBD

Over the past decades, molecular studies on the CoV S glycoprotein have shown that both the N-terminal region (NTR, domain A in β-CoV) and the C-terminal region of S_1 (CTR, comprising domain B, C, and D in β-CoV) can bind host receptors and hence function as RBDs (Fig. 2) (Li, 2015). The CTR of alpha- and betacoronaviruses appears to bind proteinaceous receptors exclusively. The α-CoV HCoV-229E, serotype II feline CoV (FCoV), TGEV, and porcine respiratory coronavirus use the human aminopeptidase N (APN) of their respective hosts as receptors (Bonavia et al., 2003; Delmas et al., 1992; Reguera et al., 2012). The HCoV-NL63 (α-CoV) and SARS-CoV (β-CoV) both utilize angiotensin-converting enzyme 2 (ACE2) as a functional receptor (Li et al., 2005b; Wu et al., 2009), whereas the β-CoVs MERS-CoV and BtCoV-HKU4 recruit dipeptidyl peptidase-4 (DPP4) as a functional receptor (Lu et al., 2013; Mou et al., 2013; Raj et al., 2013; Wang et al., 2014; Yang et al., 2014).

The receptor-binding motifs (RBMs) in the S_1 CTRs of alpha- and betacoronavirus spike proteins are presented on one or more loops extending from the β-sheet core structure. Within *alpha-* and *betacoronavirus* genera the RBD core is structurally conserved yet the RBM(s) that determine receptor specificity may vary extensively. For instance, the CTR of the α-CoVs PRCoV and HCoV-NL63 has a similar core structure suggesting common evolutionary origin but diverged in their RBMs recruiting different receptors (APN and ACE2, respectively). A similar situation is seen for the CTRs of β-CoVs SARS-CoV and MERS-CoV that bind ACE2 and DPP4, respectively (Li, 2015). Conversely, the CTRs of the α-CoV HCoV-NL63 and β-CoV SARS-CoV both recognize ACE2, yet via distinct molecular interactions (ACE2 recognition via three vs one RBM, respectively), which suggested a convergent evolution pathway for these viruses in recruiting the ACE2 receptor (Li, 2015). The core structures of the CTRs in α- and β-CoVs provide a scaffold to present RBMs from extending loop(s), which may accommodate facile receptor switching by subtle alterations in or exchange of the RBMs via mutation/recombination.

Contrary to the CTR, the NTR appears to mainly bind glycans. The NTR of the α-CoV TGEV and of the γ-CoV IBV S proteins binds to sialic acids (Promkuntod et al., 2014; Schultze et al., 1996), while the NTR of betacoronaviruses including BCoV and HCoV-OC43 was shown to bind to O-acetylated sialic acids (Künkel and Herrler, 1993; Peng et al., 2012; Schultze et al., 1991; Vlasak et al., 1988). Only the NTR of MHV (domain A) is known to interact with a protein receptor, being mCEACAM1a (Peng et al., 2011), while lacking any detectable sialic acid-binding activity (Langereis et al., 2010). However, as the NTR of MHV displays the β-sandwich fold of the galectins, a family of sugar-binding proteins, it probably has evolved from a sugar-binding domain (Li, 2012).

The presence of RBDs in different domains of the S protein that can bind either proteinaceous or glycan receptors illustrates a functional modularity of this glycoprotein in which different domains may fulfill the role of binding to cellular attachment or entry receptors. The CoV S protein is thought to have evolved from a more basic structure in which receptor recognition was confined to the CTR within S_1 (Li, 2015). The observed deletions of the NTR in some CoV species in nature are indicative of a less stringent requirement and integration of this domain with other regions of the spike trimer compared to the more C-terminally located domains of S_1 and support a scenario in which the NTR has been acquired at a later time point in CoV evolutionary history. For example, the NTR of MHV, which displays a human galectin-like fold, was suggested to originate from a cellular lectin acquired early on in CoV evolution (Peng et al., 2011). Acquisition of glycan-binding domains and fusion thereof to the ancestral S protein may have resulted in a great extension of CoV host range and may have caused an increase in CoV diversity. The general preference of the NTR and CTR to bind to, respectively, glycan or protein receptors may be related to their arrangement in the S protein trimer. In contrast to the CTR, which is located in the center of the S trimer, the NTR is more distally oriented (Fig. 1B). As protein–glycan interactions are often of low affinity, the more distal orientation of domain A may allow multivalent receptor interactions, thereby increasing avidity. Interestingly, some CoVs appear to have a dual receptor usage as they may bind via their NTR and CTR to glycan and protein receptors, respectively (Fig. 2).

3.2 CoV Protein Receptor Preference

Although the number of currently known CoV receptors is limited, receptor usage does not appear to be necessarily conserved between closely related virus species such as HCoV-229E (APN) and HCoV-NL63 (ACE2), whereas identical receptors (ACE2) can be targeted by virus species from different genera such as HCoV-NL63 and SARS-CoV. It seems that CoVs prefer certain types of host proteins as their entry receptor, with three out of four of the so far identified proteinaceous receptors being ectopeptidases (APN, ACE2, and DPP4), although enzymatic activity of these proteins was shown not to be required for infection by their respective viruses (Bosch et al., 2014). Possibly, the localization to certain membrane micro-domains and efficient internalization of two of these proteins in polarized cells (APN and DPP4) may contribute to their suitability to function as entry receptors (Aït-Slimane et al., 2009). In the case of MERS-CoV, the region of DPP4 that is bound by the S protein coincides with the binding site for its physiological ligand adenosine deaminase (Raj et al., 2014). Employment of conserved epitopes such as these may also contribute to the cross-species transmission potential of viruses (Bosch et al., 2014), as is exemplified by MERS-CoV being able to use goat, camelid, cow, sheep, horse, pig, monkey, marmoset, and human DPP4 as entry receptor (Barlan et al., 2014; Eckerle et al., 2014; Falzarano et al., 2014; Müller et al., 2012; van Doremalen et al., 2014). Similarly, this may apply for the ability of feline, canine, porcine, and human CoVs to use fAPN as entry receptor, at least in vitro (Tresnan et al., 1996).

4. S PROTEIN PROTEOLYTIC CLEAVAGE AND CONFORMATIONAL CHANGES

Coronavirus entry is a tightly regulated process that appears to be orchestrated by multiple triggers that include receptor binding and proteolytic processing of the S protein and that ultimately results in virus–cell fusion. It is initiated by virion attachment mediated through interaction of either the NTR or CTR (or both) in the S_1 subunit of the spike protein with host receptors. Upon attachment, the virus is taken up via receptor-mediated endocytosis by clathrin- or caveolin-dependent pathways (Burkard et al., 2014; Eifart et al., 2007; Inoue et al., 2007; Nomura et al., 2004) although other entry routes have also been reported (Wang et al., 2008). Prior to and/or during endocytic uptake the CoV S protein is proteolytically processed. The spike protein may contain two proteolytic cleavage sites. One of the cleavage sites is located at the boundary between the S_1 and S_2 subunits (S_1/S_2 cleavage site), while the other cleavage site is located immediately upstream of the first fusion peptide (S_2' cleavage site). Although not irrevocably proven, it is expected that all CoVs depend on proteolytic cleavage on or close to S_2' for fusion to occur. Virus–cell fusion thus not only critically depends on the conformational changes following spike–receptor engagement, and perhaps on acidification of endosomal vesicles (Eifart et al., 2007; Matsuyama and Taguchi, 2009; Zelus et al., 2003), but also on proteolytic activation of the S protein by proteases along the endocytic route (Burkard et al., 2014; Simmons et al., 2005). Indeed, inhibition of intracellular proteases has been shown to block virus entry and virus–cell fusion (Burkard et al., 2014; Frana et al., 1985; Simmons et al., 2005; Yamada and Liu, 2009). The specific proteolytic cleavage requirements of the S protein at the S_1/S_2 boundary and particularly at the S_2' site may furthermore determine the intracellular site of fusion (Burkard et al., 2014). In agreement herewith, it has become evident that the protease expression profile of host cells may form an additional determinant of the host cell tropism of coronaviruses (Millet and Whittaker, 2015).

Analysis of the CoV S prefusion conformation suggests that relocation (or shedding) of the S_1 subunits that cap the S_2 subunit is a prerequisite for the conformational changes in S_2 that ultimately result in fusion. Shedding of S_1 probably requires receptor binding as well as proteolytic processing at S_1/S_2. The cryo-EM structure indicates that the S_1/S_2 proteolytic cleavage site is accessible to proteases prior to spike–receptor interaction, and depending on the particular cleavage site present may already be processed in the cell in which the virions are produced. As indicated earlier, the conformational changes in the S protein that result in virus–cell fusion most likely also require cleavage at the S_2' site immediately upstream of the fusion peptide. Interestingly, the S_2' cleavage site is located within an α-helix exposed on the prefusion S structure which prevents efficient proteolytic cleavage (Robertson et al., 2016). This indicates the necessity for preceding conformational changes induced by receptor binding and subsequent shedding of S_1, upon which the secondary structure of the S_2' site transforms into a cleavable flexible loop. Following proteolytic cleavage activation at the S_2' site, hydrophobic interactions between the fusion peptide and the adjacent S_2 helices are disturbed which allows the four α-helices and the connecting regions that make up the HR1 region in the prefusion S protein to refold into a long trimeric coiled coil (Fig. 1D). This coiled coil forms an N-terminal extension of the central helix projecting the fusion peptide(s) toward the target membrane. Successively, the fusion peptide(s) will be inserted into the limiting membrane of the host cell endocytic compartment. Next, as a consequence of S_2 rearrangements, the two HR regions will interact to form an antiparallel energetically stable six-helix bundle (Bosch et al., 2003, 2004), enabling the close apposition and subsequent fusion of the viral and host lipid bilayers.

5. TROPISM CHANGES ASSOCIATED WITH S PROTEIN MUTATIONS

Changes in the S protein may result in an altered host, tissue, or cellular tropism of the virus. This is clearly exemplified by genomic recombination events that result in exchange of (part of) the S protein and in a concomitant change in tropism. The propensity of CoVs to undergo homologous genomic recombination has been exploited for the genetic manipulation of these viruses (de Haan et al., 2008; Haijema et al., 2003; Kuo et al., 2000). To this end, interspecies chimeric coronaviruses were generated, which carried the spike ectodomain of another CoV and which could be selected based on their altered requirement for an entry receptor. Exchange of S protein genes may also occur in vivo, resulting in altered tropism as is illustrated by the occurrence of serotype II feline infectious peritonitis virus (FIPV). This virus results from a naturally occurring recombination event between feline and canine CoVs (CCoVs) in which the feline virus acquires a CCoV spike gene (Herrewegh et al., 1995; Terada et al., 2014). As a result of the acquisition of this new S protein, the rather harmless enteric feline CoV (FECV) turns into a systemically replicating and deadly FIPV. As FECV has a strict feline tropism (Myrrha et al., 2011), while CCoV has been shown to infect feline cells (Levis et al., 1995), it is likely that serotype II FIPVs arise in cats coinfected with serotype I FECV and CCoV. Furthermore, as different recombination sites have been observed for each serotype II FIPV, while serotype II FECVs have not been observed, it appears that serotype II FIPVs exclusively result of reoccurring recombination events (Terada et al., 2014). In addition to these feline–CCoV recombinants, a chimeric porcine coronavirus with a TGEV backbone and a spike of the porcine epidemic diarrhea virus (PEDV) was recently isolated from swine fecal samples in Italy and Germany, likely also resulting from a recombination event (Akimkin et al., 2016; Boniotti et al., 2016). Moreover, the α-CoV HKU2 BtCoV probably resulted from genomic recombination as it encodes an S protein that resembles a betacoronavirus S protein except for its N-terminal region that is similar to that of alphacoronaviruses (Lau et al., 2007). Thus, such genomic recombination events are not necessarily restricted to occur between viruses of the same genus.

5.1 S_1 Receptor Interactions Determining Tropism

5.1.1 S_1 NTR Changes

Several changes in the amino-terminal domain of S_1 have been associated with changes in the tropism of the virus. For example, for several α-CoVs, loss of NTR of the S protein appears to be accompanied with a loss of enteric tropism. While the porcine CoV TGEV displays a tropism for both the gastrointestinal and respiratory tract, the closely related PRCoV, which lacks the sialic acid-binding N-terminal region (Krempl et al., 1997), only replicates in the respiratory tract. The loss of sialic acid-binding activity by four-amino acid changes in the NTR of its S protein resulted in an almost complete loss of enteric tropism (Krempl et al., 1997). Similar to TGEV, enteric serotype I FCoVs also have been reported to bind to sialic acids (Desmarets et al., 2014). Large deletions within the S_1 subunit corresponding to the N-terminal region have been found in variants of the systemically replicating FIPV (strains UU16, UU21, and C3663) after intrahost emergence from enteric FECV (Chang et al., 2012; Terada et al., 2012). Also FIPVs seem to have lost the ability to replicate in the enteric tract (Pedersen, 2014). Clinical isolates of human coronavirus 229E as well as of the related alpaca coronavirus, both of which cause respiratory infections, encode relatively short spike proteins that lack the NTR (Crossley et al., 2012; Farsani et al., 2012). In contrast, closely related bat coronaviruses with intestinal tropism contain S proteins with a NTR or sometimes even two copies of the NTR (Corman et al., 2015) (Fig. 2). Overall, these observations suggest that the alphacoronavirus spike NTR—in particular its sialic acid-binding activity—may contribute to the enteric tropism of these alphacoronaviruses, while it is not required for replication in the respiratory tract or in other extraintestinal organs. It has been hypothesized that the sialic acid-binding activity of the spike protein can allow virus binding to (i) soluble sialoglycoconjugates that may protect the virus from hostile conditions in the stomach or (ii) to mucins that may prevent the loss of viruses by intestinal peristalsis and allow the virus to pass the thick mucus barrier, thereby gaining access to the intestinal cells to initiate infection (Schwegmann-Wessels et al., 2003).

Besides deletions of entire domains of the S protein, more subtle changes consisting of amino acid substitutions in S_1 NTR may also suffice to alter the virus' tropism. For example, MHV variants have been observed that acquired the ability to use the human homologue of their murine CEACAM1a receptor to enter cells as a result of mutations in their RBD that is located in S_1 NTR (Baric et al., 1999).

5.1.2 S_1 CTR Changes

As the CTR of the S_1 subunit contains the protein RBD for most CoVs, also mutations in this part of S have been associated with changes in the virus' tropism. Perhaps the most well-known example of viral cross-species transmission involves the SARS-CoV. Studies support a transmission model in which a SARS-like CoV was transmitted from *Rhinolophus* bats to palm civets, which subsequently transmitted the palm civet-adapted virus to humans at local food markets in southern China (Li et al., 2006). According to this model, SARS-like viruses adapted to both the palm civet and human host, which was reflected in the rapid viral evolution observed for these viruses within these species (Song et al., 2005). Two-amino acid substitutions within the RBD were elucidated that are of relevance for binding to the ACE2 proteins of palm civets and humans (Li et al., 2005b, 2006; Qu et al., 2005). From these studies it appears that due to strong conservation of ACE2 between mammalian species only a few amino acid alterations within the RBD are needed to change coronavirus host species tropism. Indeed serial passage of SARS-CoVs in vitro or in vivo can rapidly lead to adaptation to new host species (Roberts et al., 2007). SARS-like viruses isolated from bats displayed major differences including a deletion in the ACE2 RBM compared to human SARS-CoV (Drexler et al., 2010; Ren et al., 2008) and as a consequence were unable of using human ACE2 as an entry receptor (Becker et al., 2008). However, recently a novel SARS-like BtCoV was identified, which could use ACE2 of *Rhinolophus* bats, palm civets as well as of humans as a functional receptor (Ge et al., 2013). These findings not only provide further evidence that bats are indeed the natural reservoir for SARS-like CoVs, but also that these bat coronaviruses can directly include human ACE2 in their receptor repertoire. The detection of sequences of SARS-CoV-like viruses in palm civets and raccoon dogs (Guan et al., 2003; Tu et al., 2004) therefore probably reflects the unusually wide host range of these viruses. A similar promiscuous receptor usage is also observed for MERS-CoV which binds to DPP4 of many species (Barlan et al., 2014; Eckerle et al., 2014; Falzarano et al., 2014; Müller et al., 2012; van Doremalen et al., 2014) as indicated earlier.

Just as SARS like and MERS-CoVs are able to use entry receptors of different host species, also several α-CoVs display promiscuity to ortho-logous receptors. For example, the feline APN molecule can be used as a receptor by feline (serotype II FIPV), canine (CCoV), porcine (TGEV), and human (HCoV-229E) α-CoVs in cell culture (Tresnan and Holmes, 1998; Tresnan et al., 1996). Conversely, serotype II FIPV can only enter cells expressing feline APN (Tresnan and Holmes, 1998). The ability of TGEV and CCoV to use feline APN as a receptor probably results from strong conservation of the viral-binding motif (VBM) among APN orthologs in combination with the RBDs recognizing APN in a similar fashion (Reguera et al., 2012). Though recruiting the same receptor, HCoV-229E binds another domain within APN, which apparently is also conserved in feline APN (Kolb et al., 1997; Tusell et al., 2007). Conserva-tion of the VBM obviates the need for large adaptations within the RBD of these viruses to orthologous receptors allowing more facile cross-species transmission.

Other mutations in the S_1 CTR associated with altered tropism have been described for the β-CoV MHV. Similar to the humanized CEACAM1a-recognizing MHV variant, serial passaging of virus-infected cells resulted in the selection of viruses with an extended host range, which were subsequently shown to be able to enter cells in a heparan sulfate-dependent and CEACAM1a-independent manner (de Haan et al., 2005; Schickli et al., 1997). Two sets of mutations in the S protein were shown to be critically required for this phenotype, both of which resulted in the occurrence of multibasic heparan sulfate-binding sites. While one heparan sulfate-binding site was located in the S_2 subunit immediately upstream of the fusion peptide, the other was located in the S_1 CTR. The presence of this latter, but not of the former, domain resulted in MHV that depended on both heparan sulfate and CEACAM1a for entry. Additional introduction of the second heparan sulfate-binding site enabled the virus to become mCEACAM1a independent (de Haan et al., 2006). In addition, a mutation of the HVR of S_1 may affect CoV tropism as was demonstrated for the MHV strain JHM (MHV-JHM). The spike protein of MHV-JHM may induce receptor-independent fusion (Gallagher et al., 1992, 1993). However, dele-tion of residues in HVR of MHV-JHM resulted in the spike protein being entirely dependent on CEACAM1a binding for fusion (Dalziel et al., 1986; Gallagher and Buchmeier, 2001; Phillips and Weiss, 2001).

5.2 Changes in Proteolytic Cleavage Site and Other S_2 Mutations Associated with Altered Tropism

5.2.1 Changes in Proteolytic Cleavage Sites

Although the S_2 subunit does not appear to contain any RBDs, several mutations in this subunit have been associated with changes in the virus' tropism. Some of these changes affect the cleavage sites in the S protein that are located at the S_1/S_2 boundary or immediately upstream of the fusion peptide (S_2' cleavage site). As these cleavages appear to be essential for virus-cell fusion, the availability of host proteases to process the S protein is of critical importance for the virus' tropism. The importance of S protein cleavage at the S_1/S_2 boundary for the tropism of the virus is exemplified by the BtCoV HKU4, which is closely related to the MERS-CoV. Although domain B of the HKU4 S protein can interact with both bat and human DPP4, it is only in the context of bat cells, but not human cells, that the virus can utilize these molecules as entry receptors (Yang et al., 2014). In contrast, MERS-CoV can enter cells of human and bat origin via both DPP4 orthologues. This difference results from host restriction factors at the level of proteolytic cleavage activation. Two-amino acid substitutions (S746R and N762A) in the S_1/S_2 boundary of the S protein were shown to be crucial for the adaptation of bat MERS-like CoV to the proteolytic environment of the human cells (Yang et al., 2015).

Although probably not directly responsible for the tropism change associated with the enterically replicating FECV evolving into the systemically replicating FIPV, loss of a furin cleavage site at S_1/S_2 junction is observed in the majority of the FIPVs, whereas this furin cleavage site is strictly conserved in the parental FECV strains (Licitra et al., 2013). Apparently, conservation of this furin cleavage site is not required for efficient systemic replication. However, as FIPV is generally not found in the feces of cats, it may well be that loss of the furin cleavage site at S_1/S_2—as well as mutations in other parts of the genome, such as the accessory genes—may prevent efficient replication of FIPV in the enteric tracts.

Besides the influence of the S_1/S_2 cleavage site, virus tropism may also depend on the S_2' cleavage site upstream of FP1. In contrast to wild-type MHV strain A59, a recombinant MHV carrying a furin cleavage site at this position was shown to no longer depend on lysosomal proteases for efficient entry to occur (Burkard et al., 2014). As a consequence, this virus was able to infect cells in which trafficking to lysosomes was inhibited. Cleavage at the S_2' site may also be important for the tropism of PEDV, which causes major damage to the biofood industry in Asia and the Americas (Lee, 2015;

Song et al., 2015). PEDV replication in cell culture is strictly dependent on trypsin-like proteases, a requirement which is expected to limit its tropism in vivo to the enteric tract. The trypsin dependency of PEDV entry was shown, however, to be lifted after introduction of a furin cleavage site at the S_2' cleavage site by a single-amino acid substitution. Such mutations may potentially affect the spread of this virus in the pig by allowing it to replicate in nonenteric tissues in the absence of trypsin-like proteases (Li et al., 2015).

5.2.2 Other S_2 Mutations Associated with Altered Tropism

Mutations in other parts of the S_2 subunit than those affecting the proteolytic cleavage sites may also influence the tropism of different CoVs. Several studies report a correlation between mutations in the HR1 region of FCoVs and the conversion of FECV into FIPV (Bank-Wolf et al., 2014; Desmarets et al., 2016; Lewis et al., 2015). Such a correlation appeared even more convincing for mutations found in the recently identified FP2 (Chang et al., 2012; Ou et al., 2016). While these correlations suggest an important role for the S protein in the transition of FECV into FIPV, the causal relationship between these mutations in S and FIP remains to be determined. It is plausible, however, that such mutations may play a role in the acquired ability of FIPVs to infect macrophages. Indeed, for serotype II FCoV, the ability to replicate in macrophages was shown to be determined by residues located in the C-terminal part of the S_2 subunit, although the responsible residues were not identified (Rottier et al., 2005).

Also for other CoVs, mutations in the S_2 subunit have been linked to changes in the virus' tropism. A serially passaged MHV-A59 virus was shown to obtain mutations (M936V, P939L, F948L, and S949I) in and adjacent to the HR1 region which conveyed host range expansion of the mutant virus to normally nonpermissive mammalian cell types in vitro (Baric et al., 1999; McRoy and Baric, 2008). Contrary, Krueger et al. reported three mutations in the S_2 subunit of MHV-JHM (V870A located upstream of the S_2' cleavage site and A994V and A1046V located in the HR1 region) all of which reduced the CEACAM1a-independent fusogenicity of this virus (Krueger et al., 2001). Many studies on MHV-JHM point to a crucial role of a leucine at amino acid position 1114 in S protein fusogenicity. The MHV S cryo-EM structure demonstrates that the L1114 residue is located in the central helix and contributes to interprotomer interactions. A L1114F substitution in the MHV-JHM S protein was observed in a mutant strain of JHM and correlated with an increased S_1–S_2 stability and the loss of the ability to induce CEACAM1a-independent fusion (Taguchi and Matsuyama, 2002), while a substitution of the same residue to an Arg (L1114R) reduced the neurotropism of this virus (Tsai et al., 2003). Mutants resistant to a monoclonal antibody (Wang et al., 1992) and soluble receptor (Saeki et al., 1997) also correlate with substitutions at this specific residue, illustrating the importance of this residue in S fusogenicity. For the MERS-CoV, mutations in HR1 have been identified that are thought to be associated with its adaptive evolution (Forni et al., 2015). Among these sites, position 1060 is particularly interesting, as it appears to correspond to substitutions found in MHV and IBV that modify the tropism of these viruses (MHV: E1035D; IBV: L857F; Navas-Martin et al., 2005; Yamada et al., 2009). Substitution E1035D in HR1 of MHV was shown to restore the hepatotropism of an otherwise non-hepatotropic MHV, the latter resulting from mutations in the S_1 NTR and the S_1/S_2 cleavage site. These studies collectively indicate that mutations in and close to the HR regions may affect CoV tropism, possibly by affecting the metastability and consequently fusogenicity of the S protein and/or the formation of the postfusion six-helix bundle.

6. CONCLUDING REMARKS

It appears that changes in the S protein associated with altered tropism can be found in several regions of the spike protein. These regions obviously include the NTR and CTR of S_1 that are involved in the interaction with attachment and/or entry receptors. Substitutions within the S_1 RBDs may convey an altered viral tropism by adaptation of the virus to new or orthologous entry receptors. In addition, the S protein cleavage sites are important for host tropism as the processing of these sites by host proteases will critically affect the removal of the S_1-mediated locking of the S_2 prefusion conformation by shedding of S_1 (S_1/S_2 cleavage site) and the release of the fusion peptide(s) (S_2' cleavage site). Finally, changes in S_2 (particularly in the HR regions) may compensate for yet suboptimal spike binding to orthologous receptors by which low relative affinity interactions suffice to induce the required conformational changes of the S protein that ultimately result in the formation of the postfusion six-helix bundle and virus–cell fusion.

The observation that the different domains of the S protein all contribute to the tropism of CoVs is indicative of a coordinated interplay between these domains. This interplay has also been inferred from several studies, which reported changes in one S protein subunit often to be accompanied by adaptations in the other subunit (Saeki et al., 1997; Wang et al., 1992). In addition, the interplay between S_1 and S_2 has also been shown to be important for changes in the tropism of the virus as indicated earlier (de Haan et al., 2006; Navas-Martin et al., 2005). The recently published cryo-EM structures of CoV spike proteins (Kirchdoerfer et al., 2016; Walls et al., 2016) now provide structural evidence for the complex interplay between the subunits and domains of the S protein.

From all these studies, a picture arises in which the S protein is progressively destabilized through receptor engagement and proteolytic activation. In this process the S_1 subunits serve as a safety pin that stabilizes the fusogenic S_2 trimer. The safety pin is discharged upon interaction with a specific receptor and processing by host cell proteases and thereby gives way to conformational changes of the instable S_2 subunit. Subsequent release of the fusion peptide may resemble the pulling of the trigger which inevitably results in fusion of viral and host membranes through interaction of the heptad repeats regions.

Based on the presented data we propose a model in which the ability of a CoV to cross the host species barrier is critically dependent on the interplay between the different regions of the S proteins. In this model, the probable low affinity of the S_1 RBD for a novel receptor must be compensated by sufficiently low S_2 metastability, which depends on both proteolytic cleavage of the S protein and the S_2 interprotomer interactions. These required S protein characteristics may be generated during naturally occurring quasispecies variation and may result in the ability of the virus to replicate in and adapt to a new host.

ACKNOWLEDGMENTS

This study is supported by TOP Project Grant (91213066) funded by ZonMW and as part of the Zoonotic Anticipation and Preparedness Initiative (ZAPI project; IMI Grant Agreement No. 115760), with the assistance and financial support of IMI and the European Commission. We thank Mark Bakkers for his help in preparing figures.

REFERENCES

Aït-Slimane, T., Galmes, R., Trugnan, G., Maurice, M., 2009. Basolateral internalization of GPI-anchored proteins occurs via a clathrin-independent flotillin-dependent pathway in polarized hepatic cells. Mol. Biol. Cell 20 (17), 3792–3800. http://dx.doi.org/10.1091/mbc.E09-04-0275.

Akimkin, V., Beer, M., Blome, S., Hanke, D., Höper, D., Jenckel, M., Pohlmann, A., 2016. New chimeric porcine coronavirus in swine feces, Germany, 2012. Emerg. Infect. Dis. 22 (7), 1314–1315. http://dx.doi.org/10.3201/eid2207.160179.

Alagaili, A.N., Briese, T., Mishra, N., Kapoor, V., Sameroff, S.C., Burbelo, P.D., et al., 2014. Middle East respiratory syndrome coronavirus infection in dromedary camels in Saudi Arabia. mBio 5 (2). http://dx.doi.org/10.1128/mBio.00884-14. e00884-14.

Annan, A., Baldwin, H.J., Corman, V.M., Klose, S.M., Owusu, M., Nkrumah, E.E., et al., 2013. Human betacoronavirus 2c EMC/2012-related viruses in bats, Ghana and Europe. Emerg. Infect. Dis. 19 (3), 456–459. http://dx.doi.org/10.3201/eid1903.121503.

Azhar, E.I., El-Kafrawy, S.A., Farraj, S.A., Hassan, A.M., Al-Saeed, M.S., Hashem, A.M., Madani, T.A., 2014. Evidence for camel-to-human transmission of MERS coronavirus. N. Engl. J. Med. 26 (26), 2499–2505. http://dx.doi.org/10.1056/NEJMoa1401505.

Baker, K.A., Dutch, R.E., Lamb, R.A., Jardetzky, T.S., 1999. Structural basis for paramyxovirus-mediated membrane fusion. Mol. Cell 3 (3), 309–319. http://dx.doi.org/10.1016/S1097-2765(00)80458-X.

Bank-Wolf, B.R., Stallkamp, I., Wiese, S., Moritz, A., Tekes, G., Thiel, H.J., 2014. Mutations of 3c and spike protein genes correlate with the occurrence of feline infectious peritonitis. Vet. Microbiol. 173 (3–4), 177–188. http://dx.doi.org/10.1016/j.vetmic.2014.07.020.

Baric, R.S., Sullivan, E., Hensley, L., Yount, B., Chen, W., 1999. Persistent infection promotes cross-species transmissibility of mouse hepatitis virus. J. Virol. 73 (1), 638–649.

Barlan, A., Zhao, J., Sarkar, M.K., Li, K., McCray, P.B., Perlman, S., Gallagher, T., 2014. Receptor variation and susceptibility to Middle East respiratory syndrome coronavirus infection. J. Virol. 88 (9), 4953–4961. http://dx.doi.org/10.1128/JVI.00161-14.

Bartesaghi, A., Merk, A., Borgnia, M.J., Milne, J.L.S., Subramaniam, S., 2013. Prefusion structure of trimeric HIV-1 envelope glycoprotein determined by cryo-electron microscopy. Nat. Struct. Mol. Biol. 20 (12), 1352–1357. http://dx.doi.org/10.1038/nsmb.2711.

Becker, M.M., Graham, R.L., Donaldson, E.F., Rockx, B., Sims, A.C., Sheahan, T., et al., 2008. Synthetic recombinant bat SARS-like coronavirus is infectious in cultured cells and in mice. Proc. Natl. Acad. Sci. U.S.A. 105 (50), 19944–19949. http://dx.doi.org/10.1073/pnas.0808116105.

Bolles, M., Donaldson, E., Baric, R., 2011. SARS-CoV and emergent coronaviruses: viral determinants of interspecies transmission. Curr. Opin. Virol. 1 (6), 624–634. http://dx.doi.org/10.1016/j.coviro.2011.10.012.

Bonavia, A., Zelus, B.D., Wentworth, D.E., Talbot, P.J., Holmes, K.V., 2003. Identification of a receptor-binding domain of the spike glycoprotein of human coronavirus HCoV-229E. J. Virol. 77 (4), 2530–2538. http://dx.doi.org/10.1128/JVI.77.4.2530-2538.2003.

Boniotti, M.B., Papetti, A., Lavazza, A., Alborali, G., Sozzi, E., Chiapponi, C., et al., 2016. Porcine epidemic diarrhea virus and discovery of a recombinant swine enteric coronavirus, Italy. Emerg. Infect. Dis. 22 (1), 83–87. http://dx.doi.org/10.3201/eid2201.

Bosch, B.J., Rottier, P.J.M., 2008. Nidovirus entry into cells. In: Perlman, S., Gallagher, T., Snijder, E. (Eds.), Nidoviruses. American Society of Microbiology, Washington, DC, pp. 157–178. http://dx.doi.org/10.1128/9781555815790.ch11.

Bosch, B.J., Van Der Zee, R., de Haan, C.A.M., Rottier, P.J.M., 2003. The coronavirus spike protein is a class I virus fusion protein: structural and functional characterization of the fusion core complex. J. Virol. 77 (16), 8801–8811. http://dx.doi.org/10.1128/JVI.77.16.8801.

Bosch, B.J., Martina, B.E.E., Van Der Zee, R., Lepault, J., Haijema, B.J., Versluis, C., et al., 2004. Severe acute respiratory syndrome coronavirus (SARS-CoV) infection inhibition using spike protein heptad repeat-derived peptides. Proc. Natl. Acad. Sci. U.S.A. 101 (22), 8455–8460. http://dx.doi.org/10.1073/pnas.0400576101.

Bosch, B.J., Smits, S.L., Haagmans, B.L., 2014. Membrane ectopeptidases targeted by human coronaviruses. Curr. Opin. Virol. 6 (1), 55–60. http://dx.doi.org/10.1016/j.coviro.2014.03.011.

Briese, T., Mishra, N., Jain, K., East, M., Syndrome, R., Quasispecies, C., et al., 2014. Dromedary camels in Saudi Arabia include homologues of human isolates revealed through whole-genome analysis etc. mBio 5 (3), 1–5. http://dx.doi.org/10.1128/mBio.01146-14. Editor.

Burkard, C., Verheije, M.H., Wicht, O., van Kasteren, S.I., van Kuppeveld, F.J., Haagmans, B.L., et al., 2014. Coronavirus cell entry occurs through the endo-/lysosomal pathway in a proteolysis-dependent manner. PLoS Pathog. 10 (11), e1004502. http://dx.doi.org/10.1371/journal.ppat.1004502.

Chambers, P., Pringle, C.R., Easton, A.J., 1990. Heptad repeat sequences are located adjacent to hydrophobic regions in several types of virus fusion glycoproteins. J. Gen. Virol. 71 (12), 3075–3080. http://dx.doi.org/10.1099/0022-1317-71-12-3075.

Chan, F.J., To, K.K., Tse, H., Jin, D.-Y., Yuen, K.-Y., 2013. Interspecies transmission and emergence of novel viruses: lessons from bats and birds. Trends Microbiol. 21 (10), 544–555. http://dx.doi.org/10.1016/j.tim.2013.05.005.

Chang, H.W., Egberink, H.F., Halpin, R., Spiro, D.J., Rottier, P.J.M., 2012. Spike protein fusion peptide and feline coronavirus virulence. Emerg. Infect. Dis. 18 (7), 1089–1095. http://dx.doi.org/10.3201/eid1807.120143.

Corman, V.M., Baldwin, H.J., Tateno, A.F., Zerbinati, R.M., Annan, A., Owusu, M., et al., 2015. Evidence for an ancestral association of human coronavirus 229E with bats. J. Virol. 89 (23), 11858–11870. http://dx.doi.org/10.1128/JVI.01755-15.

Crossley, B.M., Mock, R.E., Callison, S.A., Hietala, S.K., 2012. Identification and characterization of a novel alpaca respiratory coronavirus most closely related to the human coronavirus 229E. Viruses 4 (12), 3689–3700. http://dx.doi.org/10.3390/v4123689.

Dalziel, R.G., Lampert, P.W., Talbot, P.J., Buchmeier, M.J., 1986. Site-specific alteration of murine hepatitis virus type 4 peplomer glycoprotein E2 results in reduced neurovirulence. J. Virol. 59 (2), 463–471. Retrieved from, http://www.ncbi.nlm.nih.gov/entrez/query.fcgi?cmd=Retrieve&db=PubMed&dopt=Citation&list_uids=3016306.

de Haan, C.A.M., Li, Z., te Lintelo, E., Bosch, B.J., Haijema, B.J., Rottier, P.J.M., 2005. Murine coronavirus with an extended host range uses heparan sulfate as an entry receptor. J. Virol. 79 (22), 14451–14456. http://dx.doi.org/10.1128/JVI.79.22.14451-14456.2005.

de Haan, C.A.M., te Lintelo, E., Li, Z., Raaben, M., Wurdinger, T., Bosch, B.J., Rottier, P.J.M., 2006. Cooperative involvement of the S1 and S2 subunits of the murine coronavirus spike protein in receptor binding and extended host range. J. Virol. 80 (22), 10909–10918. http://dx.doi.org/10.1128/JVI.00950-06.

de Haan, C.A.M., Haijema, B.J., Masters, P.S., Rottier, P.J.M., 2008. Manipulation of the coronavirus genome using targeted RNA recombination with interspecies chimeric coronaviruses. Methods Mol. Biol. 454, 229–236. http://dx.doi.org/10.1007/978-1-59745-181-9_17.

Delmas, B., Gelfi, J., L'Haridon, R., Vogel, L.K., Sjöström, H., Norén, O., Laude, H., 1992. Aminopeptidase N is a major receptor for the entero-pathogenic coronavirus TGEV. Nature 357 (6377), 417–420. http://dx.doi.org/10.1038/357417a0.

Desmarets, L.M.B., Theuns, S., Roukaerts, I.D.M., Acar, D.D., Nauwynck, H.J., 2014. Role of sialic acids in feline enteric coronavirus infections. J. Gen. Virol. 95 (9), 1911–1918. http://dx.doi.org/10.1099/vir.0.064717-0.

Desmarets, L.M.B., Vermeulen, B.L., Theuns, S., Conceição-Neto, N., Zeller, M., Roukaerts, I.D.M., et al., 2016. Experimental feline enteric coronavirus infection reveals an aberrant infection pattern and shedding of mutants with impaired infectivity in enterocyte cultures. Sci. Rep. 6, 20022. http://dx.doi.org/10.1038/srep20022.

Drexler, J.F., Gloza-Rausch, F., Glende, J., Corman, V.M., Muth, D., Goettsche, M., et al., 2010. Genomic characterization of severe acute respiratory syndrome-related coronavirus in European bats and classification of coronaviruses based on partial RNA-dependent RNA polymerase gene sequences. J. Virol. 84 (21), 11336–11349. http://dx.doi.org/10.1128/JVI.00650-10.

Duquerroy, B.S., Vigouroux, A., Rottier, P.J.M., Rey, F.A., Berend, T., Bosch, J., 2005. Central ions and lateral asparagine/glutamine zippers stabilize the post-fusion hairpin conformation of the SARS coronavirus spike glycoprotein. Virology 335 (2), 276–285. http://dx.doi.org/10.1016/j.virol.2005.02.022.

Eckerle, I., Corman, V.M., Müller, M.A., Lenk, M., Ulrich, R.G., Drosten, C., 2014. Replicative capacity of MERS coronavirus in livestock cell lines. Emerg. Infect. Dis. 20 (2), 276–279. http://dx.doi.org/10.3201/eid2002.131182.

Eifart, P., Ludwig, K., Böttcher, C., de Haan, C.A.M., Rottier, P.J.M., Korte, T., Herrmann, A., 2007. Role of endocytosis and low pH in murine hepatitis virus strain A59 cell entry. J. Virol. 81 (19), 10758–10768. http://dx.doi.org/10.1128/JVI.00725-07.

Falzarano, D., de Wit, E., Feldmann, F., Rasmussen, A.L., Okumura, A., Peng, X., et al., 2014. Infection with MERS-CoV causes lethal pneumonia in the common marmoset. PLoS Pathog. 10 (8), e1004250. http://dx.doi.org/10.1371/journal.ppat.1004250.

Farsani, S.M.J., Dijkman, R., Jebbink, M.F., Goossens, H., Ieven, M., Deijs, M., et al., 2012. The first complete genome sequences of clinical isolates of human coronavirus 229E. Virus Genes 45 (3), 433–439. http://dx.doi.org/10.1007/s11262-012-0807-9.

Forni, D., Filippi, G., Cagliani, R., De Gioia, L., Pozzoli, U., Al-Daghri, N., et al., 2015. The heptad repeat region is a major selection target in MERS-CoV and related coronaviruses. Sci. Rep. 5, 14480. http://dx.doi.org/10.1038/srep14480.

Frana, M.F., Behnke, J.N., Sturman, L.S., Holmes, K.V., 1985. Proteolytic cleavage of the E2 glycoprotein of murine coronavirus: host-dependent differences in proteolytic cleavage and cell fusion. J. Virol. 56 (3), 912–920. Retrieved from, http://www.pubmedcentral.nih.gov/articlerender.fcgi?artid=252664&tool=pmcentrez&rendertype=abstract.

Gallagher, T.M., Buchmeier, M.J., 2001. Coronavirus spike proteins in viral entry and pathogenesis. Virology 279 (2), 371–374. http://dx.doi.org/10.1006/viro.2000.0757.

Gallagher, T.M., Buchmeier, M.J., Perlman, S., 1992. Cell receptor-independent infection by a neurotropic murine coronavirus. Virology 19 (1), 517–522. Retrieved from, http://www.ncbi.nlm.nih.gov/pubmed/1413526.

Gallagher, T.M., Buchmeier, M.J., Perlman, S., 1993. Dissemination of MHV4 (strain JHM) infection does not require specific coronavirus receptors. Adv. Exp. Med. Biol. 342, 279–284. Retrieved from, http://www.ncbi.nlm.nih.gov/pubmed/8209743.

Gao, J., Lu, G., Qi, J., Li, Y., Wu, Y., Deng, Y., et al., 2013. Structure of the fusion core and inhibition of fusion by a heptad repeat peptide derived from the S protein of Middle East respiratory syndrome coronavirus. J. Virol. 87 (24), 13134–13140. http://dx.doi.org/10.1128/JVI.02433-13.

Ge, X.Y., Li, J.L., Yang, X.L., Chmura, A.A., Zhu, G., Epstein, J.H., et al., 2013. Isolation and characterization of a bat SARS-like coronavirus that uses the ACE2 receptor. Nature 503 (7477), 535–538. http://dx.doi.org/10.1038/nature12711.

Guan, Y., Zheng, B.J., He, Y.Q., Liu, X.L., Zhuang, Z.X., Cheung, C.L., et al., 2003. Isolation and characterization of viruses related to the SARS coronavirus from animals in southern China. Science 302 (5643), 276–278. http://dx.doi.org/10.1126/science.1087139.

Haijema, B.J., Volders, H., Rottier, P.J.M., 2003. Switching species tropism: an effective way to manipulate the feline coronavirus genome. J. Virol. 77 (8), 4528–4538. http://dx.doi.org/10.1128/JVI.77.8.4528-4538.2003.

Herrewegh, A.A.P.M., Vennema, H., Horzinek, M.C., Rottier, P.J.M., de Groot, R.J., 1995. The molecular genetics of feline coronaviruses: comparative sequence analysis of the ORF7a/7b transcription unit of different biotypes. Virology 212 (2), 622–631.

Hofmann, H., Hattermann, K., Marzi, A., Gramberg, T., Geier, M., Krumbiegel, M., et al., 2004. S protein of severe acute respiratory syndrome-associated coronavirus mediates entry into hepatoma cell lines and is targeted by neutralizing antibodies in infected patients. J. Virol. 78 (12), 6134–6142. http://dx.doi.org/10.1128/JVI.78.12.6134-6142.2004.

Hu, B., Ge, X., Wang, L.-F., Shi, Z., 2015. Bat origin of human coronaviruses. Virol. J. 12 (1), 221. http://dx.doi.org/10.1186/s12985-015-0422-1.

Huynh, J., Li, S., Yount, B., Smith, A., Sturges, L., Olsen, J.C., et al., 2012. Evidence supporting a zoonotic origin of human coronavirus strain NL63. J. Virol. 86 (23), 12816–12825. http://dx.doi.org/10.1128/JVI.00906-12.

Inoue, Y., Tanaka, N., Tanaka, Y., Inoue, S., Morita, K., Zhuang, M., et al., 2007. Clathrin-dependent entry of severe acute respiratory syndrome coronavirus into target cells expressing ACE2 with the cytoplasmic tail deleted. J. Virol. 81 (16), 8722–8729. http://dx.doi.org/10.1128/JVI.00253-07.

Kirchdoerfer, R.N., Cottrell, C.A., Wang, N., Pallesen, J., Yassine, H.M., Turner, H.L., et al., 2016. Pre-fusion structure of a human coronavirus spike protein. Nature 531 (7592), 118–121. http://dx.doi.org/10.1038/nature17200.

Kolb, A.F., Hegyi, A., Siddell, S.G., 1997. Identification of residues critical for the human coronavirus 229E receptor function of human aminopeptidase N. J. Gen. Virol. 78 (11), 2795–2802.

Krempl, C., Schultze, B., Laude, H., 1997. Point mutations in the S protein connect the sialic acid binding activity with the enteropathogenicity of transmissible gastroenteritis coronavirus. J. Virol. 71 (4), 3285–3287.

Krueger, D.K., Kelly, S.M., Lewicki, D.N., Ruffolo, R., Gallagher, T.M., 2001. Variations in disparate regions of the murine coronavirus spike protein impact the initiation of membrane fusion. J. Virol. 75 (6), 2792–2802. http://dx.doi.org/10.1128/JVI.75.6.2792-2802.2001.

Künkel, F., Herrler, G., 1993. Structural and functional analysis of the surface protein of human coronavirus OC43. Virology 195 (1), 195–202.

Kuo, L., Godeke, G.J., Raamsman, M.J., Masters, P.S., Rottier, P.J., 2000. Retargeting of coronavirus by substitution of the spike glycoprotein ectodomain: crossing the host cell species barrier. J. Virol. 74 (3), 1393–1406. http://dx.doi.org/10.1128/JVI.74.3.1393-1406.2000.

Langereis, M.A., van Vliet, A.L.W., Boot, W., de Groot, R.J., 2010. Attachment of mouse hepatitis virus to O-acetylated sialic acid is mediated by hemagglutinin-esterase and not by the spike protein. J. Virol. 84 (17), 8970–8974. http://dx.doi.org/10.1128/JVI.00566-10.

Lau, S.K.P., Woo, P.C.Y., Li, K.S.M., Huang, Y., Wang, M., Lam, C.S.F., et al., 2007. Complete genome sequence of bat coronavirus HKU2 from Chinese horseshoe bats revealed a much smaller spike gene with a different evolutionary lineage from the rest of the genome. Virology 367 (2), 428–439. http://dx.doi.org/10.1016/j.virol.2007.06.009.

Lee, C., 2015. Porcine epidemic diarrhea virus: an emerging and re-emerging epizootic swine virus. Virol. J. 12 (1), 193. http://dx.doi.org/10.1186/s12985-015-0421-2.

Lee, J.E., Fusco, M.L., Hessell, A.J., Oswald, W.B., Burton, D.R., Saphire, E.O., 2008. Structure of the ebola virus glycoprotein bound to an antibody from a human survivor. Nature 454 (7201), 177–182. http://dx.doi.org/10.1038/nature07082.

Levis, R., Cardellichio, C.B., Scanga, C.A., Compton, S.R., Holmes, K.V., 1995. Multiple receptor-dependent steps determine the species specificity of HCV-229E infection. Adv. Exp. Med. Biol. 380, 337–343. Retrieved from, http://www.ncbi.nlm.nih.gov/pubmed/8830504.

Lewis, C.S., Porter, E., Matthews, D., Kipar, A., Tasker, S., Helps, C.R., Siddell, S.G., 2015. Genotyping coronaviruses associated with feline infectious peritonitis. J. Gen. Virol. 96 (Pt. 6), 1358–1368. http://dx.doi.org/10.1099/vir.0.000084.

Li, F., 2012. Evidence for a common evolutionary origin of coronavirus spike protein receptor-binding subunits. J. Virol. 86 (5), 2856–2858. http://dx.doi.org/10.1128/JVI.06882-11.

Li, F., 2015. Receptor recognition mechanisms of coronaviruses: a decade of structural studies. J. Virol. 89 (4), 1954–1964. http://dx.doi.org/10.1128/JVI.02615-14.

Li, F., Li, W., Farzan, M., Harrison, S.C., 2005a. Structure of SARS coronavirus spike receptor-binding domain complexed with receptor. Science 309 (5742), 1864–1868. http://dx.doi.org/10.1126/science.1116480.

Li, W., Zhang, C., Sui, J., Kuhn, J.H., Moore, M.J., Luo, S., et al., 2005b. Receptor and viral determinants of SARS-coronavirus adaptation to human ACE2. EMBO J. 24 (8), 1634–1643. http://dx.doi.org/10.1038/sj.emboj.7600640.

Li, W., Wong, S.-K., Li, F., Kuhn, J.H., Huang, I.-C., Choe, H., Farzan, M., 2006. Animal origins of the severe acute respiratory syndrome coronavirus: insight from ACE2-S-protein interactions. J. Virol. 80 (9), 4211–4219. http://dx.doi.org/10.1128/JVI.80.9.4211-4219.2006.

Li, W., Wicht, O., van Kuppeveld, F.J.M., He, Q., Rottier, P.J.M., Bosch, B.-J., 2015. A single point mutation creating a furin cleavage site in the spike protein renders porcine epidemic diarrhea coronavirus trypsin-independent for cell entry and fusion. J. Virol. 89 (15), 8077–8081. http://dx.doi.org/10.1128/JVI.00356-15.

Licitra, B.N., Millet, J.K., Regan, A.D., Hamilton, B.S., Rinaldi, V.D., Duhamel, G.E., Whittaker, G.R., 2013. Mutation in spike protein cleavage site and pathogenesis of feline coronavirus. Emerg. Infect. Dis. 19 (7), 1066–1073. http://dx.doi.org/10.3201/eid1907.121094.

Lin, X., Eddy, N.R., Noel, J.K., Whitford, P.C., Wang, Q., Ma, J., Onuchic, J.N., 2014. Order and disorder control the functional rearrangement of influenza hemagglutinin. Proc. Natl. Acad. Sci. U.S.A. 111, 12049–12054. http://dx.doi.org/10.1073/pnas.1412849111.

Lu, G., Hu, Y., Wang, Q., Qi, J., Gao, F., Li, Y., et al., 2013. Molecular basis of binding between novel human coronavirus MERS-CoV and its receptor CD26. Nature 500 (7461), 227–231. http://dx.doi.org/10.1038/nature12328.

Lu, L., Liu, Q., Zhu, Y., Chan, K.-H., Qin, L., Li, Y., et al., 2014. Structure-based discovery of Middle East respiratory syndrome coronavirus fusion inhibitor. Nat. Commun. 5, 3067. http://dx.doi.org/10.1038/ncomms4067.

Matsuyama, S., Taguchi, F., 2009. Two-step conformational changes in a coronavirus envelope glycoprotein mediated by receptor binding and proteolysis. J. Virol. 83 (21), 11133–11141. http://dx.doi.org/10.1128/JVI.00959-09.

McRoy, W.C., Baric, R.S., 2008. Amino acid substitutions in the S2 subunit of mouse hepatitis virus variant V51 encode determinants of host range expansion. J. Virol. 82 (3), 1414–1424. http://dx.doi.org/10.1128/JVI.01674-07.

Millet, J.K., Whittaker, G.R., 2015. Host cell proteases: critical determinants of coronavirus tropism and pathogenesis. Virus Res. 202, 120–134. http://dx.doi.org/10.1016/j.virusres.2014.11.021.

Mou, H., Raj, V.S., van Kuppeveld, F.J.M., Rottier, P.J.M., Haagmans, B.L., Bosch, B.J., 2013. The receptor binding domain of the new Middle East respiratory syndrome coronavirus maps to a 231-residue region in the spike protein that efficiently elicits neutralizing antibodies. J. Virol. 87 (16), 9379–9383. http://dx.doi.org/10.1128/JVI.01277-13.

Müller, M.A., Raj, V.S., Muth, D., Meyer, B., Kallies, S., Smits, S.L., et al., 2012. Human coronavirus EMC does not require the SARS-coronavirus receptor and maintains broad replicative capability in mammalian cell lines. mBio 3 (6). http://dx.doi.org/10.1128/mBio.00515-12. e00515-12.

Myrrha, L.W., Silva, F.M.F., de Oliveira Peternelli, E.F., Junior, A.S., Resende, M., de Almeida, M.R., 2011. The paradox of feline coronavirus pathogenesis: a review. Adv. Virol. 2011, 109849. http://dx.doi.org/10.1155/2011/109849.

Navas-Martin, S., Hingley, S.T., Weiss, S.R., 2005. Murine coronavirus evolution in vivo: functional compensation of a detrimental amino acid substitution in the receptor binding domain of the spike glycoprotein. J. Virol. 79 (12), 7629–7640. http://dx.doi.org/10.1128/JVI.79.12.7629-7640.2005.

Nomura, R., Kiyota, A., Suzaki, E., Kataoka, K., Ohe, Y., Miyamoto, K., et al., 2004. Human coronavirus 229E binds to CD13 in rafts and enters the cell through caveolae. J. Virol. 78 (16), 8701–8708. http://dx.doi.org/10.1128/JVI.78.16.8701-8708.2004. 78/16/8701 [pii].

Ou, X., Zheng, W., Shan, Y., Mu, Z., Dominguez, S.R., Holmes, K.V., Qian, Z., 2016. Identification of the fusion peptide-containing region in betacoronavirus spike glyco-proteins. J. Virol. 90 (12), 5586–5600. http://dx.doi.org/10.1128/JVI.00015-16. JVI. 00015-16.

Pedersen, N.C., 2014. An update on feline infectious peritonitis: virology and im-munopathogenesis. Vet. J. 201 (2), 123–132. http://dx.doi.org/10.1016/j.tvjl. 2014.04.017.

Peng, G., Sun, D., Rajashankar, K.R., Qian, Z., Holmes, K.V., Li, F., 2011. Crystal structure of mouse coronavirus receptor-binding domain complexed with its murine receptor. Proc. Natl. Acad. Sci. U.S.A. 108 (26), 10696–10701. http://dx.doi.org/10.1073/pnas.1104306108.

Peng, G., Xu, L., Lin, Y.L., Chen, L., Pasquarella, J.R., Holmes, K.V., Li, F., 2012. Crystal structure of bovine coronavirus spike protein lectin domain. J. Biol. Chem. 287 (50), 41931–41938. http://dx.doi.org/10.1074/jbc.M112.418210.

Phillips, J.J., Weiss, S.R., 2001. MHV neuropathogenesis: the study of chimeric S genes and mutations in the hypervariable region. Adv. Exp. Med. Biol. 494, 115–119. Retrieved from, http://www.ncbi.nlm.nih.gov/pubmed/11774454.

Promkuntod, N., van Eijndhoven, R., de Vrieze, G., Gröne, A., Verheije, M., 2014. Map-ping of the receptor-binding domain and amino acids critical for attachment in the spike protein of avian coronavirus infectious bronchitis virus. Virology 448, 26–32. http://dx. doi.org/10.1016/j.virol.2013.09.018.

Qu, X.X., Hao, P., Song, X.J., Jiang, S.M., Liu, Y.X., Wang, P.G., et al., 2005. Identifica-tion of two critical amino acid residues of the severe acute respiratory syndrome coro-navirus spike protein for its variation in zoonotic tropism transition via a double substitution strategy. J. Biol. Chem. 280 (33), 29588–29595. http://dx.doi.org/10.1074/jbc.M500662200.

Racaniello, V.R., Skalka, A.M., Flint, J., Rall, G.F., 2015. Principles of Virology, Bundle. American Society of Microbiology, Washington, DC. http://dx.doi.org/10.1128/9781555819521.

Raj, V.S., Mou, H., Smits, S.L., Dekkers, D.H.W., Müller, M.A., Dijkman, R., et al., 2013. Dipeptidyl peptidase 4 is a functional receptor for the emerging human coronavirus-EMC. Nature 495, 251–254. http://dx.doi.org/10.1038/nature12005.

Raj, V.S., Smits, S.L., Provacia, L.B., van den Brand, J.M.A., Wiersma, L., Ouwendijk, W.J.D., et al., 2014. Adenosine deaminase acts as a natural antagonist for dipeptidyl peptidase 4-mediated entry of the Middle East respiratory syndrome corona-virus. J. Virol. 88 (3), 1834–1838. http://dx.doi.org/10.1128/JVI.02935-13.

Reguera, J., Santiago, C., Mudgal, G., Ordoño, D., Enjuanes, L., Casasnovas, J.M., 2012. Structural bases of coronavirus attachment to host aminopeptidase N and its inhibition by neutralizing antibodies. PLoS Pathog. 8 (8), e1002859. http://dx.doi.org/10.1371/journal.ppat.1002859.

Ren, W., Qu, X., Li, W., Han, Z., Yu, M., Zhou, P., et al., 2008. Difference in receptor usage between severe acute respiratory syndrome (SARS) coronavirus and SARS-like coronavirus of bat origin. J. Virol. 82 (4), 1899–1907. http://dx.doi.org/10.1128/JVI.01085-07.

Reusken, C.B., Haagmans, B.L., Müller, M.A., Gutierrez, C., Godeke, G.-J., Meyer, B., et al., 2013. Middle East respiratory syndrome coronavirus neutralising serum antibodies in dromedary camels: a comparative serological study. Lancet Infect. Dis. 13 (10), 859–866. http://dx.doi.org/10.1016/S1473-3099(13)70164-6.

Roberts, A., Deming, D., Paddock, C.D., Cheng, A., Yount, B., Vogel, L., et al., 2007. A mouse-adapted SARS-coronavirus causes disease and mortality in BALB/c mice. PLoS Pathog. 3 (1), e5. http://dx.doi.org/10.1371/journal.ppat.0030005.

Robertson, A.L., Headey, S.J., Ng, N.M., Wijeyewickrema, L.C., Scanlon, M.J., Pike, R.N., Bottomley, S.P., 2016. Protein unfolding is essential for cleavage within the α-helix of a model protein substrate by the serine protease, thrombin. Biochimie 122, 227–234. http://dx.doi.org/10.1016/j.biochi.2015.09.021.

Rottier, P.J.M., Nakamura, K., Schellen, P., Volders, H., Haijema, B.J., 2005. Acquisition of macrophage tropism during the pathogenesis of feline infectious peritonitis is determined by mutations in the feline coronavirus spike protein. J. Virol. 79 (22), 14122–14130. http://dx.doi.org/10.1128/JVI.79.22.14122-14130.2005.

Saeki, K., Ohtsuka, N., Taguchi, F., 1997. Identification of spike protein residues of murine coronavirus responsible for receptor-binding activity by use of soluble receptor-resistant mutants. J. Virol. 71 (12), 9024–9031. Retrieved from, http://www.ncbi.nlm.nih.gov/pubmed/9371559.

Schickli, J.H., Zelus, B.D., Wentworth, D.E., Sawicki, S.G., Holmes, K.V., 1997. The murine coronavirus mouse hepatitis virus strain A59 from persistently infected murine cells exhibits an extended host range. J. Virol. 71 (12), 9499–9507. Retrieved from, http://www.ncbi.nlm.nih.gov/pubmed/9371612.

Schultze, B., Gross, H.-J., Brossmer, R., Herrler, G., 1991. The S protein of bovine coronavirus is a hemagglutinin recognizing 9-0-acetylated sialic acid as a receptor determinant. J. Virol. 65 (11), 6232–6237.

Schultze, B., Krempl, C., Ballesteros, M.L., Shaw, L., Schauer, R., Enjuanes, L., Herrler, G., 1996. Transmissible gastroenteritis coronavirus, but not the related porcine respiratory coronavirus, has a sialic acid (N-glycolylneuraminic acid) binding activity. J. Virol. 70 (8), 5634–5637. Retrieved from, http://www.ncbi.nlm.nih.gov/pubmed/8764078.

Schwegmann-Wessels, C., Zimmer, G., Schroder, B., Breves, G., Herrler, G., 2003. Binding of transmissible gastroenteritis coronavirus to brush border membrane sialoglycoproteins. J. Virol. 77 (21), 11846–11848. http://dx.doi.org/10.1128/JVI.77.21.11846.

Simmons, G., Gosalia, D.N., Rennekamp, A.J., Reeves, J.D., Diamond, S.L., Bates, P., 2005. Inhibitors of cathepsin L prevent severe acute respiratory syndrome coronavirus entry. Proc. Natl. Acad. Sci. U.S.A. 102 (33), 11876–11881.

Song, H.-D., Tu, C.-C., Zhang, G.-W., Wang, S.-Y., Zheng, K., Lei, L.-C., et al., 2005. Cross-host evolution of severe acute respiratory syndrome coronavirus in palm civet and human. Proc. Natl. Acad. Sci. U.S.A. 102 (7), 2430–2435. http://dx.doi.org/10.1073/pnas.0409608102.

Song, D., Moon, H., Kang, B., 2015. Porcine epidemic diarrhea: a review of current epidemiology and available vaccines. Clin. Exp. Vaccine Res. 4 (2), 166–176. http://dx.doi.org/10.7774/cevr.2015.4.2.166.

Supekar, V.M., Bruckmann, C., Ingallinella, P., Bianchi, E., Pessi, A., Carfi, A., 2004. Structure of a proteolytically resistant core from the severe acute respiratory syndrome coronavirus S2 fusion protein. Proc. Natl. Acad. Sci. U.S.A. 101 (52), 17958–17963. http://dx.doi.org/10.1073/pnas.0406128102.

Taguchi, F., Matsuyama, S., 2002. Soluble receptor potentiates receptor-independent infection by murine coronavirus. J. Virol. 76 (3), 950–958. Retrieved from, http://www.ncbi.nlm.nih.gov/pubmed/11773370.

Terada, Y., Shiozaki, Y., Shimoda, H., Mahmoud, H.Y.A.H., Noguchi, K., Nagao, Y., et al., 2012. Feline infectious peritonitis virus with a large deletion in the 5'-terminal region of the spike gene retains its virulence for cats. J. Gen. Virol. 93 (Pt. 9), 1930–1934. http://dx.doi.org/10.1099/vir.0.043992-0.

Terada, Y., Matsui, N., Noguchi, K., Kuwata, R., Shimoda, H., Soma, T., et al., 2014. Emergence of pathogenic coronaviruses in cats by homologous recombination between feline and canine coronaviruses. PLoS One 9 (9), e106534. http://dx.doi.org/10.1371/journal.pone.0106534.

Tresnan, D.B., Holmes, K.V., 1998. Feline aminopeptidase N is a receptor for all group I coronaviruses. Adv. Exp. Med. Biol. 440, 69–75. Retrieved from, http://www.ncbi.nlm.nih.gov/pubmed/9782266.

Tresnan, D.B., Levis, R., Holmes, K.V., 1996. Feline aminopeptidase N serves as a receptor for feline, canine, porcine, and human coronaviruses in serogroup I. J. Virol. 70 (12), 8669–8674. Retrieved from, http://www.pubmedcentral.nih.gov/articlerender.fcgi?artid=190961&tool=pmcentrez&rendertype=abstract.

Tsai, J.C., De Groot, L., Pinon, J.D., Iacono, K.T., Phillips, J.J., Seo, S.H., et al., 2003. Amino acid substitutions within the heptad repeat domain 1 of murine coronavirus spike protein restrict viral antigen spread in the central nervous system. Virology 312 (2), 369–380. http://dx.doi.org/10.1016/S0042-6822(03)00248-4.

Tu, C., Crameri, G., Kong, X., Chen, J., Sun, Y., Yu, M., et al., 2004. Antibodies to SARS coronavirus in civets. Emerg. Infect. Dis. 10 (12), 2244–2248. http://dx.doi.org/10.3201/eid1012.040520.

Tusell, S.M., Schittone, S.A., Holmes, K.V., 2007. Mutational analysis of aminopeptidase N, a receptor for several group 1 coronaviruses, identifies key determinants of viral host range. J. Virol. 81 (3), 1261–1273. http://dx.doi.org/10.1128/JVI.01510-06.

van Doremalen, N., Miazgowicz, K.L., Milne-Price, S., Bushmaker, T., Robertson, S., Scott, D., et al., 2014. Host species restriction of Middle East respiratory syndrome coronavirus through its receptor, dipeptidyl peptidase 4. J. Virol. 88 (16), 9220–9232. http://dx.doi.org/10.1128/JVI.00676-14.

Vijgen, L., Keyaerts, E., Lemey, P., Maes, P., Van Reeth, K., Nauwynck, H., et al., 2006. Evolutionary history of the closely related group 2 coronaviruses: porcine hemagglutinating encephalomyelitis virus, bovine coronavirus, and human coronavirus OC43. J. Virol. 80 (14), 7270–7274. http://dx.doi.org/10.1128/JVI.02675-05.

Vlasak, R., Luytjes, W., Spaan, W., Palese, P., 1988. Human and bovine coronaviruses recognize sialic acid-containing receptors similar to those of influenza C viruses. Proc. Natl. Acad. Sci. U.S.A. 85 (12), 4526–4529. http://dx.doi.org/10.1073/pnas.85.12.4526.

Walls, A.C., Tortorici, M.A., Bosch, B.-J., Frenz, B., Rottier, P.J.M., DiMaio, F., et al., 2016. Cryo-electron microscopy structure of a coronavirus spike glycoprotein trimer. Nature 531 (7592), 114–117. http://dx.doi.org/10.1038/nature16988.

Wang, F.I., Fleming, J.O., Lai, M.M., 1992. Sequence analysis of the spike protein gene of murine coronavirus variants: study of genetic sites affecting neuropathogenicity. Virology 186 (2), 742–749. Retrieved from, http://www.ncbi.nlm.nih.gov/pubmed/1310195.

Wang, H., Yang, P., Liu, K., Guo, F., Zhang, Y., Zhang, G., Jiang, C., 2008. SARS coronavirus entry into host cells through a novel clathrin- and caveolae-independent endocytic pathway. Cell Res. 18 (2), 290–301. http://dx.doi.org/10.1038/cr.2008.15.

Wang, N., Shi, X., Jiang, L., Zhang, S., Wang, D., Tong, P., et al., 2013. Structure of MERS-CoV spike receptor-binding domain complexed with human receptor DPP4. Cell Res. 23 (8), 986–993. http://dx.doi.org/10.1038/cr.2013.92.

Wang, Q., Qi, J., Yuan, Y., Xuan, Y., Han, P., Wan, Y., et al., 2014. Bat origins of MERS-CoV supported by bat coronavirus HKU4 usage of human receptor CD26. Cell Host Microbe 16 (3), 328–337. http://dx.doi.org/10.1016/j.chom.2014.08.009.

WHO | Middle East respiratory syndrome coronavirus (MERS-CoV) – Saudi Arabia, 2016. WHO.

Widagdo, W., Raj, V.S., Schipper, D., Kolijn, K., van Leenders, G.J.L.H., Bosch, B.J., Bensaid, A., 2016. Differential expression of the Middle East respiratory syndrome coronavirus receptor in the upper respiratory tracts of humans and dromedary camels. J. Virol. 90 (9), 4838–4842. http://dx.doi.org/10.1128/JVI.02994-15. Editor.

Woo, P.C.Y., Lau, S.K.P., Lam, C.S.F., Lau, C.C.Y., Tsang, A.K.L., Lau, J.H.N., et al., 2012. Discovery of seven novel mammalian and avian coronaviruses in the genus deltacoronavirus supports bat coronaviruses as the gene source of alphacoronavirus and betacoronavirus and avian coronaviruses as the gene source of gammacoronavirus and deltacoronavirus. J. Virol. 86 (7), 3995–4008. http://dx.doi.org/10.1128/JVI.06540-11.

Wu, K., Li, W., Peng, G., Li, F., 2009. Crystal structure of NL63 respiratory coronavirus receptor-binding domain complexed with its human receptor. Proc. Natl. Acad. Sci. U.S.A. 106 (47), 19970–19974. http://dx.doi.org/10.1073/pnas.0908837106.

Xu, Y., Lou, Z., Liu, Y., Pang, H., Tien, P., Gao, G.F., Rao, Z., 2004. Crystal structure of severe acute respiratory syndrome coronavirus spike protein fusion core. J. Biol. Chem. 279 (47), 49414–49419. http://dx.doi.org/10.1074/jbc.M408782200.

Yamada, Y., Liu, D.X., 2009. Proteolytic activation of the spike protein at a novel RRRR/S motif is implicated in furin-dependent entry, syncytium formation, and infectivity of coronavirus infectious bronchitis virus in cultured cells. J. Virol. 83 (17), 8744–8758. http://dx.doi.org/10.1128/JVI.00613-09.

Yamada, Y., Liu, X.B., Fang, S.G., Tay, F.P.L., Liu, D.X., 2009. Acquisition of cell-cell fusion activity by amino acid substitutions in spike protein determines the infectivity of a coronavirus in cultured cells. PLoS One 4 (7), e6130. http://dx.doi.org/10.1371/journal.pone.0006130.

Yang, Y., Du, L., Liu, C., Wang, L., Ma, C., Tang, J., et al., 2014. Receptor usage and cell entry of bat coronavirus HKU4 provide insight into bat-to-human transmission of MERS coronavirus. Proc. Natl. Acad. Sci. U.S.A. 111 (34), 12516–12521. http://dx.doi.org/10.1073/pnas.1405889111.

Yang, Y., Liu, C., Du, L., Jiang, S., Shi, Z., Baric, R.S., Li, F., 2015. Two mutations were critical for bat-to-human transmission of Middle East respiratory syndrome coronavirus. J. Virol. 89 (17), 9119–9123. http://dx.doi.org/10.1128/JVI.01279-15.

Zelus, B.D., Schickli, J.H., Blau, D.M., Weiss, S.R., Holmes, K.V., 2003. Conformational changes in the spike glycoprotein of murine coronavirus are induced at 37 degrees C either by soluble murine CEACAM1 receptors or by pH 8. J. Virol. 77 (2), 830–840. http://dx.doi.org/10.1128/JVI.77.2.830-840.2003.

CHAPTER THREE

The Nonstructural Proteins Directing Coronavirus RNA Synthesis and Processing

E.J. Snijder*,[1], E. Decroly[†,‡], J. Ziebuhr[§,1]

*Leiden University Medical Center, Leiden, The Netherlands
†Aix-Marseille Université, AFMB UMR 7257, Marseille, France
‡CNRS, AFMB UMR 7257, Marseille, France
§Institute of Medical Virology, Justus Liebig University Giessen, Giessen, Germany
[1]Corresponding authors: e-mail address: e.j.snijder@lumc.nl; john.ziebuhr@viro.med.uni-giessen.de

Contents

Advances in Virus Research, Volume 96
ISSN 0065-3527
http://dx.doi.org/10.1016/bs.aivir.2016.08.008

59

Abstract

Coronaviruses are animal and human pathogens that can cause lethal zoonotic infections like SARS and MERS. They have polycistronic plus-stranded RNA genomes and belong to the order *Nidovirales*, a diverse group of viruses for which common ancestry was inferred from the common principles underlying their genome organization and expression, and from the conservation of an array of core replicase domains, including key RNA-synthesizing enzymes. Coronavirus genomes (~26–32 kilobases) are the largest RNA genomes known to date and their expansion was likely enabled by acquiring enzyme functions that counter the commonly high error frequency of viral RNA polymerases. The primary functions that direct coronavirus RNA synthesis and processing reside in nonstructural protein (nsp) 7 to nsp16, which are cleavage products of two large replicase polyproteins translated from the coronavirus genome. Significant progress has now been made regarding their structural and functional characterization, stimulated by technical advances like improved methods for bioinformatics and structural biology, in vitro enzyme characterization, and site-directed mutagenesis of coronavirus genomes. Coronavirus replicase functions include more or less universal activities of plus-stranded RNA viruses, like an RNA polymerase (nsp12) and helicase (nsp13), but also a number of rare or even unique domains involved in mRNA capping (nsp14, nsp16) and fidelity control (nsp14). Several smaller subunits (nsp7–nsp10) act as crucial cofactors of these enzymes and contribute to the emerging "nsp interactome." Understanding the structure, function, and interactions of the RNA-synthesizing machinery of coronaviruses will be key to rationalizing their evolutionary success and the development of improved control strategies.

1. INTRODUCTION

Coronaviruses (CoVs) are the best-known and best-studied clade of the order *Nidovirales*, which is comprised of enveloped plus-stranded (+RNA) viruses and currently also comprises the *Arteriviridae*, *Roniviridae*, and *Mesoniviridae* families (de Groot et al., 2012a,b; Lauber et al., 2012). In addition to including various highly pathogenic CoVs of livestock (Saif, 2004) and four "established" human CoVs causing a large number of common colds (Pyrc et al., 2007), CoVs have attracted abundant attention due to their potential to cause lethal zoonotic infections (Graham et al., 2013). This was exemplified by the 2003 outbreak of severe acute respiratory syndrome-coronavirus (SARS-CoV) in Southeast Asia and the ongoing transmission—since 2012—of the Middle East respiratory syndrome-coronavirus (MERS-CoV), which causes ~35% mortality among patients seeking medical attention. Both these viruses are closely related to CoVs that are circulating in bats (Ge et al., 2013; Menachery et al., 2015) and other potential reservoir species. They may be transmitted to humans either directly or through intermediate hosts, like civet cats for SARS-CoV (Song et al., 2005) and dromedary camels for MERS-CoV (Reusken et al., 2013). Formally, the family *Coronaviridae* now includes about 30 species, divided into the subfamilies *Torovirinae* and *Coronavirinae*, the latter being further subdivided in the genera *Alpha-*, *Beta-*, *Gamma-*, and *Deltacoronavirus*. SARS-CoV and MERS-CoV are betacoronaviruses, and the same holds true for one of the best-characterized animal CoV models, murine hepatitis virus (MHV). This explains why the bulk of our current knowledge of CoV molecular biology is betacoronavirus based, even more so for the replicative proteins that are the central theme of this review, which will mainly summarize data obtained studying SARS-CoV proteins.

Despite their unification in the same virus order, nidoviruses cover an unusually broad range of genome sizes, ranging from ∼13–16 kilobases (kb) for arteriviruses, via ∼20 kb for mesoniviruses, to ∼26–32 kb for CoVs (Nga et al., 2011). Together with the genomes of roniviruses, which infect invertebrate hosts, CoV genomes are the largest RNA genomes known to date (Gorbalenya et al., 2006). The common ancestry of these extremely diverse virus lineages was inferred from their polycistronic genome structure, the common principles underlying the expression of these genomes, and—most importantly—the conservation of an array of "core replicase domains," including key enzymes required for RNA synthesis. While retaining this conserved genomic and proteomic blueprint, nidovirus genomes are thought to have expanded gradually by gene duplication and acquisition of novel genes (Lauber et al., 2013), most likely by RNA recombination. In addition to the high mutation rate that characterizes all RNA viruses, these genomic innovations appear to have enabled nidoviruses to explore an unprecedented evolutionary space and adapt to a wide variety of host organisms, including mammals, birds, reptiles, fish, crustaceans, and insects. Whereas the poor replication fidelity generally restricts RNA virus genome sizes, it has been postulated that nidovirus genome expansion was enabled by the acquisition of specific replicative functions that counter the error rate of the RNA polymerase (Deng et al., 2014; Eckerle et al., 2010; Snijder et al., 2003) (discussed in more detail later).

As in all nidoviruses, at least two-thirds of the CoV genome capacity is occupied by the two large open reading frames (ORFs) that together constitute the replicase gene, ORF1a and ORF1b (Fig. 1). These ORFs overlap by a few dozen nucleotides and are both translated from the viral genome, with expression of ORF1b requiring a -1 ribosomal frameshift to occur just

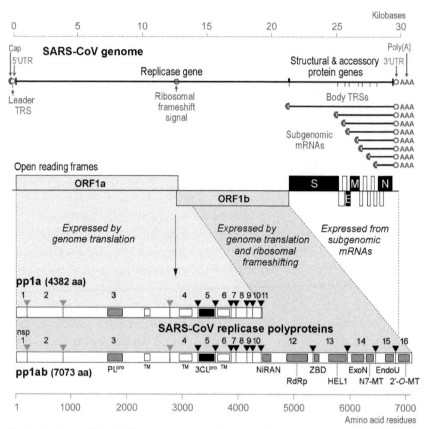

Fig. 1 Outline of the CoV genome organization and expression strategy, based on SARS-CoV. The *top panel* depicts the SARS-CoV genome, including various regulatory RNA elements, and the 5′- and 3′-coterminal nested set of subgenomic mRNAs used to express the genes downstream of the replicase gene. *UTR*, untranslated region; *TRS*, transcription-regulatory sequence. Below the RNAs, the 14 open reading frames in the genome are indicated, i.e., the replicase ORFs 1a and 1b, the four common CoV structural protein genes (S, E, M, and N) and the ORFs encoding "accessory proteins." The *bottom panel* explains the organization and proteolytic processing of the pp1a and pp1ab replicase polyproteins, the latter being produced by -1 ribosomal frameshifting. The nsp3 (PL^pro) and nsp5 (3CL^pro) proteases and their cleavage sites are indicated in matching colors. The resulting 16 cleavage products (nonstructural proteins (nsps)) are indicated, as are the conserved replicase domains that are relevant for this review. Domain abbreviations and corresponding nsp numbers: *PL^pro*, papain-like proteinase (nsp3); *3CL^pro*, 3C-like proteinase (nsp5); *TM*, transmembrane domain (nsp3, nsp4, and nsp6); *NiRAN*, nidovirus RdRp-associated nucleotidyl transferase (nsp12); *RdRp*, RNA-dependent RNA polymerase (nsp12); *ZBD*, zinc-binding domain (nsp13); *HEL1*, superfamily 1 helicase (nsp13); *ExoN*, exoribonuclease (nsp14); *N7-MT*, N7-methyl transferase (nsp14); *endoU*, uridylate-specific endoribonuclease (nsp15); *2′-O-MT*, 2′-O-methyl transferase (nsp16).

upstream of the ORF1a termination codon (Brierley et al., 1989). The efficiency of this highly conserved frameshift event, which may approach 50% in the case of CoVs (Irigoyen et al., 2016), is promoted by specific primary and higher-order RNA structures. As a result, in CoV-infected cells, the replicase subunits encoded in ORF1a are overexpressed in a fixed ratio relative to the proteins encoded in ORF1b. The primary translation products of the CoV replicase are two huge polyproteins, the ORF1a-encoded pp1a and the C-terminally extended pp1ab frameshift product (Fig. 1). The former is roughly 4000–4500 amino acids long, depending on the CoV species analyzed. The size of the ORF1b-encoded extension is more conserved (around 2700 residues), resulting in pp1ab sizes in the range of 6700–7200 amino acids. Probably already during their synthesis, either two or three ORF1a-encoded proteases initiate the proteolytic cleavage of pp1a and pp1ab to release (sometimes) 15 or (mostly) 16 functional nonstructural proteins (nsps; Fig. 1). The highly conserved nsp5 protease has a chymotrypsin-like fold (3C-like protease, 3CL^{pro}) (Anand et al., 2002, 2003; Gorbalenya et al., 1989) and is the viral "main protease" (therefore sometimes also referred to as M^{pro}). The 3CL^{pro} cleaves the nsp4–nsp11 part of pp1a and the nsp4–nsp16 part of pp1ab at 7 and 11 conserved sites, respectively. These sites can be summarized with the P4-P2$'$ consensus motif (small)-X-(L/I/V/F/M)-Q↓(S/A/G), where X is any amino acid and ↓ represents the cleavage. The processing of three sites in the nsp1–nsp4 region is performed by one or two papain-like proteases (PL^{pro}) residing in the very large nsp3 subunit (Mielech et al., 2014). Whereas alphacoronaviruses and most betacoronaviruses (though not SARS-CoV and MERS-CoV) have two PL^{pro} domains in their nsp3, presumably the result of an ancient duplication event, gamma- and deltacoronaviruses have only a single PL^{pro}. The cleavage sites (LXGG↓ or similar) resemble the C-terminal LRGG↓ motif of ubiquitin, which explains why CoV PL^{pro} domains were found capable to also act as deubiquitinases (Ratia et al., 2006). This secondary function has been implicated in the disruption of host innate immune signaling by removing ubiquitin from certain cellular substrates. More than any other CoV-encoded enzyme, the CoV 3CL^{pro} and PL^{pro} domains have been characterized in exquisite structural and biochemical detail, both in their capacity of critical regulators of nsp synthesis and as two of the primary drug targets for this virus family. Space limitations unfortunately prevent us from summarizing these studies in more detail, but a variety of excellent reviews is available to compensate for this omission (Baez-Santos et al., 2015; Hilgenfeld, 2014; Mielech et al., 2014; Steuber and Hilgenfeld, 2010).

Once released from pp1a and pp1ab, most CoVs nsps studied thus far assemble into a membrane-bound ribonucleoprotein complex that drives the synthesis of different forms of viral RNA (see later) and is sometimes referred to as the replication and transcription complex (RTC). While viral RNA production takes off, peculiar convoluted membrane structures, spherules tethered to zippered endoplasmic reticulum, and double-membrane vesicles begin to accumulate in CoV-infected cells (Gosert et al., 2002; Knoops et al., 2008; Maier et al., 2013). As for other +RNA viruses, they have been postulated to serve as scaffolds, or perhaps even suitable microenvironments, for viral RNA synthesis. Nevertheless, many questions on their biogenesis and function remain to be answered, and the exact location of the metabolically active RTC still has to be pinpointed "beyond reasonable doubt" for CoVs and other nidoviruses (Hagemeijer et al., 2012; Neuman et al., 2014a; van der Hoeven et al., 2016). Three ORF1a-encoded replicase subunits containing transmembrane domains (nsp3, nsp4, and nsp6; Fig. 1) have been implicated in the formation of the membrane structures that are induced upon CoV infection and with which the RTC is thought to be associated (Angelini et al., 2013; Hagemeijer et al., 2014). In addition to actively engaging in host membrane remodeling, they may serve as membrane anchors for the RTC by binding the nsps that lack hydrophobic domains, like all of the ORF1b-encoded enzymes. For more details, the reader is referred to the numerous recent reviews of the "replication organelles" of CoVs and other +RNA viruses (den Boon and Ahlquist, 2010; Hagemeijer et al., 2012; Neuman et al., 2014a; Romero-Brey and Bartenschlager, 2016; van der Hoeven et al., 2016; Xu and Nagy, 2014).

The common ancestry of nidovirus replicases is not only reflected in their conserved core replicase domains but also in the synthesis of subgenomic (sg) mRNAs that are used to express the genes located downstream of ORF1b (Fig. 1) (Gorbalenya et al., 2006). Although some nidoviruses (e.g., roni- and mesoniviruses) have only a few of these genes, they are much more numerous in arteriviruses and CoVs, their number going up to about a dozen ORFs for some CoVs. In addition to the standard set of four CoV structural protein genes (encoding the spike (S), envelope (E), membrane (M), and nucleocapsid (N) protein), genomes in different CoV clusters contain varying numbers of ORFs encoding so-called "accessory proteins" (Liu et al., 2014; Narayanan et al., 2008). The proteins they encode are often dispensable for the basic replicative cycle in cultured cells, but highly relevant for CoV viability and pathogenesis in vivo, for example, because they enable the virus to interfere with the host's immune response. Most of the genes downstream of ORF1b are made accessible to ribosomes by positioning them at the 5′ end of their own sg transcript. Occasionally, two or even three genes are expressed from the same sg mRNA, usually by employing ribosomal "leaky scanning" during translation initiation.

Nidoviral sg mRNAs are 3′-coterminal with the viral genome, but in most nidovirus taxa, including CoVs, the sg transcripts also carry common 5′ leader sequences (~65–95 nucleotides in CoVs), which are identical to the 5′-terminal sequence of the viral genome (Fig. 1) (Pasternak et al., 2006; Sawicki et al., 2007; Sola et al., 2011). The joining of common leader and different sg RNA "body" sequences occurs during minus-strand RNA synthesis (Sawicki and Sawicki, 1995; Sethna et al., 1989). This step can be either continuous, to produce the full-length minus strand required for genome replication, or interrupted (discontinuous) to produce a subgenome-length minus-strand RNA that can subsequently serve as the template for the synthesis of one of the sg mRNAs. The polymerase jumping that is the basis for leader-to-body joining occurs at specific "transcription-regulatory sequences" (TRSs). These conserved sequence motifs are comprised of up to a dozen nucleotides, and are found in the genome at the 3′ end of the leader sequence and at the 5′ end of each of the sg mRNA bodies. Quite likely, also higher-order RNA structure and transcription-specific protein factors play a role in the interruption of minus-strand RNA synthesis at a body TRS, after which the nascent minus strand (with a body TRS complement at its 3′ end) is translocated to the 5′-proximal part of the genomic template. Guided by a base-pairing interaction with the leader TRS, the synthesis of the subgenome-length minus-strand RNA is resumed and completed with the addition of the complement of the genomic leader sequence. In this manner, a nested set of subgenome-length templates for sg mRNA synthesis is produced, providing a mechanism to regulate the abundance of the different viral proteins by fine-tuning the level at which the corresponding sg mRNA is generated (Nedialkova et al., 2010). The CoV transcription strategy allows the RTC to use the same 3′-terminal recognition/initiation signals in both full- and subgenome-length templates of either polarity. Moreover, the presence of the common 5′ leader sequence may be important for mRNA capping or other translation-related features.

During the past two decades, studies on the CoV enzyme complex that controls this elegant replication and transcription mechanism have been accelerated by four important developments. First, using bioinformatics, expression systems, and virus-infected cells, the replicase polyprotein processing scheme and the proteases involved were elucidated, thus defining the boundaries of the 16 mature nsps (Fig. 1) that are working together during CoV replication (Ziebuhr et al., 2000). Second, using this information and promoted by rapidly advancing methods in structural biology, X-ray or NMR structures were obtained for numerous (recombinant) full-length CoV nsps or domains thereof, in particular for SARS-CoV (Neuman et al., 2014b). Third, multiple techniques for the targeted mutagenesis of CoV genomes were developed and refined, which was a specific technical challenge due to the exceptionally large size of the CoV RNA genome (Almazan et al., 2014). By launching engineered mutant genomes in susceptible cells, the RNA and protein players in the CoV replication cycle can now be interrogated directly, to reveal their importance, function(s) and/or interactions in vivo. Finally, in vitro biochemical assays were developed for a variety of CoV replicative enzymes, including many of those involved in RNA synthesis and processing. For the purpose of this review, we have chosen to focus on these latter functions, as performed by the CoV nsp7 to nsp16 products (Gorbalenya et al., 2006; Nga et al., 2011; Sevajol et al., 2014; Subissi et al., 2014a). These subunits include several replicative enzymes that are more or less universal among + RNA viruses, such as RNA polymerase (nsp12) and helicase (nsp13), but also a number of rare or even unique domains involved in, e.g., mRNA capping, cap modification, and promoting the fidelity of CoV RNA synthesis. Several smaller subunits, in particular nsp7 to nsp10, have been identified as crucial cofactors of these enzymes and contribute to the emerging CoV "nsp interactome," which will likely need to be advanced considerably to achieve a more complete understanding of the intricacies of CoV RNA synthesis. Making that step will obviously be key to understanding the evolutionary success of CoVs, and nidoviruses at large. Moreover, this knowledge will lay the foundation for the development of improved strategies to combat current and future emerging CoVs, including targeted antiviral drug development.

2. CORONAVIRUS nsp7–10: SMALL BUT CRITICAL REGULATORY SUBUNITS?

The 3'-terminal part of ORF1a, the approximately 1.7 kb separating the nsp6-coding sequence and the ORF1a/1b ribosomal frameshift site, encodes a set of four small replicase subunits, named nsp7 to nsp10 (Fig. 1). Although highly conserved among *Coronavirinae*, these proteins seem to lack enzymatic functions. Instead, they have emerged as (putative) interaction partners and modulators of ORF1b-encoded core enzymes like nsp12 (RNA-dependent RNA polymerase, RdRp), nsp14 (exoribonuclease, ExoN), and nsp16 (ribose $2'$-O-methyl transferase, $2'$-O-MTase). Furthermore, several of them have been predicted or shown to interact with RNA. Additionally, a fifth, very small cleavage product is assumed to be released from this region of pp1a: the nsp11 peptide resulting from cleavage of pp1a at the nsp10/11 junction (Fig. 1). In the pp1ab frameshift product, the N-terminal sequence of nsp11 (encoded between the nsp10/11 junction and ORF1a/1b frameshift site) equals the N-terminal part of the nsp12 subunit. Depending on the CoV species, nsp11 consists of 13–23 residues and its actual release, function (if any), or fate in CoV-infected cells have not been established. In cell culture models, for some (infectious bronchitis virus (IBV)) but not other (MHV) CoVs, the nsp10/11 and nsp10/12 cleavages were found to be dispensable for virus replication (Deming et al., 2007; Fang et al., 2008), even though the conservation of this cleavage site suggests that it is generally required for full replicase functionality.

Processing of the nsp7–nsp10 region of pp1a/pp1ab has been studied in some detail for MHV (Bost et al., 2000; Deming et al., 2007), human CoV 229E (HCoV-229E) (Ziebuhr and Siddell, 1999), and IBV (Ng et al., 2001), confirming the release of these subunits in infected cells and the use of the predicted 3CL^{pro} cleavage sites. Processing at these sites was found to be critical for MHV replication, the exception being inactivation of the nsp9/10 cleavage site, which yielded a crippled mutant virus. Depending on antibody availability, the subcellular localization of nsp7 to nsp10 has been studied for several CoVs using immunofluorescence microscopy. Without exception, and in line with their role as interaction partner of key replicative enzymes, these subunits localize to the perinuclear region of infected cells (Bost et al., 2000), where the membranous replication organelles of CoVs accumulate (Gosert et al., 2002; Knoops et al., 2008; Maier et al., 2013). It should be noted, however, that these labeling techniques cannot distinguish between fully processed nsps and polyprotein precursors or processing intermediates.

2.1 Coronavirus nsp7

The structure of the 83-amino acid SARS-CoV nsp7 was determined using both NMR (Peti et al., 2005) and X-ray crystallography (Zhai et al., 2005), with the latter study resolving the structure of a hexadecameric

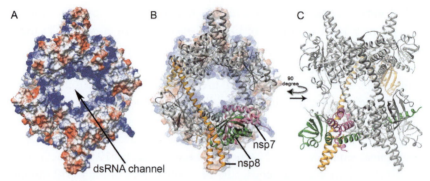

Fig. 2 Crystal structure of the SARS-CoV nsp7–nsp8 hexadecamer (pdb 2AHM) (Zhai et al., 2005). Purified recombinant SARS-CoV nsp7 and nsp8 were found to self-assemble into a supercomplex of which the structure was determined at 2.4 Å resolution. (A) The complex forms a *doughnut-shaped* hollow structure of which the central channel is lined with positively charged side chains (in *blue*) and was postulated to mediate double-stranded RNA binding. The outside of the structure is predominantly negatively charged (*red*) surface shading. (B and C) SARS-CoV nsp8 resembles a "golf club"-like shape that can adopt two conformations, as presented here in *orange* and *green*. These nsp8 conformations are integrated into a much larger, hexadecameric structure that is composed of eight nsp8 subunits and eight nsp7 subunits, of which one is shaded *pink*. In (B), the hexadecamer is depicted against the background of the surface plot presented in (A).

supercomplex consisting of recombinant nsp7 and nsp8 (see later; Fig. 2). In both structures, the nsp7-fold includes four helices, but their position and spatial orientation is quite different, suggesting that the protein's conformation is strongly affected by the interaction with nsp8, in particular, where it concerns helix α4 (Johnson et al., 2010). Reverse-genetics studies targeting specific residues in SARS-CoV nsp7 confirmed the protein's importance for virus replication (Subissi et al., 2014b), although the impact of single point mutations was smaller than anticipated on the basis of the biochemical characterization of the RNA-binding properties of nsp7-containing protein complexes in vitro (see later).

2.2 Coronavirus nsp8 and nsp7–nsp8 Complexes

The ~200-amino-acid-long nsp8 subunit initially took center stage due to two studies, the first describing a fascinating hexadecameric structure consisting of eight copies each of nsp7 and nsp8 (Fig. 2) (Zhai et al., 2005), and the second reporting an nsp8-specific "secondary" RNA polymerase activity (Imbert et al., 2006) that was implicated in the mechanism of initiation of CoV RNA synthesis. This template-dependent activity was reported to depend on the presence of Mn^{2+} or Mg^{2+} and to typically generate products of up to six nucleotides (for more details, see Section 3.2). Around the same time, purified recombinant SARS-CoV nsp7 and nsp8 were found to self-assemble into the hexadecameric supercomplex of which the structure was determined at 2.4 Å resolution (Zhai et al., 2005). The complex was described, and also visualized by electron microscopy, as a doughnut-shaped hollow structure of which the central channel is lined with positively charged side chains (Fig. 2A). A combination of structural modeling, RNA-binding studies, and site-directed mutagenesis led to the hypothesis that the complex may slide along the replicating viral RNA together with other viral proteins, possibly as a processivity factor for the RdRp (nsp12; see later). Within the nsp7–nsp8 hexadecamer, SARS-CoV nsp8 was found to adopt two different conformations (Fig. 2B and C). These were named "golf club" and "golf club with a bent shaft" (Zhai et al., 2005), with the globular head of the golf club being considered a new fold. Although the structures of feline coronavirus (FCoV) nsp7 and nsp8 were found to resemble their SARS-CoV equivalents, they were found to assemble into a quite different higher-order complex, with two copies of nsp7 and a single copy of nsp8 forming a heterotrimer (Xiao et al., 2012).

Biochemical and reverse-genetics studies pointed toward an important role in RNA synthesis for SARS-CoV nsp8 residues K58, P183, and R190, whose replacement was lethal to SARS-CoV. Of these residues, P183 and R190 were postulated to be involved in interactions with nsp12, whereas K58 may be critical for nsp8–RNA interactions (Subissi et al., 2014b). Reverse-genetics studies targeting the 3′-proximal RNA replication signals in the MHV genome provided strong evidence for an interaction between nsp8 and these RNA structures (a so-called "bulged stem-loop" and RNA pseudoknot). When making a particular 6-nucleotide insertion in the RNA pseudoknot, which strongly affected MHV replication, multiple suppressor mutations evolved, of which several mapped to the genomic region encoding nsp8 and nsp9 (Züst et al., 2008). These interactions were postulated to be part of a molecular switch that controls minus-strand RNA synthesis, or its initiation from the 3′ end of the viral genome (te Velthuis et al., 2012; Züst et al., 2008). Using screening approaches based on yeast two-hybrid and glutathione S-transferase (GST) pull-down assays, SARS-CoV nsp8 was reported to be an interaction partner of many other viral proteins (including nsp2, nsp3, and nsp5 to nsp16), although most of these interactions remain to be verified in the infected cell (von Brunn et al., 2007).

2.3 Coronavirus nsp9

The CoV nsp9 subunit is about 110 amino acids long and was the second replicase cleavage product, after nsp5, for which crystal structures were obtained (Egloff et al., 2004; Sutton et al., 2004). The biologically active form of the protein is believed to be a dimer that is capable of binding nucleic acids in a nonsequence-specific manner, with an apparent preference for single-stranded RNA (Egloff et al., 2004; Ponnusamy et al., 2008; Sutton et al., 2004). Several nsp9 point mutations that block CoV replication have now been described (Chen et al., 2009a; Miknis et al., 2009), but the protein's exact function has remained enigmatic thus far.

The nsp9 monomer consists of a β-barrel, composed of seven β-strands, and a C-terminal domain formed by a single α-helix. The latter domain plays a key role in the formation of the parallel helix-helix dimer conformation that—based on sequence conservation, structural considerations, and experimental data (Miknis et al., 2009)—is thought to be the biologically most relevant state of SARS-CoV nsp9. Nevertheless, multiple alternative structures were described, including a SARS-CoV form that is stabilized by β-sheet interactions (Sutton et al., 2004) and, for HCoV-229E nsp9, an antiparallel helix–helix dimer that is stabilized by a disulfide bond (Ponnusamy et al., 2008). Replacement of the HCoV-229E Cys residue involved in dimerization (Cys-69) resulted in conversion to the parallel helix–helix dimer described for SARS-CoV nsp9. Whereas wild-type HCoV-229E nsp9 is organized as a trimer of dimers, the Cys-69 → Ala mutant and SARS-CoV nsp9 both form rod-like polymers (Ponnusamy et al., 2008). Disulfide bonding of the latter protein could not be detected (Miknis et al., 2009). Although SARS-CoV and other betacoronaviruses do contain an equivalent Cys residue, the feature is not conserved in alphacoronaviruses that are much more closely related to HCoV-229E. Thus, it cannot be excluded that the disulfide-bonded form of HCoV-229E nsp9 is an artifact of recombinant protein purification and crystallization, although it was suggested that oxidative stress due to viral infection may favor its formation in CoV-infected cells (Ponnusamy et al., 2008). We are not aware of experiments directly addressing the existence of such a disulfide-linked nsp9 dimer in CoV-infected cells.

The importance of nsp9 dimerization for SARS-CoV and IBV viability was demonstrated in reverse-genetics studies (Chen et al., 2009a; Miknis et al., 2009) that also independently confirmed the importance of dimerization of the α-helical domain and in particular a putative GxxxG protein–protein interaction motif. Although RNA binding in vitro was not disrupted in dimerization-incompetent SARS-CoV nsp9 variants, their affinity for ssRNA 20-mers was reduced by 5- to 12-fold compared to the wild-type protein (Miknis et al., 2009). Replacement of some of the basic residues (e.g., Lys-10, Lys-51, and Lys-90) in the β-barrel domain of IBV nsp9 also significantly reduced the protein's capability to bind RNA in vitro, but these mutations only modestly affected virus replication upon reverse engineering (Chen et al., 2009a). It remains to be studied how nsp9 dimerization and mutagenesis may affect interactions with other replicase subunits, like nsp8 and nsp12-RdRp. These proteins were identified as nsp9 interaction partners using different technical approaches (Brockway et al., 2003; Sutton et al., 2004; von Brunn et al., 2007) and colocalize with nsp9 on the membranous replication organelles (Bost et al., 2000). At present, the available data suggest that, for efficient CoV replication, nsp9 homodimerization is a more critical feature than the protein's affinity for RNA per se. Alternatively, the correct positioning of RNA on larger protein complexes consisting of (or containing) nsp9 may be important for the protein's correct functioning in viral RNA synthesis (Miknis et al., 2009). Currently, the fact that suppressor mutations arose in MHV nsp9 (and nsp8) after mutagenesis of 3′-proximal MHV replication signals (see earlier) is the most compelling evidence for the involvement of nsp9–RNA interactions in a critical step of CoV replication. The protein may be part of a molecular switch (Züst et al., 2008) and/or possess features that are relevant to viral pathogenesis, as mutations in nsp9 were found to contribute to increased SARS-CoV pathogenesis in an animal model employing young mice infected with a mouse-adapted virus strain (MA-15) (Frieman et al., 2012).

2.4 Coronavirus nsp10

The small nsp10 subunit (139 residues in the case of SARS-CoV) is among the more conserved CoV proteins and is thought to serve as an important multifunctional cofactor in replication. Using yeast two-hybrid assays, nsp10 was shown to interact with itself, as well as with nsp1, nsp7, nsp14, and nsp16. These interactions were confirmed by coimmunoprecipitation and/or GST pull-down assays (Brockway et al., 2004; Imbert et al.,

2008; Pan et al., 2008; von Brunn et al., 2007). The important role of nsp10 in replication was first inferred from the phenotype of temperature-sensitive mutants of MHV in which an nsp10 mutation was responsible for a defect in minus-strand RNA synthesis (Sawicki et al., 2005). In addition, the protein was implicated in the regulation of polyprotein processing since an engineered MHV nsp10 double mutant (Asp-47 and His-48 to Ala) was partially impaired in the processing of the nsp4–nsp11 region (Donaldson et al., 2007).

When nsp10 was characterized in biochemical and structural studies, the protein was found to bind two Zn^{2+} ions with high affinity, suggesting the presence of two zinc-finger motifs (Matthes et al., 2006). Additionally, in in vitro assays, nsp10 displayed a weak affinity for single- and double-stranded RNA and DNA, although no obvious sequence specificity could be established, suggesting that the protein may function as part of a larger RNA-binding complex. Crystal structures of monomeric and dodecameric forms of SARS-CoV nsp10 were solved by different laboratories, but obvious structural rearrangements between the two forms were not detected (Joseph et al., 2007; Su et al., 2006). The structures revealed a new fold in which the Zn^{2+} ions are coordinated in a unique conformation and in which a cluster of basic residues on the protein's surface probably contributes to the RNA-binding properties of nsp10. More recent biochemical studies revealed that nsp10 interacts with nsp14 and nsp16 and regulates their respective ExoN and ribose-2′-O-MTase (2′-O-MTase) activities (Bouvet et al., 2010, 2012). Both these cofactor functions will be discussed in more detail later, in Section 5.

3. CORONAVIRUS nsp12: A MULTIDOMAIN RNA POLYMERASE

Although a virus-encoded RdRp is at the hub of the replication of all RNA viruses, special properties have long been attributed to the CoV RdRp. These ideas find their origin in a combination of CoV features, like the exceptionally long RNA genome (Gorbalenya et al., 2006), the complex mechanism underlying subgenomic RNA synthesis (Gorbalenya et al., 2006; Pasternak et al., 2006; Sawicki et al., 2007; Sola et al., 2011), the reported high RNA recombination frequency (Graham and Baric, 2010; Lai and Cavanagh, 1997), and the size and positioning of the RdRp-containing subunit, nsp12, within the replicase polyprotein. It remains to be elucidated to which extent features like polymerase processivity, fidelity, and template switching (during either genomic recombination or subgenome-length negative-strand RNA synthesis) are determined by the properties of the nsp12-RdRp subunit itself or by some of its protein cofactors, such as nsp7 and nsp8 (see earlier). In fact, some cofactors have been studied more extensively than nsp12 itself, and the same holds true for some of the specific RNA signals employed by the RdRp during, e.g., replication and subgenomic mRNA synthesis. Protein subunits of the larger RNA-synthesizing complex, like nsp7–nsp8, the nsp13-helicase, and the nsp14-ExoN, likely exert a strong influence on RdRp behavior and performance. On the other hand, a recent study employing homology modeling and reverse genetics of the MHV RdRp domain described the first two nsp12 mutations that can induce resistance to a mutagen and reduce the MHV RdRp error rate during virus passaging (Sexton et al., 2016). So, not unexpectedly, also features within nsp12 itself contribute to properties like nucleotide selectivity and fidelity regulation. All of the currently identified nsp12 cofactors, and most other CoV nsps, assemble into membrane-associated enzyme complexes (see earlier). The large number of viral subunits in these complexes (Subissi et al., 2014a), the likely requirement for host factors (van Hemert et al., 2008), and the concept of RNA synthesis occurring in a dedicated microenvironment in the infected cell (Knoops et al., 2008; V'Kovski et al., 2015) complicate the straightforward characterization of the CoV RdRp. To reconstitute the enzyme's activities in vitro, purified recombinant nsp12 is a key reagent but, for many years, such studies were hampered by poor nsp12 expression in *Escherichia coli*. The first in vitro activity assays have only been developed recently (Subissi et al., 2014b; te Velthuis et al., 2010), and the same technical issues with protein production explain the current lack of an nsp12 crystal structure. Consequently, structural information is restricted to sequence comparisons and some homology-based structure models of the C-terminal RdRp domain of the ~930-residue-long nsp12 (Xu et al., 2003). Moreover, most of what we have learned so far is based on the characterization of a single nsp12 homolog only, that of the SARS-CoV.

3.1 The nsp12 RdRp Domain

The nsp12-coding sequence includes the ORF1a/1b ribosomal frameshift site and a programmed -1 frameshifting event directs ORF1b translation to yield the pp1ab polyprotein that includes nsp12. The 3CL^{pro}-driven cleavage required to release the N-terminus of nsp12 is the same that separates nsp10 and nsp11. About 925–940 amino acids downstream (932 in the case of SARS-CoV), the nsp12/nsp13 cleavage site separates the CoV RdRp subunit from the helicase-containing cleavage product, which— uniquely among +RNA viruses—resides downstream of the RdRp domain for reasons that are poorly understood thus far (Gorbalenya et al., 2006).

Nsp12 consists of at least two domains, the recently described N–terminal "nidovirus-wide conserved domain with nucleotidyl transferase activity" (nidovirus RdRp-associated nucleotidyltransferase (NiRAN); see later) (Lehmann et al., 2015a) and the C-terminal canonical RdRp domain (Gorbalenya et al., 1989). The latter possesses the common motifs and structural features found in other RNA polymerases, which are often summarized as a "cupped right hand" with subdomains called fingers, palm, and thumb each playing specific roles in binding of templates and NTPs, initiation, and elongation (te Velthuis, 2014; Xu et al., 2003). In simplified form, the reaction catalyzed by the RdRp comes down to selecting the appropriate NTP to match with the template and the formation of a phosphodiester bond to extend the 3′ end of the nascent RNA chain with this incoming nucleotide (Ng et al., 2008; van Dijk et al., 2004). Reconstituting these activities in vitro using a purified RdRp preparation can be relatively straightforward, but sometimes is a huge technical challenge depending—among other factors—on the efficiency of recombinant RdRp expression and purification, the existence of specific template requirements (e.g., recognition signals), and the need for protein cofactors.

3.2 The Initiation Mechanism of the nsp12 RdRp

The initiation mechanism of the CoV RdRp, primer dependent or de novo, continues to be a much-debated issue, with important implications for the question of how CoVs maintain the integrity of the crucial terminal sequences of their genome. Compared to a de novo–initiating RdRp, the enzyme's active site, which is enclosed by the thumb and fingers domains, needs to be more accessible when a primer-template duplex has to be accommodated. De novo initiation, on the other hand, requires specific structural elements (so-called "priming loops") that serve to properly position the initiating NTPs for catalysis, thus creating an initiation platform for RNA synthesis. Bioinformatics analyses grouped the CoV RdRp with primer-dependent RdRps, as found in, e.g., picornaviruses and caliciviruses, in part based on the identification of a specific sequence motif (motif G) that is thought to mediate primer recognition (Fig. 3A) (Beerens et al., 2007;

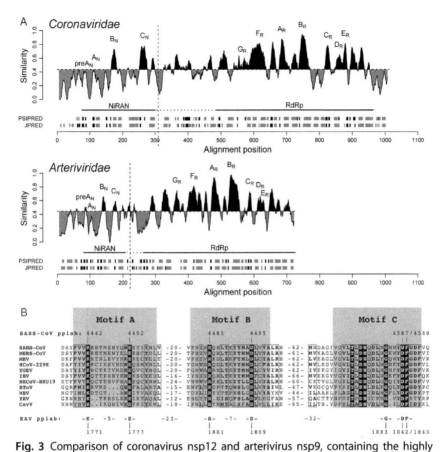

Fig. 3 Comparison of coronavirus nsp12 and arterivirus nsp9, containing the highly conserved NiRAN and RdRp domains. (A) Similarity density plot derived from a multiple sequence alignment including RdRp subunits from all nidovirus lineages. To highlight local deviations from the average, areas displaying conservation above and below the mean similarity are shaded in black and gray, respectively. Conserved sequence motifs of NiRAN (*subscript* N; see also B) and RdRp (*subscript* R) are labeled. Domain boundaries used for bioinformatics analyses and uncertainty with respect to the NiRAN/RdRp domain boundary are indicated with *vertical* and by *dashed horizontal lines*, respectively. Below each plot, the predicted secondary structure elements are presented in *gray* for α-helices and black for β-strands. (B) Multiple sequence alignment showing the three conserved motifs of the NiRAN domain from representative species across the *Nidovirales* order. Conserved residues in this alignment are shown in *white font*, while partially conserved residues are *boxed*. The *bottom line* depicts residues also conserved in the arterivirus EAV, which was used for a first experimental analysis of the NiRAN domain (Lehmann et al., 2015a). Abbreviations not explained in the main text: *NHCoV*, night-heron coronavirus HKU19 (genus *Deltacoronavirus*); *BToV*, bovine torovirus (family *Coronaviridae*, subfamily *Torovirinae*, genus *Torovirus*); *WBV*, white bream virus (family *Coronaviridae*, subfamily *Torovirinae*, genus *Bafinivirus*); *YHV*, yellow head virus (family *Roniviridae*, genus *Okavirus*); *CavV*, Cavally virus (family *Mesoniviridae*, genus *Alphamesonivirus*). (A) *Modified with permission from Lehmann, K.C., Gulyaeva, A., Zevenhoven-Dobbe, J.C., Janssen, G.M., Ruben, M., Overkleeft, H.S. et al., 2015. Discovery of an essential nucleotidylating activity associated with a newly delineated conserved domain in the RNA polymerase-containing protein of all nidoviruses. Nucleic Acids Res., 43, 8416–8434.*

Gorbalenya et al., 2002; Xu et al., 2003). This prediction appeared to be further supported by the identification of SARS-CoV nsp8 as a de novo-initiating second RNA polymerase (see earlier), capable of synthesizing products of up to six nucleotides in length that could serve to prime RNA synthesis by the nsp12-RdRp (Imbert et al., 2006). Support for a direct interaction between nsp8 and nsp12 was obtained using different technical approaches (Imbert et al., 2008; Subissi et al., 2014b; von Brunn et al., 2007). However, although a similar primer-independent RdRp activity was reported for the FCoV nsp8 (Xiao et al., 2012), other studies have called into question this concept of a primase–main RdRp (i.e., nsp8/nsp12) tandem working in concert to achieve initiation of processive CoV RNA synthesis (see later).

Using recombinant SARS-CoV nsp12, preliminary evidence for primer-dependent RdRp activity on poly(A) templates was first obtained using a GST–nsp12 fusion protein, although these efforts were hampered by protein instability, which also led to the conclusion that the N-terminal domain of nsp12 is required for activity (Cheng et al., 2005). Subsequently, a C-terminally His_6-tagged SARS-CoV nsp12 was found to mediate homo-polymeric RNA synthesis in a primer-dependent manner (te Velthuis et al., 2010). Both these activities must probably be considered relatively weak and nonprocessive compared to the activity observed when a SARS-CoV nsp12 RdRp assay was supplemented with nsp7 and nsp8 (Subissi et al., 2014b). However, at the same time, this study reinvigorated the debate on the initiation mechanism of the coronavirus RdRp, as the nsp7–8–12 tripartite complex displayed both primer-dependent and de novo initiation of RNA synthesis, whereas no de novo-initiating RdRp activity could be detected for nsp8 or the nsp7–nsp8 complex alone (Subissi et al., 2014b). To add to the confusion, other studies reported de novo initiation by SARS-CoV nsp12 alone (Ahn et al., 2012) and primer-dependent RdRp activity of SARS-CoV nsp8, when expressed without affinity tags commonly used to facilitate purification (te Velthuis et al., 2012). Technical differences between these studies and those summarized earlier (e.g., regarding expression constructs and templates used) may have contributed to the contradictory results obtained on the RdRp activities of nsp8 and nsp12. Thus far, five different laboratories addressed the two (putative) coronavirus RdRps in seven independent studies, none of which succeeded in exactly reproducing the results of any of the other studies (Ahn et al., 2012; Cheng et al., 2005; Imbert et al., 2006; Subissi et al., 2014b; te Velthuis et al., 2010, 2012; Xiao et al., 2012). Nidovirus RdRps appear to be technically challenging and sensitive proteins that may respond to minute changes in purification protocols or assay conditions. Clearly, both the role of nsp8 (primase or processivity factor?) and the initiation mechanism employed by the nsp12-RdRp require further study. Although the bioinformatics-based prediction that nsp12 uses a primer-dependent initiation mechanism is compelling, it lacks the direct support of an nsp12 crystal structure. At the same time, the question of the nature and source of the primer that would be used by nsp12 seems to be wide open again.

3.3 Inhibitors of the nsp12 RdRp

As for other RNA viruses, the nsp12-RdRp of CoVs is a primary drug target that may, in principle, be inhibited without major toxic side effects for the host cell. Nucleoside analogs constitute an important class of antiviral drug candidates that can target viral RdRps, but efforts to use them to inhibit CoV replication were not very successful thus far (Chu et al., 2006; Ikejiri et al., 2007). Moreover, it remains to be established that their target in the infected cell is indeed the nsp12-RdRp. The mismatch repair capabilities attributed to the nsp14-ExoN domain (see later) (Bouvet et al., 2012) may pose an additional hurdle, as the efficacy of a nucleoside analogue with anticoronavirus activity may be determined by the balance between its propensity to be incorporated by the nsp12-RdRp and its tendency to resist excision by the mismatch repair mechanism mediated by nsp14-ExoN.

Similar considerations apply to ribavirin, a guanosine analog with broad-spectrum antiviral activity that is used to treat patients infected with a variety of RNA viruses. Its mechanism of action appears to differ on a case-by-case basis, but may include the induction of lethal mutagenesis by increasing the RdRp error rate, inhibition of viral mRNA capping, and reduction of viral RNA synthesis by inhibition of the cellular enzyme inosine monophosphate dehydrogenase (IMPDH), which decreases the availability of intracellular GTP (Crotty et al., 2000, 2002; Smith et al., 2013, 2014). Although ribavirin was used to treat small numbers of SARS and MERS patients, high doses were used and the benefits of the treatment remained essentially unclear (Zumla et al., 2016). Experiments with different CoVs in animal models (Barnard et al., 2006; Falzarano et al., 2013) and infected cell cultures (Ikejiri et al., 2007; Pyrc et al., 2006) also established its poor activity and strongly suggested that ribavirin does not target the CoV RdRp directly or is targeted (itself) by the nsp14-ExoN activity (Smith et al., 2013). Innovative nucleoside inhibitors continue to be identified or developed (Peters et al., 2015; Warren et al., 2014) and the recently described in vitro RdRp assay (Subissi et al., 2014b) may prove very useful for establishing their mechanism of inhibition more precisely. A better understanding of nsp12-RdRp structure and function will also be required to design strategies that minimize the impact of drug resistance-inducing mutations, which are a common problem when targeting enzymes of rapidly evolving RNA viruses.

3.4 The nsp12 NiRAN Domain

Since the delineation of the borders of the CoV RdRp-containing replicase cleavage product (Boursnell et al., 1987; Gorbalenya et al., 1989), which is now known as nsp12, it had been clear that the protein must be a multi-domain subunit, with the canonical RdRp domain roughly occupying its C-terminal half (Fig. 3A). Only recently, first clues to some of the properties and possible functions of the N-terminal part of nsp12 were obtained (Lehmann et al., 2015a). A renewed bioinformatics analysis across the (still expanding) order Nidovirales revealed that the nidoviral RdRp-containing replicase subunit contains a conserved N-terminal domain of 200–300 residues (~225 residues in CoV nsp12; Fig. 3B). In CoV nsp12, about 175 residues separate the NiRAN and RdRp domains, leaving space for the presence of an additional domain between the two.

Based mainly on biochemical data obtained with the arterivirus homolog (see later), the N-terminal domain was concluded to possess an essential nucleotidylation activity and hence it was coined nidovirus RdRp-associated nucleotidyltransferase (NiRAN) (Lehmann et al., 2015a). NiRAN conservation was found to be lower than that of the downstream RdRp domain (Fig. 3A), but the analysis suggested that the evolutionary constraints on NiRAN have been similar in different nidovirus lineages, which would be in line with a conserved function. Gorbalenya and colleagues identified three key NiRAN motifs (A–B–C) containing seven invariant residues (Fig. 3B), with domains B and C being most conserved (Lehmann et al., 2015a). The identification of the NiRAN domain was further supported by the conservation of its predicted secondary structure elements in different nidovirus families (Fig. 3A). Extensive database searches did not reveal potential NiRAN homologs in either the viral or the cellular world, although it cannot be excluded that the domain has diverged from cellular ancestors to a level that prevents their identification with the currently available sequences and tools. Nevertheless, its unique presence in nidoviruses and its association with the important RdRp domain suggest that NiRAN may be a crucial regulator or interaction partner of the downstream RdRp domain that must have been acquired before the currently known nidovirus lineages diverged. NiRAN and the zinc-binding domain (ZBD) that is associated with the nsp13-helicase protein (see later) are the only unique genetic markers of the order *Nidovirales* identified thus far.

Mainly due to the lack of sufficient amounts of recombinant CoV nsp12, the preliminary biochemical characterization of NiRAN was restricted to its arterivirus homolog, using recombinant nsp9 of equine arteritis virus (EAV) (Lehmann et al., 2015a). For both EAV and SARS-CoV, it could be shown that replacement of conserved NiRAN residues can cripple or completely block virus replication in cultured cells. A combination of biochemical assays revealed that in vitro the NiRAN domain exhibits a specific, Mn^{2+}-dependent enzymatic activity that results in the self-nucleotidylation of EAV nsp9. The activity was abolished upon mutagenesis of conserved key residues in NiRAN motifs A, B, and C. Although UTP was found to be the preferred substrate for NiRAN's in vitro nucleotidylation activity, also GTP could be used, albeit less efficiently. The conserved lysine residue in motif A (the EAV equivalent of Lys-73 in SARS-CoV nsp12) was concluded to be the most likely target residue for nucleotidylation via formation of a phosphoamide bond.

Although the importance of the NiRAN domains of arterivirus nsp9 and coronavirus nps12 was supported by the outcome of reverse-genetics studies (Lehmann et al., 2015a), the role of the produced protein–nucleoside adducts in viral replication remains unclear at present. In fact, the unique dual specificity for UTP and GTP seems to argue against two initially considered potential NiRAN functions (Lehmann et al., 2015a). The first of these was a role as an RNA ligase, a type of activity however that commonly is ATP dependent. The second was its involvement in synthesizing mRNA cap structures. One of the four enzymes required for this process, the crucial guanylyl transferase (GTase), still remains to be identified for CoVs (see later). However, NiRAN's substrate preference for UTP over GTP is difficult to reconcile with this hypothesis and has not been observed for other GTases involved in mRNA capping. The third hypothesis that was put forward links back to the open question of the initiation of coronavirus RNA synthesis, which presumably is a primer-dependent step (see earlier). Nsp12 nucleotidylation could be envisioned to play a role in protein-primed RNA synthesis, a strategy used by, e.g., picornaviruses and their relatives, which covalently attach an oligonucleotide to a viral protein (called VPg in the case of picornaviruses) that subsequently mediates the initiation of RNA synthesis (Paul et al., 2000). The first step in the synthesis of the "protein primer" is a nucleotidylation step during which a nucleotide monophosphate is covalently attached to the VPg. NiRAN could be involved in a similar mechanism either directly or indirectly, by transferring the bound nucleotide to another protein player. Although such a mechanism would definitely revolutionize the concept of the initiation of CoV RNA synthesis, it is clearly not very compatible with some of the currently available data, such as the reported presence of a $5'$ cap structure (rather than a VPg–like molecule) on CoV mRNAs. Evidently, the further in–depth characterization of NiRAN is needed to fill the current knowledge gaps, starting with the biochemical characterization of a CoV NiRAN domain, which may confirm and extend the features now deduced from the analysis of its distantly related arterivirus homolog.

4. CORONAVIRUS nsp13: A MULTIFUNCTIONAL AND HIGHLY CONSERVED HELICASE SUBUNIT

Helicases are versatile NTP–dependent motor proteins that play a role in cellular nucleic acid metabolism in the broadest possible sense, including processes like DNA replication, recombination and repair, transcription, translation, as well as RNA processing. Helicases are also encoded by all +RNA viruses with a genome size exceeding 7 kb, suggesting they are required for the efficient replication of +RNA viral genomes above this size threshold. Given the large size of the genomes of CoVs and related nidoviruses, they may depend on the function(s) of a replicative helicase even more than other +RNA virus taxa. However, despite their abundance and conservation, the specific role of helicases in +RNA virus replication remains poorly understood. For an extensive recent review of nidovirus helicases, the reader is referred to Lehmann et al. (2015c).

Currently, helicases are classified into six superfamilies (SFs) (Singleton et al., 2007), with +RNA viral helicases belonging to SF1 (e.g., alphaviruses and nidoviruses), SF2 (e.g., flaviviruses), or SF3 (e.g., picornaviruses). The presence of a SF1 helicase (HEL1) domain in the CoV replicase polyprotein was discovered upon the early in–depth analysis of the first full–length CoV genome sequence that became available (IBV) (Gorbalenya et al., 1989). The HEL1 domain maps to the C-terminal part of the replicase cleavage product that is now known as nsp13, which is about 600 residues long. The CoV HEL1 domain contains all characteristic sequence motifs of the

SF1 superfamily. The N-terminal part of nsp13 is formed by a multinuclear ZBD, one of the most conserved domains across the order *Nidovirales* (Gorbalenya, 2001; Nga et al., 2011). This qualification also applies to the helicase-containing subunit as a whole, despite considerable size differences between, e.g., CoV nsp13 and its arterivirus homolog (designated nsp10) (Lehmann et al., 2015c). The ZBD and HEL1 domains occupy a conserved position downstream of the RdRp domain in all nidovirus replicase polyproteins studied so far.

4.1 The Coronavirus nsp13 SF1 Helicase (HEL1)

SF1 helicases contain at least a dozen conserved motifs that direct the binding of NTPs and nucleic acids. Of these, motifs I and II (also known as the Walker A and B boxes) are common to helicases of all SFs as well as NTPases. Structurally, the catalytic core of SF1 helicases like the CoV HEL1 domain is formed by two RecA-like domains, designated 1A and 2A (Fig. 4), that bind to nucleic acids through stacking interactions of aromatic residues with the bases of their nucleic acid substrates (Velankar et al., 1999). Cyclic conformational changes of the RecA-like domains mediate the conversion of the energy from hydrolysis of the phosphodiester bonds of NTPs into directional movement along the nucleic acid substrate, with the so-called "inchworm" model now widely being considered as best supported by the available experimental data (Lehmann et al., 2015c; Velankar et al., 1999; Yarranton and Gefter, 1979). Additional domains, located up- or downstream of 1A and 2A, or inserted internally, can mediate supplemental protein–protein and protein–nucleic acid interactions or enzymatic activities, thus contributing to the functional versatility and specificity of the enzyme (Lehmann et al., 2015c; Singleton et al., 2007).

Within helicase SF1, the CoV HEL1 domain belongs to the Upf1-like family (SF1B) which is characterized by moving in the 5′-to-3′ direction along the nucleic acid strand to which they bind. Upf1-like helicases may unwind either DNA or RNA and, in some cases, also both substrates without a clear preference, as was readily observed during the in vitro characterization of different nidovirus helicases. The CoV HEL1 activity was first demonstrated in vitro using recombinant HCoV-229E nsp13 (Seybert et al., 2000a). Bacterially expressed nsp13 from HCoV-229E and SARS-CoV, and also the homologous helicase (nsp10) of the arterivirus EAV, displayed 5′-to-3′ unwinding activity on double-stranded RNA or DNA substrates containing single-stranded 5′ overhangs (Ivanov and Ziebuhr,

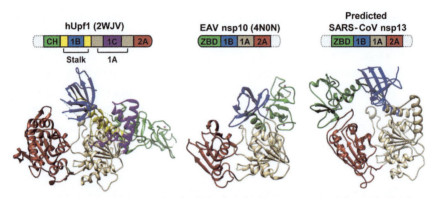

Fig. 4 Three-dimensional models of cellular hUpf1 (the prototype of the Upf1-like family of SF1 helicases), the EAV nsp10 helicase (Deng et al., 2014), and the predicted structure of SARS-CoV nsp13. Based on sequence and structural comparisons, nidovirus helicases are classified into the Upf1-like family. Domain colors in the structures correspond to those used in the domain organization depicted above each structure, in which domain sizes are not drawn to scale. *Dashed* domains represent parts that could not be modeled. Zn^{2+} ions bound to the respective N-terminal domains are depicted as *pink spheres*. The identical coloring of domains other than 1A and 2A does not imply an evolutionary relationship. PDB accession numbers are listed in brackets. *Modified with permission from Lehmann, K.C., Snijder, E.J., Posthuma, C.C., Gorbalenya, A.E., 2015. What we know but do not understand about nidovirus helicases. Virus Res., 202, 12–32.*

2004; Ivanov et al., 2004b; Seybert et al., 2000a,b; Tanner et al., 2003). Following the biochemical characterization of SARS-CoV nsp13, it was calculated that unwinding occurs in discrete steps of 9.3 base pairs each, with a catalytic rate of 30 steps per second (Adedeji et al., 2012a). The nsp13 NTPase activity can use all four natural ribonucleotides and nucleotides as substrate, with ATP, dATP, and GTP being hydrolyzed most efficiently, and UTP being the least preferred substrate (Ivanov and Ziebuhr, 2004; Ivanov et al., 2004b; Tanner et al., 2003). Replacement of a conserved Lys in motif I, the Walker A box (Walker et al., 1982), kills the in vitro NTPase activity of all nidovirus helicases tested thus far and, when introduced by reverse genetics, this mutation also abolished replication of the arterivirus EAV (Seybert et al., 2000b).

The substrate preferences summarized earlier support a three-dimensional model of the SARS-CoV HEL1 core domains (1A and 2A) that was based on structural information available for multiple cellular helicases (Hoffmann et al., 2006). The model predicts both the existence of multiple hydrogen bonding interactions with the β- and γ-phosphates of the NTP and a lack of specific interactions with the nucleobase. Thus, the mere presence of a $5'$ triphosphate group appears to be the main determinant for NTP/dNTP binding. Since nidovirus helicases are presumed to unwind double-stranded RNA intermediates that are formed during viral replication, considerable attention was given to the in vitro characterization of their nucleic acid substrate preferences (Seybert et al., 2000a,b). The HCoV-229E and EAV helicases could not unwind substrates with $3'$ single-stranded tails or blunt-ended substrates. In contrast, RNA and DNA substrates with one or two $5'$ single-stranded regions were unwound efficiently, suggesting that the nidovirus helicase must bind to a single-stranded region before initiating unwinding in the $5'$-to-$3'$ direction. However, the in vitro assays did not yield any clear indications for the preferred recognition of specific sequences or higher-order structures in the substrate (Lehmann et al., 2015c). Also a more in-depth biochemical characterization, performed with SARS-CoV nsp13, confirmed that the CoV helicase does not discriminate between RNA and DNA substrates (Adedeji et al., 2012a). Consequently, it cannot be excluded that the enzyme, in addition to being engaged in viral RNA synthesis, may also target host DNA. Nuclear translocation of nidovirus helicases has not been reported thus far, but the light microscopy techniques used to study the protein's subcellular distribution would not suffice to detect the nuclear import of only a small fraction of the protein.

As a final *caveat* it should be stressed that the biochemical properties summarized earlier are all derived from in vitro studies using recombinant helicases, expressed in different systems and sometimes containing substantial foreign sequences. The in situ characterization of the helicase as one of the key enzymes of the nidovirus RNA-synthesizing machinery remains to be addressed. In that context, sequence specificity, for example, could be conveyed by other subunits of the replicase complex, which may target the helicase to, e.g., the initiation sites for viral genome or antigenome synthesis, or to signals controlling the production of subgenomic mRNAs. As summarized by Lehmann et al. (2015c), other important helicase features that could be dramatically different in the setting of the infected cell are (the need for) helicase oligomerization, cooperativity between multiple helicase molecules binding to the same substrate, and—consequently—the overall processivity of the enzyme, which in vitro appeared to be quite low given the large CoV genome size (Adedeji et al., 2012a).

4.2 The Helicase-Associated ZBD

The nidovirus helicase subunit domain is unique among its +RNA virus homologs in having a conserved N-terminal domain of 80–100 residues that contains 12 or 13 conserved Cys/His residues (den Boon et al., 1991; van Dinten et al., 2000). The domain was recognized as a potential ZBD (Gorbalenya et al., 1989) and early in vitro studies with the recombinant HCoV-229E and EAV helicases confirmed that Zn^{2+} ions are essential for retaining the protein's enzymatic activities, suggesting that ZBD modulates nidovirus helicase function (Seybert et al., 2005). A recent structural study of the arterivirus nsp10-helicase (Deng et al., 2014) will be discussed in more detail later. This first nidovirus helicase structure confirmed the binding of three zinc ions by the ZBD, which adopts a unique fold that combines a RING-like module with a so-called "treble-clef" zinc finger.

ZBD and HEL1 interact extensively (Deng et al., 2014) but, in the helicase primary structure, they are separated by a variable and uncharacterized domain that essentially explains the size difference of about 130 residues between CoV and arterivirus helicase subunits (Seybert et al., 2005). Using the arterivirus prototype EAV, the functional importance of ZBD was probed extensively by combining biochemistry and reverse genetics (Seybert et al., 2005; van Dinten et al., 2000). This yielded a variety of phenotypes for nsp10-ZBD mutants, the most striking being mutants deficient in subgenomic mRNA synthesis while remaining capable of (and even enhancing) viral genome replication (van Dinten et al., 1997, 2000) (see later). Most replacements of conserved ZBD Cys and His residues profoundly impacted the helicase activity of EAV nsp10, even when performed in a semiconservative manner that could preserve zinc binding. In reverse-genetics studies, most of these ZBD mutations rendered the virus nonviable. Recently, the impact of these mutations on ZBD integrity and ZBD–HEL1 interactions could be rationalized with the help of the nsp10 crystal structure (Deng et al., 2014).

4.3 Nidovirus Helicase Structural Biology

Despite its importance as a potential drug target, a CoV nsp13 or HEL1 crystal structure has not been obtained thus far due to technical complications with recombinant protein production and crystallization. Instead, several CoV helicase models have been described, mainly based on cellular helicase structures (Bernini et al., 2006; Hoffmann et al., 2006; Lehmann et al., 2015c). Given this limitation, and despite the large evolutionary distance between the two enzymes, it is interesting to have a closer look at the recently published EAV nsp10-helicase structure (Deng et al., 2014).

The overall structure of EAV nsp10 (Fig. 4) consists of the N-terminal ZBD, a new domain designated 1B, the two recA-like HEL1 domains

(1A and 2A), and a short C-terminal domain, which is not conserved among nidoviruses and needed to be deleted to allow nsp10 crystallization. This 65-amino-acid C-terminal truncation did not affect the helicase core domains and only modestly influenced levels of ATPase and helicase activity compared to the full-length protein (Deng et al., 2014). Compared to cellular representatives of the SF1 helicase superfamily, nsp10 is most similar to Upf1 and its close homolog Ighmbp2, which also contain an N-terminal ZBD. Moreover, the location and orientation of the newly discovered 1B domain of nsp10 (residues 83–137) resembles that of the domain with the same name found in the Upf1-like helicase subfamily.

The nsp10 ZBD uses 12 conserved Cys and His residues to coordinate three zinc ions and folds into two zinc-binding modules that are connected by a disordered region. The N-terminal RING-like structure of nsp10 coordinates two zinc ions and the closest similarity that was found for this module was with the N-terminal zinc-binding CH-domain of Upf1. Both proteins have a second zinc-binding module downstream (a so-called treble-clef zinc finger in the case of nsp10), but these are structurally different, suggesting that the nidoviral ZBD represents a new kind of complex zinc-binding element. The previous suggestion of ZBD codetermining HEL1 function was strengthened by the presence of an extensive interface of 1019 $\mathring{A}^2$ that was proposed to be involved in intramolecular signaling (Deng et al., 2014). A second crystal structure was obtained for nsp10 in complex with a partially double-stranded DNA substrate, revealing possible nucleic acid-binding clefts at the protein's surface that are formed by the ZBD + 1B and ZBD + 1A domains. Although the exact path of the nucleic acid strands could not yet be determined, it became clear that the positively charged ZBD, and in particular its N-terminal RING-like module, must be involved in nucleic acid binding. In line with the biochemical data summarized earlier, most of the nsp10-substrate contacts identified are not base-specific and occur with the nucleic acid backbone. Whereas the HEL1 core domains were found to be quite similar in the absence or presence of bound substrate, a remarkable 29 degree rotation was observed for domain 1B, enlarging the dimensions of the nucleic acid substrate channel formed by domains 1A and 1B, but not allowing it to accommodate a duplex substrate. Consequently, it was postulated that an element near the entrance of the substrate channel may destabilize the duplex and facilitate the entrance of one of the strands into the channel. Since the double-stranded region of the duplex could not be modeled, additional studies are needed to verify the existence and molecular details of this proposed unwinding mechanism

(Deng et al., 2014; Lehmann et al., 2015c). Likewise, direct structural information on CoV nsp13 is needed to be able to assess to which extent the structural observations made for arterivirus nsp10 can be translated to distantly related (and larger) nidovirus helicases (Fig. 4). In general, however, the analysis of nsp10 provided a clear basis for a model in which the function of the common RecA-like core domains of nidovirus helicases is modulated by specific extension domains, presumably to facilitate the involvement of the nidovirus helicase in multiple processes in the infected cell (see later).

4.4 Functional Characterization of the Nidovirus Helicase

As outlined earlier, the biochemical characterization of purified recombinant nidovirus helicases, the functional probing of (in particular) the EAV nsp10-helicase using reverse genetics, the EAV nsp10 structure, and advanced bioinformatics analyses together have painted a picture of an enzyme that is involved in multiple critical steps of the viral replicative cycle. Space limitations do not allow an in-depth discussion of all of these (putative) functions, which—based on EAV nsp10 studies—may include a poorly characterized role in virion biogenesis (van Dinten et al., 2000), not unlike what was uncovered for, e.g., the helicase-containing NS3 protein of flaviviruses (Liu et al., 2002; Ma et al., 2008). Likewise, we will not discuss the first reports on possible interactions with host proteins, such as the Ddx5 helicase (for SARS-CoV nsp13; Chen et al., 2009b) and polymerase δ (for IBV nsp13; Xu et al., 2011). Instead, we will focus on the most significant findings related to nidovirus helicase functions in RNA synthesis and processing, specifically (i) genome replication, (ii) transcription of subgenomic mRNAs, (iii) mRNA capping, and (iv) posttranscriptional mRNA quality control.

The presumed "default" function of +RNA viral helicases is to cooperate with the viral RdRp to achieve the efficient amplification of the genome. In this context, helicases are presumed to promote RdRp processivity by opening up the double-stranded RNA intermediates of viral replication, and possibly also by removing RNA secondary structures in single-stranded template strands (Kadare and Haenni, 1997). In this light, reports on molecular interactions between the CoV RdRp (nsp12) and helicase (nsp13) were not unexpected (Imbert et al., 2008; von Brunn et al., 2007). The same holds true for the observation that SARS-CoV nsp12 can stimulate the in vitro helicase activity of nsp13 (Adedeji et al., 2012a) and for the fact that both nsp12 and nsp13 (like most other CoV replicase cleavage products) associate with the membranous replication organelles in nidovirus-infected cells (Denison et al., 1999; Ivanov et al., 2004b; Knoops et al., 2008). In spite of all these indications for RdRp-helicase interplay, the polarity of nidovirus helicase-mediated unwinding (5′-to-3′) remains a major conundrum, as it is opposite to the polarities of the RdRp and many other +RNA helicases, which move in a 3′-to-5′ direction on the RNA strand they initially bind to (Seybert et al., 2000a). This strongly suggests that the two enzymes cannot simply operate as a tandem that moves along the same template strand while copying it, a consideration that also applies to the SF1B helicase employed by alphaviruses. This problem could be resolved if RdRp and helicase would move along different strands of the same RNA duplex, which might allow the helicase to separate the two strands and provide a single-stranded template to be copied by the RdRp (Lehmann et al., 2015c). Also, it is tempting to speculate that the helicase, by trailing along the nascent strand (following the RdRp at a certain and, possibly, somewhat flexible distance), provides (i.e., leaves behind) a single-stranded template, thus facilitating initiation and elongation of RNA synthesis by the next RTC. This, for example, could occur in cases when multiple RTCs act simultaneously/consecutively on the same template to produce multiple plus-strand RNAs from the same minus-strand template, a process that is generally thought to add to the large excess of plus- over minus-strand RNAs in nidovirus-infected cells. Clearly, significantly more work is needed to explore this possibility.

Nidovirus sg mRNAs are each produced from their own subgenome-length minus-strand template (see Section 1). In the case of arteri- and coronaviruses, these derive from a process of discontinuous minus-strand RNA synthesis and this unique mechanism likely requires specific functional interactions between (among others) RdRp and helicase. These interactions may contribute to maintaining a proper balance between continuous and discontinuous minus-strand RNA synthesis, and thus between the production of new genomes and sg mRNAs. The serendipitous identification of an arterivirus nsp10-ZBD mutation (Ser-59 → Pro) that essentially inactivated transcription while leaving replication unaffected was an early indication for the involvement of the nidovirus helicase in the control of sg mRNA synthesis (van Dinten et al., 1997). Such control functions could also be related to the recognition of TRSs (Fig. 1), the frequency with which each of the TRSs is used to produce a subgenome-length minus strand, or mechanistic aspects of the stalling and reinitiation of RNA synthesis or the transfer of the nascent strand to an upstream position on the template (see earlier), which must occur during the discontinuous step in sg RNA synthesis (Lehmann et al., 2015c). Recently, the EAV nsp10 Ser-59→Pro mutation, which selectively reduces transcription of all subgenomic mRNAs to below 1% of their normal levels, was reanalyzed in the context of the nsp10 crystal structure. As postulated when this virus mutant was first described, its phenotype appears to be based on the special structural properties of the Pro residue in combination with the position of residue 59 in a "hinge" region that connects ZBD to the rest of the protein (Deng et al., 2014). Although residue 59, located just downstream of the second zinc-binding module of the ZBD, is fairly distant from the RNA-binding surface, it resides in a region that connects the ZBD treble-clef zinc finger to a helix that interacts with domains 1A and 1B and with the nucleic acid. Thus, specific mutations affecting the flexibility of this hinge region may drastically influence the long-distance signaling within nsp10, apparently preventing the RNA-synthesizing machinery to work in "transcription mode" and dedicating it exclusively to full-length minus-strand RNA synthesis and genome amplification. Since this kind of mutations barely affected nsp10s in vitro NTPase and helicase activities (Seybert et al., 2005), it may well be that changed interactions with specific protein partners will turn out to be the key to explaining the transcription-negative phenotype of the corresponding virus mutants (Lehmann et al., 2015c). For the coronavirus IBV, a point mutation in a somewhat comparable position of nsp13 (Arg-132→Pro; just downstream of ZBD) was reported to cause a similar block of sg mRNA synthesis (Fang et al., 2007) but, thus far, this observation has not been followed up in more detail for IBV or confirmed for nsp13 of another CoV.

In addition to its role in RNA synthesis, the nidovirus helicase is also assumed to be involved in the capping pathway of viral mRNAs by exhibiting an RNA 5′-triphosphatase (RTPase) activity that can remove the 5′-terminal triphosphate from the RNA substrate. For SARS-CoV and HCoV-229E nsp13, this first step in viral cap synthesis was shown to rely on the same NTPase active site that provides the energy for the protein's helicase activity (Ivanov and Ziebuhr, 2004; Ivanov et al., 2004b). The CoV capping pathway is discussed in Section 5 of this review.

Finally, the remarkable similarities between EAV nsp10 and the cellular helicase Upf1 (Deng et al., 2014) have given rise to the intriguing but still speculative hypothesis that the nidovirus helicase may be involved in the posttranscriptional quality control of viral RNA. Common features of the two helicases include their 5′-to-3′ polarity of unwinding, their lack of substrate specificity and striking similarities in terms of domain organization and fold (Fig. 4) (Lehmann et al., 2015c). Using several pathways, including nonsense-mediated mRNA decay, Upf1 mediates RNA quality control in eukaryotic cells (Cheng et al., 2007; Clerici et al., 2009), while its activity can be modulated through interactions of its N-terminal ZBD. It was postulated that a similar function in nidovirus replication, e.g., detection and elimination of defective viral mRNAs (including the genome), could explain the conservation of the unique ZBD across the nidovirus order, which stands out for containing members with very large + RNA genomes. Such a function would prevent the synthesis of defective viral polypeptides, which might interfere with the proper functioning of full-length viral proteins. In this manner, not unlike the nsp14-ExoN domain (see earlier), the nidovirus helicase may have contributed to genome expansion by providing a form of compensation for the relatively poor fidelity of genome replication by the nidoviral RdRp (Deng et al., 2014). Clearly, this is just one of the scenarios for the involvement of the helicase in the posttranscriptional fate of viral RNA products. Further experimental work will be needed to explore these possibilities in more detail, as they are compatible with the much broader realization that the functions of RNA helicases can extend far beyond merely the unwinding of RNA structure.

4.5 The Coronavirus Helicase as Drug Target

Due to its multifunctionality and involvement in several key processes in viral RNA synthesis and processing, the nidovirus helicase is an important target for antiviral drug development, which was mainly explored for CoVs following the SARS-CoV outbreak. The highly conserved nature of the helicase offers the interesting perspective of developing inhibitors with a potential broad-spectrum activity. On the other hand, avoiding toxicity resulting from inhibition of the abundant cellular NTPases/helicases poses a serious challenge, which is why—as in the case of the RdRp—obtaining a crystal structure for a CoV helicase should be considered a research priority. In theory, a variety of helicase properties may be targeted with specific inhibitors, ranging from the active and nucleic acid-binding sites of the enzyme to interaction surfaces for multimerization and modulation by protein cofactors (Kwong et al., 2005). Several compound families were found to target the ATPase of nsp13, and thus also its helicase activity. These include naturally occurring flavonoids (Yu et al., 2012), chromones (Kim et al., 2011), and bananins (Kesel, 2005; Tanner et al., 2005), all exhibiting in vitro IC_{50} values in the low-micromolar range. Other compounds appear to target the helicase of nsp13 activity more specifically, like the triazole SSYA10-001, which was found to inhibit the replication of multiple beta-CoVs (SARS-CoV, MERS-CoV, and MHV) in cell culture-based infection models (Adedeji et al., 2012b, 2014). The IC_{50} value in in vitro helicase assays was about 5.5 μM, whereas EC_{50} values in cell cultures assays were in the range of 7–25 μM, depending on the CoV analyzed, suggesting that broad-spectrum activity may indeed be achieved. To our knowledge, the antiviral potential and toxicity of SSYA10-001 in animal models remain to be tested. Another interesting group of helicase-directed antiviral hits are bismuth complexes, which were postulated to inhibit the NTPase and helicase functions by competing for zinc ions with the ZBD. In SARS-CoV–infected cell cultures the determined EC_{50} and CC_{50} values were 6 μM and 5 mM, respectively (Yang et al., 2007).

5. THE CORONAVIRUS CAPPING MACHINERY: nsp10–13–14–16

Cap structures consists of a 7-methylguanosine (^{m7}G) linked to the first nucleotide of the RNA transcript through a $5'$–$5'$ triphosphate bridge (for a review, see Decroly et al., 2012). In eukaryotic cells, the synthesis of the cap structure is a multistep process that occurs in the nucleus and is coupled to RNA pol II–driven transcription (Shatkin, 1976). Capping begins with the hydrolysis of the $5'$ γ-phosphate of the nascent RNA transcript by an RNA $5'$-triphosphatase (RTPase). Subsequently, a GMP molecule is transferred to the $5'$-diphosphate of the RNA by a GTase, leading to the formation of GpppN-RNA. The cap structure is methylated at the N7 position of the guanosine by an (AdoMet)-dependent N7-MTase, yielding cap-0 ($^{m7}GpppN$). The cap-0 structure is then converted into cap-1 ($^{m7}GpppN_{2'-Om}$) or cap-2 by an AdoMet-dependent $2'$-O-MTase that methylates the $2'$-O position of the ribose of the first or first and second RNA nucleotide, respectively.

Due to the cytoplasmic localization of their mRNA synthesis, nidoviruses, and all other +RNA viruses of eukaryotes, cannot rely on the standard capping pathway outlined earlier, which is executed by host cell enzymes in the nucleus. Cap structures can protect viral mRNAs from degradation by the cellular $5'$-to-$3'$ exoribonucleases involved in RNA turnover (Liu and Kiledjian, 2006). Cap methylation is critical for mRNA recognition by translation initiation factor eIF4E (Filipowicz et al., 1976; Ohlmann et al., 1996), and thus for viral translation and replication as a whole (Ferron et al., 2012). In addition to promoting mRNA stability and securing their recognition by host cell ribosomes, capping of viral mRNAs also promotes escape from certain antiviral responses of the host cell. The retinoic acid-inducible gene 1 and melanoma differentiation-associated protein 5 (Mda5) were shown to detect either uncapped triphosphorylated RNA or cap-0-containing RNA (Devarkar et al., 2016; Hyde and Diamond, 2015; Hyde et al., 2014; Schuberth-Wagner et al., 2015; Züst et al., 2011), resulting in the expression of antiviral interferon-stimulated genes (ISGs) in infected and neighboring cells. Interferon-induced proteins with a tetratricopeptide repeat 1 (IFIT1/56) and IFIT2/54 (IFIT2) have been shown to recognize miscapped RNAs, in order to restrict viral translation (Daffis et al., 2010; Pichlmair et al., 2011). A subsequent study identified IFIT1 as the only interferon-induced protein whose RNA-binding affinity was affected by the ribose-$2'$-O methylation state of the $5'$ cap structure (cap-0/cap-1). The data support a model in which IFIT1 efficiently binds and sequesters capped mRNA that lacks a ribose-$2'$-O-methyl group. Consistent with this, viral mRNA translation was shown to be reduced in cells infected with $2'$-O-MTase-deficient CoVs (Habjan et al., 2013). Other studies suggested that $2'$-O methylation of cap structures prevents or delays the Mda5-dependent recognition of viral mRNAs as "nonself." This mRNA cap modification thus limits the antiviral response launched upon infection, thereby affecting viral pathogenesis (Daffis et al., 2010; Schuberth-Wagner et al., 2015; Züst et al., 2011).

The genomic and sg mRNAs of nidoviruses are presumed to be capped at their $5'$ end and polyadenylated at their $3'$ end (Fig. 1). The presence of a cap structure was first suggested based on studies using ^{32}P-labeled MHV RNA that was digested with RNase T1 and T2 and subjected to DEAE-cellulose chromatography (Lai and Stohlman, 1981). The presence of a cap was further substantiated by immunoprecipitation experiments using a cap-specific monoclonal antibody that was shown to specifically bind to equine torovirus mRNAs (van Vliet et al., 2002). Although the (presumed) capping machinery of arteriviruses has remained essentially uncharacterized thus far (Lehmann et al., 2015b), three conserved putative capping enzymes were identified in the conserved ORF1b-encoded part of the replicase of *Coronaviridae*, *Roniviridae*, and *Mesoniviridae*, which all have substantially larger genomes. These enzymatic activities, which were proposed to participate in the synthesis of a cap-1 structure (^{N7m}GpppN$_{2'Om}$), are the following: (i) the nsp13 helicase/RTPase (Ivanov and Ziebuhr, 2004), (ii) the nsp14 N7-MTase (Chen et al., 2009c; Ma et al., 2015), and (iii) the nsp16 2′-O-MTase (Bouvet et al., 2010; Decroly et al., 2008; Snijder et al., 2003; von Grotthuss et al., 2003; Zeng et al., 2016).

At present, structural and functional studies, which were mainly performed with purified recombinant SARS-CoV nsps, suggest that CoVs follow a capping pathway that is very similar to that of eukaryotic cells. Cap synthesis is presumed to start by hydrolysis of the 5′ end of a nascent RNA by the RTPase function of nsp13 to yield pp-RNA (Ivanov and Ziebuhr, 2004). Then, a still elusive GTase must transfer a GMP molecule onto the pp-RNA to yield Gppp-RNA. The cap structure is then methylated at the N7 position by the N7-MTase domain of the bifunctional nsp14 (Bouvet et al., 2010; Chen et al., 2009c). Cap modification is completed by the conversion of the cap-0 into a cap-1 structure, which involves the nsp10/nsp16 complex (Bouvet et al., 2010; Chen et al., 2009c) in which nsp16 possesses 2′-O-MTase activity. In the following paragraphs, we will describe the different CoV enzymes involved in mRNA capping in more detail.

5.1 The nsp13 RNA 5′ Triphosphatase

As described earlier, the nsp13 helicase domain is thought to unwind double-stranded RNA in a 5′-to-3′ direction, an activity that energetically depends on the nucleotide triphosphatase (NTPase) function of the enzyme (Ivanov and Ziebuhr, 2004; Ivanov et al., 2004b; Seybert et al., 2000a). Additionally, using the same active site, the protein was found to exhibit RTPase activity (Ivanov and Ziebuhr, 2004), which was found to be abolished if the conserved active-site Lys residue of the Walker A box motif was replaced with Ala (Ivanov and Ziebuhr, 2004; Ivanov et al., 2004b). Specific RTPase activity on viral mRNA species would require specific recruitment of nsp13 to the 5′ end of viral mRNAs, which has not been demonstrated for CoV helicases but may involve yet other factors. In this context, it remains to be studied if the common leader sequence present on CoV mRNAs contributes to the recruitment of nsp13 and/or other proteins involved in 5′ capping reactions.

Several other +RNA virus helicases were shown to possess an activity that can target the phosphodiester bond between the β and γ phosphate groups of the 5′-terminal NTP of diverse substrate RNAs, suggesting that this dual function of helicase and capping RTPase is a common feature in this group of viruses (Decroly et al., 2012; Ivanov and Ziebuhr, 2004). Even though experimental evidence has been obtained to suggest an nsp13-associated 5′

RTPase activity, coronavirus nsp13 homologs proved to be unable to *trans*-complement the yeast strain YBS20, which lacks the *CET1* locus encoding the yeast RTPase involved in mRNA capping (Chen et al., 2009c). The lack of *trans*-complementation by nsp13 could be due to technical reasons, such as inappropriate subcellular localization of nsp13, misfolding of the protein, or a functional mismatch with other players of the distantly related yeast capping machinery. Alternatively, it could indicate specific substrate requirements for coronavirus nsp13-associated RTPase activities. In any case, a possible involvement of nsp13 in the first step of CoV mRNA capping remains to be corroborated in further studies, for example, by in vitro reconstitution experiments of the entire CoV capping pathway.

5.2 The Elusive RNA GTase

The CoV nsp involved in the second step of RNA capping, GTase, remains to be identified. Bioinformatics analysis of CoV genome sequences failed to identify replicase gene-encoded domains that may perform this activity. Eukaryotic RNA capping enzymes belong to the ligase family and have been shown to form a GTase-GMP adduct upon incubation with GTP (Decroly et al., 2012). A substantial number of SARS-CoV nsps were expressed and purified (nsp7, nsp8, nsp10, and nsp12–16), but covalent linkage of GMP to any these proteins could not be demonstrated (Jin et al., 2013). In addition, nsps were screened for GTase activity in a yeast *trans*-complementation system using the YBS2 strain lacking the gene (*ceg1*) encoding the yeast GTase (Chen et al., 2009c), but also this powerful approach failed to identify the CoV GTase. Consequently, several hypotheses remain to be explored. First, it is possible that the N-terminal NiRAN domain of nsp12 (see earlier) forms a covalent adduct with GTP, as observed for its arterivirus homolog nsp9 (Lehmann et al., 2015a). Another possibility is that the CoV capping pathway is unconventional and, for example, resembles that of alphaviruses in which the GTP molecule needs to be methylated at its N7 position before the GTase-^{m}GMP adduct can be formed (Ahola and Ahlquist, 1999). Interestingly, this second possibility might explain nsp14s capability to convert GTP into ^{m}GTP (see later) (Jin et al., 2013). Finally, the involvement of a host GTase remains an interesting possibility, in particular since cytoplasmic forms of cellular capping enzymes have been described recently (Mukherjee et al., 2012; Schoenberg and Maquat, 2009). Further work is needed to explore these various hypotheses and resolve this important question.

5.3 The nsp14 N7-Methyl Transferase

Coronavirus nsp14 is a bifunctional protein that plays a crucial role in viral RNA synthesis. Its N-terminal exonuclease (ExoN) domain (Minskaia et al., 2006; Snijder et al., 2003), which is thought to promote the fidelity of CoV RNA synthesis, will be discussed in more detail in Section 6. The C-terminal part of nsp14 carries an AdoMet-dependent guanosine N7-MTase activity. Interestingly, as in the case of the GTase (see earlier), bioinformatics analyses of CoV genome sequences failed to identify proteins or protein domains related to cellular and/or viral N7-MTases, again illustrating the significant divergence of nidoviruses from other viral and cellular systems. However, using a *trans*-complementation assay and a yeast strain lacking the *abd1* (N7-MTase) gene, Guo et al. discovered a SARS-CoV nsp14-associated N7-MTase activity (Chen et al., 2009c) by demonstrating that the protein was able to restore the growth of the Δ*abd1* yeast mutant. They also showed that a range of alphacoronavirus nsp14 homologs were able to complement the defect of the Δ*abd1* yeast strain. The N7-MTase activity of nsp14 was subsequently confirmed and characterized using purified recombinant enzymes (Bouvet et al., 2012; Chen et al., 2009c). The SARS-CoV N7-MTase was shown to methylate 5′ cap structures in a sequence-independent manner using a range of RNAs and it also proved to be active on cap analogues and GTP (Jin et al., 2013), corroborating the *trans*-complementation experiments in yeast in which rescue required efficient methylation of a wide range of cellular RNAs. In contrast to the ExoN activity, the in vitro N7-MTase activity was not found to be affected by interactions with nsp10 (Bouvet et al., 2010).

The N7-MTase domain was further characterized by alanine scanning mutagenesis and key residues for enzymatic activity were identified (Chen et al., 2009c) including 10 crucial residues distributed throughout the domain and two clusters of residues essential for MTase activity (Fig. 5). The first cluster (nsp14 residues 331–336) corresponds to the DXGXPXA motif of the AdoMet-binding site. In cross-linking experiments, mutations in this motif strongly decreased the binding of ^{3}H-labeled AdoMet. The role of the second cluster, between residues 414 and 428, was revealed by X-ray structure analysis of a SARS-CoV nsp10/nsp14 complex expressed in E. *coli* (Fig. 6) (Ma et al., 2015). These residues form a constricted pocket that holds the cap structure (GpppA) between two β-strands (β1 and β2) and helix 1, placing the N7 position of the guanine in close proximity of AdoMet and ready for methyl transfer using an in-line mechanism.

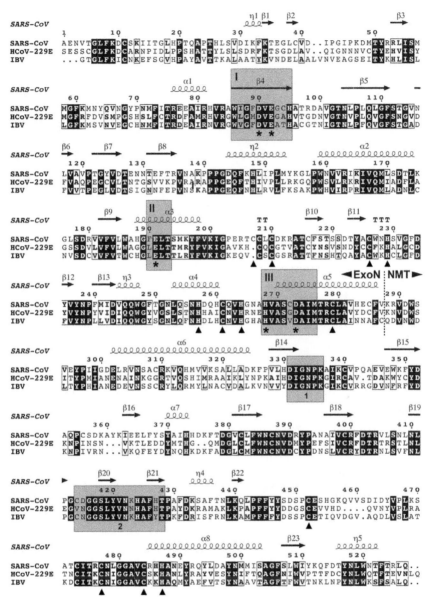

Fig. 5 Sequence alignment of coronavirus nsp14 homologs representing three genera of the *Coronavirinae* subfamily: SARS-CoV (genus *Betacoronavirus*), HCoV-229E (genus *Alphacoronavirus*), and IBV (genus *Gammacoronavirus*). The alignment was generated using Clustal Omega (Sievers et al., 2011) and rendered using ESPript version 3.0 (Robert and Gouet, 2014). Conserved ExoN motifs I, II, and III and clusters of residues involved in SAM binding and N7-MTase activity (1 and 2) are highlighted in *gray*. Catalytic residues of ExoN and residues involved in the formation of zinc fingers are indicated by *asterisks* and *arrowheads*, respectively. Also shown are the secondary structure elements of SARS-CoV nsp14 (pdb 5C8S) and the border between the N-terminal ExoN and C-terminal N7-MT (NMT) domains.

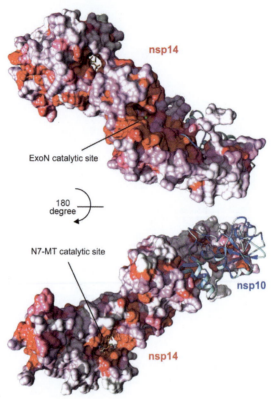

Fig. 6 Surface representation of the three-dimensional structure of the nsp10/nsp14 complex (pdb 5C8S). The nsp10 ribbon structure is shown with conserved residues colored in *blue* (using a scale from *dark* to *light blue*). The coloring of the nsp14 surface is based on the conservation of the respective residues among CoVs (using a scale from *dark* to *light red*). The *upper panel* shows the surface containing the ExoN catalytic site with one Mg^{2+} ion bound in the active site (*green sphere*). The *lower panel* shows the opposite side of the structure with the N7-MTase active site. The cap analog GpppA and SAH are shown in *stick* representation. The figures were generated using UCSF Chimera (Pettersen et al., 2004). The degree of conservation of specific residues was determined using an alignment of nsp10 and nsp14 sequences of eight coronaviruses representing the four genera of the *Coronavirinae* subfamily (SARS-CoV, MERS-CoV, MHV, TGEV, FCoV, HCoV-229E, IBV, and bulbul coronavirus HKU11).

The structure also revealed that the nsp14 N7-MTase domain exhibits a noncanonical MTase fold. Whereas the canonical fold contains a 7-strand β-sheet that is commonly present in the Rossman fold, nsp14 contains only a 5-strand β-sheet and an insertion of a 3-strand antiparallel β-sheet between β5 and β6. In line with previous mutagenesis data (Chen et al., 2009c), the nsp14 X-ray structure revealed functionally relevant interactions between the N-terminal (ExoN) and C-terminal (N7-MTase) domains, with three α-helices of ExoN stabilizing the base of the N7-MTase substrate-binding pocket.

The specific role of N7-MTase activity in virus replication was supported by reverse genetics. Nsp14 mutations were introduced in SARS-CoV RNA replicons expressing a luciferase reporter gene. A D331A mutation in the AdoMet-binding site, which blocks the N7-MTase activity of nsp14, was shown to reduce the luciferase expression (by 90%), indicating that viral RNA replication and/or transcription were impaired in this in vitro system (Chen et al., 2009c).

The nsp14 N7-MTase is an attractive target for antiviral strategies, especially because it exhibits a range of features that are distinct from host cell MTases (Ferron et al., 2012). The druggability of the enzyme was explored using a small set of previously documented MTase inhibitors. These in vitro assays revealed that AdoHcy (the coproduct of the methylation reaction), sinefungin (another AdoMet analog), and ATA efficiently inhibited nsp14 N7-MTase activity with IC_{50} values of 12 μM, 39.5 nM, and 2.1 μM, respectively (Bouvet et al., 2010). ATA was also shown to limit SARS-CoV replication in infected cells (He et al., 2004). In the yeast-ΔMTase trans-complementation assay mentioned earlier, micromolar concentrations of sinefungin were reported to effectively suppress the nsp14 N7-MTase activity of SARS-CoV, MHV, transmissible gastroenteritis virus (TGEV), and IBV (Sun et al., 2014). However, other compounds, such as ATA and AdoHcy, did not exert an inhibitory effect in the context of yeast cells. These discrepancies may reflect differences in cell penetration of the compounds between yeast and (virus-infected) mammalian cells. The yeast system was also applied to screen a library of 3000 natural product extracts, and three hits were obtained displaying potent inhibitory effects on the CoV N7-MTase (Sun et al., 2014). Further work is needed to optimize these hits and test their inhibitory activities in assays using CoV-infected cells.

5.4 The nsp16 2′-O-Methyl Transferase

The presence of a 2′-O-methyl transferase (2′-O-MTase) domain in CoV nsp16 was first inferred using bioinformatics (Snijder et al., 2003; von Grotthuss et al., 2003). Computational threading produced a model containing a conserved K–D–K–E catalytic tetrad that is characteristic of AdoMet-dependent 2′-O-MTases and a conserved AdoMet-binding site

(Fig. 7) (von Grotthuss et al., 2003). The 2′-O-MTase activity was then confirmed by in vitro biochemical assays using purified FCoV nsp16 (Decroly et al., 2008). The recombinant protein was shown to specifically recognize short, cap-0-containing RNAs and to transfer a methyl group from AdoMet to the 2′-O position of the first nucleotide of the N7-methylated substrate. In contrast, a recombinant form of the SARS-CoV nsp16 homolog was inactive under similar experimental conditions. Since yeast two-hybrid experiments and GST pull-down assays had revealed that SARS-CoV nsp16 strongly interacts with nsp10 (Imbert et al., 2008; Pan et al., 2008), a possible involvement of the latter in regulating or supporting the 2′-O-MTase activity was tested. It was shown that purified nsp10 interacts with nsp16 in vitro and thereby triggers a robust 2′-O-MTase activity (Bouvet et al., 2010). Effective methyl transfer was demonstrated using synthetic capped N7-methylated RNA and longer RNA mimicking the 5′ end of the SARS-CoV genome (Bouvet et al., 2010). In contrast, RNA with a nonmethylated cap structure (Gppp-RNA) was not bound by the nsp10/nsp16 complex and, consequently, could not serve as a substrate. These observations suggest that SARS-CoV mRNA capping follows a strict order of reaction steps: after GTP transfer by the still elusive GTase, the cap is methylated by the nsp14 N7-MTase at the guanosine-N7 position to produce a cap-0 structure that, in a subsequent reaction, is bound by the nsp10/nsp16 complex and converted to a cap-1 structure employing the 2′-O-MTase activity of nsp16.

Mutagenesis of SARS-CoV nsp10 and nsp16 confirmed the importance of the catalytic tetrad of the nsp16 2′-O-MTase (Decroly et al., 2011) and showed that the nsp10–nsp16 interaction is absolutely required for activity (Decroly et al., 2011; Lugari et al., 2010). Crystallographic studies of the nsp10/nsp16 complex revealed the molecular basis for the stimulation of the nsp16 2′-O-MTase activity by nsp10 (Fig. 7) (Chen et al., 2011; Decroly et al., 2011). The CoV 2′-O-MTase belongs to the RrmJ/fibrillarin superfamily of ribose 2′-O-methyl transferases (Feder et al., 2003) which have a number of viral orthologs in flaviviruses, alphaviruses, and other nidoviruses. As mentioned earlier, this family contains a conserved K–D–K–E catalytic tetrad (Fig. 7) that is located in close proximity to the substrate-binding pocket accommodating the RNA substrate. The structure revealed that nsp10 is an allosteric regulator that stabilizes nsp16. Moreover, structural and biochemical analyses indicated that nsp10 binding extends and narrows the RNA-binding groove that accommodates the RNA substrate, thereby promoting the RNA- and AdoMet-binding capabilities of nsp16 (Fig. 7).

A

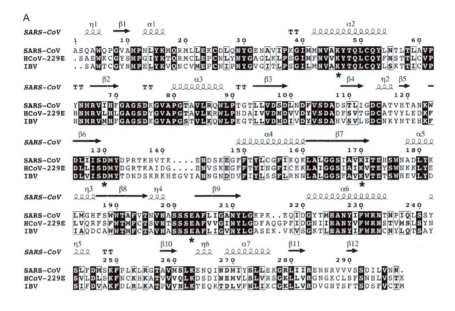

```
            η1      β1    α1                                    TT   α2
SARS-CoV   ℓℓℓ    ━━►   ℓℓℓℓ                                    TT   ℓℓℓℓℓℓℓℓℓℓℓℓℓℓℓ
            1       10      20        30        40        50        60
SARS-CoV   ASQAWQPGVAMPNLYKMQRMLLEKCDLQNYGENAVIPKGIMMNVAKYTQLCQYLNTLTLAVP
HCoV-229E  .SAEWKCGYSMPGIYKTQRMCLEPCNLYNYGAGLKLPSGIMFNVVKYTQLCQYFNSTTLCVP
IBV        ..SAWTCGYNMPELYKVQNCVMEPCNIPNYGVGITLPSGILMNVAKYTQLCQYLSKTTICVP

            β2       β3            α3            TT      β3          *   β4      η2  β5
SARS-CoV   TT ━━►    TT      ℓℓℓℓℓℓℓℓℓ    TT    ━━►                       ━━►    ℓℓℓ ━━►  -
            70       80        90       100       110       120
SARS-CoV   YNMRVIHFGAGSDKGVAPGTAVLRQWLPTGTLLVDSDLNDFVSDADSTLIGDCATVHTANKW
HCoV-229E  HNMRVLHLGAGSDVGVAPGTAVLKRWLPHDAIVVDNDVVDYVSDADFSVTGDCATVYLEDKF
IBV        HNMRVMHFGAGSDRGVAPGSTVLKQWLPEGTLLVDNDIVDYVSDAHVSVLSDCNKYNTEHKF

            β6                  α4              β7             α5
SARS-CoV   ━━►                ℓℓℓℓℓℓℓℓℓℓℓℓ   ━━►            ℓℓℓℓℓ
            130       140       150       160       170       180
SARS-CoV   DLIISDMYDPRTKHVTK....ENDSKEGFFTYLCGFIKQKLALGGSIAVKITEHSWNADLYK
HCoV-229E  DLLISDMYDGRTKAIDG....ENVSKEGFFTYINGFICEKLAIGGSIAIKVTEYSWNKKLYE
IBV        DLVISDMYTDNDSKRKHEGVIANNGNDDVFIYLSSFLRNNLALGGSFAVKVTETSWHEVLYD

            η3     *   β8    η4       β9                      α6                    *
SARS-CoV   ℓℓℓℓ      ━━►  ℓℓℓ     ━━►                      ℓℓℓℓℓℓℓℓℓℓℓℓℓ              ℓ
            190       200       210       220       230       240
SARS-CoV   LMGHFSWWTAFVTNVNASSSEAFLIGANYLGKPK..EQIDGYTMHANYIFWRNTNPIQLSSY
HCoV-229E  LVQRFSFWTMFCTSVNTSSSEAFVVGINYLGDFAQGPFIDGNIIHANYVFWRNSTVMSLSYN
IBV        IAQDCAWWTMFCTAVNASSSEAFLIGVNYLGASEK.VKVSGKTLHANYIFWRNCNYLQTSAY

            η5      TT      β10        η6    α7        β11         β12
SARS-CoV   ℓℓℓ     TT      ━━►       ℓℓℓ   ℓℓℓℓℓℓℓℓ   ━━►          ━━►
            250       260       270       280       290
SARS-CoV   SLFDMSKFPLKLRGTAVMSLKENQINDMIYSLLEKGRLIIRENNRVVVSSDILVNN.
HCoV-229E  SVLDLSKFNCKHKATVVVQLKDSDINEMVLSLVRSGKLVRGNGKCLSFSNHLVSTK
IBV        SIFDVAKFDLRLKATPVVNLKTEQKTDLVFNLIKCCGKLLVRDVGNTSFTSDSFVCTM
```

B

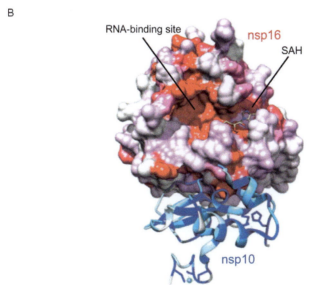

Fig. 7 Coronavirus nsp16 and its interaction with nsp10. (A) Sequence alignment of nsp16 homologs of SARS-CoV (genus *Betacoronavirus*), HCoV-229E (genus *Alphacoronavirus*), and IBV (genus *Gammacoronavirus*). The alignment was generated using Clustal Omega (Sievers et al., 2011) and rendered using ESPript version 3.0 (Robert and Gouet, 2014). Residues of the catalytic tetrad K–D–K–E are indicated by *asterisks* and secondary structure elements of SARS-CoV nsp16 (pdb 2XYV) are shown. (B) Surface representation of the three-dimensional structure of the nsp10/nsp16

(Continued)

The functional relevance of 2′-O-MTase regulation by nsp10 for virus replication is not yet understood. The nsp16-associated 2′-O-MTase activity (itself) is highly conserved across the *Coronaviridae* family and its functional relevance has been supported by reverse-genetics studies using genetically engineered alpha- and betacoronavirus mutants that lack 2′-O-MTase activity (Devarkar et al., 2016; Hyde and Diamond, 2015; Schuberth-Wagner et al., 2015; Züst et al., 2011). Furthermore, the phenotype of temperature-sensitive MHV mutants suggested nsp16 to play a role in RNA synthesis or in the stability of plus-strand RNA (Sawicki et al., 2005). An early SARS-CoV study was able to show that the insertion of a stop codon immediately upstream of the nsp16-coding sequence blocked viral RNA synthesis (Almazan et al., 2006). Subsequently, for HCoV-229E, MHV, and SARS-CoV, several mutants with reduced or ablated 2′-O-MTase activity were described that generally retained robust viral replication in cell culture (Menachery et al., 2014; Züst et al., 2011). The studies also revealed that the impact of nsp16 mutations may depend on the cell types used and that the lack of 2′-O-MTase activity causes more profound effects in primary cells and immune cells. The SARS-CoV nsp16 mutants were further characterized in infected mice and showed a robust attenuation as judged by viral titers, weight loss, lung histology, and respiratory function. The nsp16 mutants also displayed increased sensitivity to treatment with type I interferon. This was also observed for the corresponding MHV and HCoV-229E nsp16 mutants (Züst et al., 2011). However, in contrast to the latter study, the SARS-CoV nsp16 mutant was not found to induce type I interferons, either in vitro or in vivo (Menachery et al., 2014). This observation suggests that SARS-CoV may have a larger repertoire of functions for preventing induction of type I interferons following cellular sensing of "nonself" RNAs, such as viral RNAs with incompletely methylated 5′ cap structures. Together, these data established that the highly conserved nsp16 2′-O-MTase plays an important role in limiting the detection of viral

---

Fig. 7—Cont'd complex (pdb 2XYV). Nsp10 is shown in ribbon representation with conserved residues colored in *dark* to *light blue* according to their conservation among CoVs. Zinc molecules are shown as *spheres* and zinc-coordinating residues are shown in *stick* representation. The surface of nsp16 is colored in *dark* to *light red* according to the conservation of the respective residues among coronaviruses. SAH is depicted in a *stick* model. The figure was generated using UCSF Chimera (Pettersen et al., 2004). The degree of conservation of specific residues was determined using an alignment of nsp10 and nsp16 sequences of eight coronaviruses (see Fig. 6).

RNA by the host's antiviral sensors, but that the specific role of this activity in escaping host innate immune responses may differ to some extent among CoVs. The mechanism underlying the induction of the antiviral immune response following infection with $2'$-O-MTase knockout mutants was studied using specific knockout mice in which viral replication was found to be restored. The absence of Mda5, a cap-0 sensor, restored MHV replication, whereas the nuclear translocation of IRF3 and interferon induction were reduced (Züst et al., 2011). This suggested Mda5 to function in the primary recognition of the RNA produced by CoV $2'$-O-MTase mutants and to initiate the ISG cascade that restricts viral replication. Among the ISG products, the IFIT family of proteins was shown to be critically involved in reducing the replication of CoV nsp16 mutants. The replication of MHV and HCoV-229E nsp16 mutants and wild-type controls was similar in IFIT1 $-/-$ knockout mice (Habjan et al., 2013; Züst et al., 2011). Likewise, replication of a SARS-CoV Δnsp16 mutant was increased in both IFIT1 and IFIT2 knockdown mice (Menachery et al., 2014), suggesting that IFIT family proteins mediate the primary attenuation of SARS-CoV $2'$-O-MTase knockout mutants.

The earlier results indicate that the nsp16 $2'$-O-MTase constitutes a new and attractive target for the development of antiviral drugs against SARS-CoV and HCoV-229E, as well as newly emerging CoVs like MERS-CoV and porcine epidemic diarrhea virus (PEDV). For example, the nsp10–nsp16 interface may be targeted to limit viral $2'$-O-MTase activity and thus restore the antiviral responses mediated by Mda5 and IFIT1 (Menachery and Baric, 2013). Interestingly, the nsp10 residues involved in the nsp10/nsp16 interaction are quite conserved within the CoV family and it was recently demonstrated that nsp10 of different CoVs (FCoV, MHV, SARS-CoV, MERS-CoV) is functionally interchangeable in the stimulation of nsp16 $2'$-O-MTase activity (Wang et al., 2015). Thus, molecules or peptides blocking this interface may have broad-spectrum anti-CoV effects, a concept that was explored and supported using synthetic peptides that mimic the nsp10 interface and suppress nsp16 $2'$-O-MTase activity in vitro (Ke et al., 2012; Wang et al., 2015). The antiviral effect of the MHV TP29 peptide, for example, was first demonstrated in MHV-infected cells and was subsequently confirmed to limit MHV replication in mice and to enhance the type I interferon response (Wang et al., 2015). The same peptide was also effective in blocking the replication of a SARS-CoV replicon.

6. CORONAVIRUS nsp14 ExoN: KEY TO A UNIQUE MISMATCH REPAIR MECHANISM THAT PROMOTES FIDELITY

The CoV ExoN domain was identified on the basis of comparative sequence analyses (Snijder et al., 2003) that suggested a distant relationship of the nsp14 N-terminal domain (and equivalent polyprotein regions of other nidoviruses) with cellular DEDD exonucleases, a large protein superfamily containing RNA and DNA exonucleases from all kingdoms of life (Zuo and Deutscher, 2001). The designation "DEDD" alludes to four invariant Asp/Glu residues that are part of three sequence motifs, I–III, that are conserved in members of this superfamily (Fig. 5). The DEDD superfamily is also referred to as DnaQ-like family because it includes DnaQ, the ε subunit of *E. coli* DNA polymerase III, a well-characterized proofreading enzyme (Echols et al., 1983; Scheuermann et al., 1983). Conservation of a fifth residue, His, located four positions upstream of the conserved Asp in motif III identifies ExoN as a member of the DEDDh subgroup, while members of the DEDDy exonucleases contain Tyr at the equivalent position. The acidic residues are required to form two metal-binding sites. Based on catalytic models initially developed for cellular exonucleases and catalytic RNA (Beese and Steitz, 1991; Steitz and Steitz, 1993), the conserved His and the site A metal ion are thought to activate a water molecule that launches a nucleophilic attack on the phosphorus group of the $3'$-terminal phosphodiester, while the site B metal ion is thought to stabilize the transition state (Derbyshire et al., 1991).

ExoN is conserved in CoVs and all other known nidoviruses with genome sizes of >20 kb (Gorbalenya et al., 2006; Minskaia et al., 2006; Nga et al., 2011; Snijder et al., 2003; Zirkel et al., 2011). The correlation between genome size and ExoN conservation suggests that, in nidoviruses with medium- and large-size genomes, the correct nucleotide selection and recognition of properly formed base pairs by the RdRp is not enough to accomplish the necessary replication fidelity and, therefore, requires additional functions suitable to detect and remove misincorporated nucleotides. Recently, biochemical evidence has been provided to suggest that ExoN may have exactly this function (Bouvet et al., 2012). CoV mutants that lack ExoN activity provided additional evidence for ExoN being involved in mechanisms that keep the CoV mutation rate at a relatively low level ($<10^{-6}$ mutations per site per round of replication for MHV and SARS-CoV) (Eckerle et al., 2007, 2010), while other RNA viruses have much higher mutation rates, ranging from 10^{-3} to 10^{-5} mutations per site per round of replication (Drake and Holland, 1999; Sanjuan et al., 2010). ExoN knockout mutants of SARS-CoV and MHV were shown to display a mutator phenotype with significantly increased mutation frequencies approaching those of other RNA viruses (Eckerle et al., 2007, 2010). Considering that these studies were performed under selection pressure favoring genotypes with high replication efficiency, the total number of mutations in RNAs produced by ExoN-deficient viruses may be even higher than calculated in that study, especially in genome regions that are not subject to selection in in vitro cell culture systems. While inactivation of ExoN activity was tolerated by MHV and SARS-CoV (albeit with reductions in replication efficiency), stable ExoN-deficient mutants of the alphacoronaviruses HCoV-229E and TGEV could not be recovered (Becares et al., 2016; Minskaia et al., 2006), supporting the critical role of this activity in CoV replication. Taken together, the available information provides compelling evidence for ExoN playing a key role in high-fidelity replication of CoVs. Consistent with this hypothesis, genetically engineered ExoN knockout mutants were shown to be significantly more sensitive to RNA mutagens such as ribavirin and 5-fluorouracil (up to 300-fold). Furthermore, compared to wild-type virus, the ExoN knockout mutants were shown to accumulate a much higher number of mutations when propagated in the presence of mutagens (Smith et al., 2013). The lack of ExoN activity was also shown to have profound effects on viral replication and pathogenesis in vivo (Graham et al., 2012). ExoN-negative mutants displayed a stable mutator phenotype in a number of mouse models of human SARS, providing a promising approach for the stable attenuation of highly pathogenic CoVs with important implications for vaccine development (Graham et al., 2012).

Coronavirus nsp14 is a bifunctional protein comprised of an N-terminal ExoN domain and a C-terminal N7-MTase domain. Surprisingly, the latter domain is not conserved in the *Torovirinae* (genera *Torovirus* and *Bafinivirus*), representing the other subfamily of the *Coronaviridae*, but in other (more distantly related) nidovirus branches (Lauber et al., 2013; Nga et al., 2011). This conservation pattern suggests that a common ancestor of the *Corona-*, *Mesoni-*, and *Roniviridae* contained the two-domain ExoN/N7-MTase structure while some lineages lost the N7-MTase domain at a later stage. Although nsp14 does not require other proteins for activity (Chen et al., 2007; Minskaia et al., 2006), its ribonucleolytic (but not the N7-MTase) activity was shown to be stimulated significantly in the presence nsp10. In line with this, nsp10 variants carrying amino acid substitutions that prevent the interaction with nsp14 failed to stimulate ExoN activity (Bouvet et al., 2012, 2014). To date, there is no evidence to suggest a direct role for nsp10 in catalysis. Most likely, interactions between the nsp10 and the N-terminal domain of nsp14 stabilize the ExoN active site in a catalytically competent conformation. Mutagenesis data and a recent X-ray structure analysis of a SARS-CoV nsp10/nsp14 complex (see Fig. 6) revealed that the nsp10 surface required for interaction with nsp14 overlaps with the surface involved in the interaction and activation of the nsp16 2′-O-MTase activity (see earlier) (Bouvet et al., 2014; Ma et al., 2015).

Coronavirus ExoN activities were first demonstrated using recombinant forms of SARS-CoV nsp14 expressed in *E. coli* (Minskaia et al., 2006). The protein was shown to require Mg^{2+} or Mn^{2+} ions for activity and to degrade a range of single-stranded (ss) synthetic RNAs with 3′-to-5′ directionality to yield reaction products of about 8–12 nucleotides. The data further suggested that RNA secondary structure affects ExoN activity (Minskaia et al., 2006). Mutational analysis of predicted active-site (Asp/Glu) residues confirmed their critical involvement in catalysis (Minskaia et al., 2006). In a subsequent study, using nsp10/nsp14 complexes, the substrate specificity was characterized in more detail and revealed dsRNA with a terminal mismatch to be the preferred substrate for ExoN activity. Excision efficiencies using different mismatched base pairs (A:G, A:A, A:C, U:G, U:C, U:U) were found to be similar, suggesting that the mismatch rather than the nature of the nucleotide misincorporated at the 3′ end determines ExoN activity (Bouvet et al., 2012).

In a recent study, the structures of unliganded, SAM-bound, and SAH-GpppA-bound SARS-CoV nsp10–nsp14 complexes were determined by X-ray crystallography (Ma et al., 2015). The structures provide important insight into the two-subdomain structure of nsp14, the two catalytic sites of ExoN and N7-MTase, critical substrate-binding residues, the contribution of nsp10 to enhancing ExoN activity, and the roles of as many as three zinc fingers present in nsp14 (Fig. 6). In the structure, one molecule of nsp14 was found to bind one molecule of nsp10. Given that nsp10 tends to form multimers (Su et al., 2006), it is tempting to speculate that, in infected cells, the nsp10–nsp14 complex may form even larger complexes, for example, by interacting with nsp10–nsp16 complexes that might be stabilized by nsp10–nsp10 interactions. In this way, consecutive methylation reactions of the 5′ cap structure could be spatially coordinated. Comparison of the nsp10 surfaces that interact with nsp14 and nsp16, respectively, revealed a significant overlap (Bouvet et al., 2014), with a substantially larger surface being involved in interactions between nsp10 and nsp14. Buried solvent accessible areas for interactions with nsp14 and nsp16 were determined to be 2236 and 938 Å^2, respectively (Decroly et al., 2011; Ma et al., 2015). The structure of the nsp10–nsp14 complex also helps to explain the observed stimulation of ExoN activity by nsp10 (see earlier). Two regions of nsp10 interact extensively with different structural elements of the nsp14 N-terminal domain, most likely to maintain the structural integrity of the ExoN domain. Interestingly, the observed interaction of N-terminal residues of nsp10 with nsp14 led to interpretable electron density for these residues that had not been observed in previous structures of nsp10 (Joseph et al., 2007; Su et al., 2006) or the nsp10–nsp16 complex (Chen et al., 2011; Decroly et al., 2011), consistent with the proposed role of the N-terminal loop of nsp10 in stabilizing interactions with nsp14.

The structure of the ExoN domain is essentially comprised of a twisted β-sheet that is formed by five β-strands and flanked by α-helices on either side. It resembles that of other DEDD superfamily exonucleases, such as the ε subunit of *E. coli* DNA polymerase III, but also has unique features. These include two segments (residues 1–76 and 119–145) that are involved in the interaction with nsp10 and two zinc fingers in the ExoN domain (see Fig. 5). The second zinc finger was found in close proximity to the catalytic site. Both its position in the structure and mutagenesis data support a role for this zinc finger in catalysis (Ma et al., 2015). The other zinc finger appears to be required to maintain the structural integrity of nsp14. Consistent with this, nsp14 variants containing substitutions in zinc finger 1 proved to be insoluble when expressed in *E. coli*.

Five residues predicted to coordinate two Mg^{2+} ions, Asp-90, Glu-92, Glu-191, His-268, and Asp-273, were found in the catalytic site, with one Mg^{2+} ion being coordinated by Asp-90 (ExoN motif I) and Glu-191 (motif II) (Fig. 5). The second Mg^{2+} ion expected to be involved in the two-metal-ion-assisted catalytic mechanism of ExoN (Beese and Steitz, 1991; Chen et al., 2007; Ulferts and Ziebuhr, 2011) was not identified, presumably due to the lack of an RNA substrate and/or product in this structure. With one exception, metal ion-coordinating residues identified in the active site corresponded well to those identified in related cellular proofreading exonucleases (Beese et al., 1993; Hamdan et al., 2002) and previous predictions for nidovirus homologs (Snijder et al., 2003). The structure clearly revealed Glu-191 (instead of Asp-243) to be involved in catalysis, thus revising previous predictions on the identity of ExoN motif II in the CoV nsp14 primary structure and making nsp14 a "DE͟ED" outlier" in the DEDD superfamily of exonucleases.

The combined structural and functional information obtained for nsp14 including its subdomains and the nsp10 cofactor provides an excellent basis for studies using even larger multisubunit complexes to obtain insight into the coordinated action of key replicative enzymes involved in RNA synthesis, quality control, capping, methylation, and other functions (Subissi et al., 2014a).

7. CORONAVIRUS nsp15: A REMARKABLE ENDORIBONUCLEASE WITH ELUSIVE FUNCTIONS

The nsp15-associated endoU domain is one of the most conserved proteins among CoVs and related viruses (Fig. 8), suggesting important functions in the viral replicative cycle. Already in the first sequence analyses of torovirus and arterivirus replicase genes published more than 25 years ago (den Boon et al., 1991; Snijder et al., 1990), the identification of a conserved sequence in the 3′-terminal ORF1b region, including the (at the time unknown) endoU domain, was key to establishing phylogenetic relationships between corona- and toroviruses and, subsequently, also arteriviruses (Cavanagh and Horzinek, 1993; Snijder et al., 1993). Outside the *Nidovirales*, no viral homologs of endoU have been identified to date. Together with the helicase-associated ZBD (see earlier), the nidoviral endoU has therefore been proposed to be a unique and universally conserved genetic marker common to all nidoviruses (Ivanov et al., 2004a; Snijder et al., 2003). Only recently, with the identification of the first nidoviruses in insects (now classified in the family *Mesoniviridae*) and reanalysis of the ronivirus replicase gene, it was found that endoU is not conserved in those nidovirus branches that replicate in invertebrate hosts (*Mesoniviridae, Roniviridae*) (Lauber et al., 2012, 2013; Nga et al., 2011; Zirkel et al., 2013), suggesting specific roles in vertebrate hosts. To date, these functions and, more specifically, the biologically relevant substrates of endoU have not been identified. Characterization of MHV endoU knockout mutants revealed only minor effects on viral RNA synthesis in infected cells, with all RNA species being equally affected, and caused a slight reduction in virus titers, which was most evident at later time points post infection (Kang et al., 2007). For the arterivirus EAV, substitutions of several conserved residues in the endoU domain were found to cause more

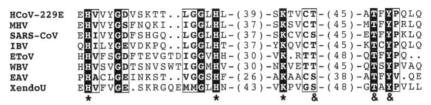

Fig. 8 Alignment of nidovirus endoU domains and XendoU from *Xenopus laevis*. Residues involved in catalysis (*) and substrate binding (&) are indicated. Abbreviations not explained in the main text: *EToV*, Equine torovirus (subfamily *Torovirinae*, genus *Torovirus*); *WBV*, white bream virus (subfamily *Torovirinae*, genus *Bafinivirus*).

profound effects, both in virus reproduction and viral RNA accumulation, with sg RNA being more affected than genome replication in several mutants. In some cases, reduction in virus titers by up to 5 log was observed (Posthuma et al., 2006). The (limited) information obtained in these studies suggests that endoU activity is not strictly required for nidovirus RNA synthesis, at least in cell culture. However, the strong conservation clearly suggests an important in vivo function that remains to be identified in suitable model systems. Initial evidence for specific functions of nidovirus endoU domains in infected cells was obtained in studies showing that SARS-CoV nsp15, but surprisingly not the homologous proteins from HCoV-NL63 and HCoV-HKU1, counteracts MAVS-induced apoptosis (Lei et al., 2009). Further experiments will be necessary to confirm and assess the significance of these functions for virus replication and/or virus–host interactions.

Nidovirus-encoded endoU activities have been characterized using recombinant forms of CoV nsp15 (SARS-CoV, HCoV-229E, MHV-A59, IBV) and arterivirus nsp11 (EAV, PRRSV) (Bhardwaj et al., 2004; Ivanov et al., 2004a; Kang et al., 2007; Nedialkova et al., 2009). In a number of studies, recombinant nidovirus endoUs were shown (i) to have endonucleolytic activity, (ii) to cleave 3′ of pyrimidines, preferring uridine over cytidine, and (iii) to release reaction products with 2′,3′-cyclic phosphate and 5′-OH ends. Using suitable test substrates, RNA structural features were shown to affect endoU cleavage efficiency, with unpaired pyrimidines being processed more efficiently (Bhardwaj et al., 2006; Nedialkova et al., 2009). The role of metal ions in endoU activity is not entirely clear, with somewhat contradictory data being reported for different homologs. While the activities of cellular and CoV homologs were shown to require (or to be significantly stimulated by) Mn^{2+} ions (Bhardwaj et al., 2004; Ivanov et al., 2004a; Laneve et al., 2003, 2008), arterivirus endoU activities proved to be less dependent on metal ions. Low concentrations of Mn^{2+} were found to stimulate only marginally the arterivirus endoU activities, whereas higher concentrations (previously shown to be required for optimal nucleolytic activities in CoV and cellular homologs) inhibited the activities of EAV and PRRSV endoUs (Nedialkova et al., 2009). Metal ion requirements are commonly used to distinguish between the two basic catalytic mechanisms employed by ribonucleases: the metal-independent mechanism that, for example, is employed by RNase A and results in products with $2',3'$-cyclic phosphate ends (as described earlier for endoU), and the metal-dependent mechanism in which catalysis is aided by two divalent cations coordinated by conserved acidic residues and generates products with $3'$-OH and $5'$-phosphate ends (as described earlier for ExoN). The critical (or supportive) role of metal ions observed for several cellular and viral endoU homologs is inconsistent with an RNase A-like (metal-independent) reaction mechanism. Also, metal ions were not detected in any of the structures determined for coronavirus nsp15s or XendoU, arguing against a direct role of metal ions in catalysis (Renzi et al., 2006; Ricagno et al., 2006). However, Mn^{2+} ions were found to change the intrinsic tryptophan fluorescence of SARS-CoV nsp15, suggesting conformational changes that, potentially, may affect activity and were shown to be unrelated to protein multimerization (Bhardwaj et al., 2004; Guarino et al., 2005). Also, regarding the role of Mn^{2+} in RNA binding, contradicting data have been reported, with RNA binding by SARS-CoV nsp15 being enhanced in the presence of Mn^{2+} or not affected by metal ions in the case of XendoU (Bhardwaj et al., 2006; Gioia et al., 2005).

Structural information has been obtained by X-ray crystallography and cryoelectron microscopy studies for several CoV and cellular endoU homologs (Bhardwaj et al., 2008; Renzi et al., 2006; Ricagno et al., 2006; Xu et al., 2006). SARS-CoV and MHV nsp15s were shown to form homohexamers comprised of a dimer of trimers. The nsp15 monomers have an $\alpha + \beta$ structure with three subdomains, a small N-terminal, an intermediate-sized middle, and a large C-terminal domain, the latter basically representing the "conserved domain" of the CoV-like superfamily (see earlier). One side of this domain contains two β-sheets that line the positively charged active-site groove, while the other side is formed by five α-helices. The structures of the catalytic domains are largely conserved among CoV endoUs and XendoU, supporting the proposed common phylogeny of viral cellular members of this large endoribonuclease family (Bhardwaj et al., 2008; Renzi et al., 2006; Snijder et al., 2003).

The endoU hexamer reported for SARS-CoV nsp15 (Ricagno et al., 2006) has dimensions of 80–96 × 110 Å, forming a three-petal-shaped surface that surrounds a small, predominantly negatively charged central channel with an inner diameter of ~15 Å. The hexamer has six-independent active sites located on the surface of the molecule. Interactions between individual protomers are predominantly mediated by residues located in the N-terminal and middle domains. Database searches failed to identify closely related structural homologs, suggesting that endoUs diverged profoundly from other ribonucleases (Renzi et al., 2006; Ricagno et al., 2006). Nevertheless, the presumed endoU catalytic residues (His/His/Lys, Fig. 8) could be superimposed with the catalytic His/His/Lys residues of bovine RNase A, the prototype of a large superfamily of pyrimidine-specific ribonucleases (Ricagno et al., 2006; Ulferts and Ziebuhr, 2011). The superposition also includes a number of conserved substrate-binding residues. A comparison of viral and cellular endoU structures with that of RNase A and related nucleases using the PDBefold server revealed similarities between the structural cores of these enzymes that may be described as "interrelated by topological permutation," providing initial evidence for a common ancestry of the two endonuclease families (Ulferts and Ziebuhr, 2011). Further studies are required to substantiate this hypothesis (see later).

The hexameric form is thought to be the fully active form of CoV nsp15. This is supported by the exponential increase of activity with increased protein concentrations, the reduced activities determined for protein variants that do not multimerize and the increased RNA-binding activities observed for hexameric forms of endoU (Bhardwaj et al., 2006; Guarino et al., 2005; Xu et al., 2006). Consistent with this, hexamerization has been confirmed for different CoV endoU homologs and residues confirmed to be involved in intersubunit interactions are highly conserved among CoVs.

In the structure of a truncated, monomeric form of SARS-CoV nsp15 that lacks 28 N-terminal and 11 C-terminal residues, two loops of the catalytic domain were found to be displaced compared to their location in the hexamer, resulting in the destruction of the active site (Joseph et al., 2007). In hexameric structures, the two loops pack against each other and are stabilized by intermonomer interactions, suggesting that hexamerization may induce an allosteric switch. Furthermore, cross-linking and cryoelectron microscopy studies support a specific role of hexamerization in RNA binding (Bhardwaj et al., 2006, 2008).

In the structure model, the proposed active-site His and Lys residues (Fig. 8) identified by comparative sequence analysis and site-directed mutagenesis (Bhardwaj et al., 2004; Gioia et al., 2005; Guarino et al., 2005; Ivanov et al., 2004a; Kang et al., 2007; Snijder et al., 2003) were found to be embedded in a positively charged groove of the catalytic domain. Other residues identified in the active site and proposed to be involved in binding to the substrate phosphate include the side chain of a conserved Thr (Fig. 8) and the main chain amide of a conserved Gly (Fig. 1) (Bhardwaj et al., 2008; Ricagno et al., 2006; Xu et al., 2006). The proposed functional role of the latter residues has also been corroborated by mutagenesis data for several endoU homologs (Kang et al., 2007; Renzi et al., 2006; Ricagno et al., 2006).

As mentioned earlier, endoU and RNase A share a number of features in their active sites. In RNase A, pyrimidine binding primarily involves Thr-45 and Phe-120. While Thr-45 forms hydrogen bonds with the pyrimidine base, Phe-120 interacts with the base through stacking interactions (Raines, 1998). The Ser-293/Tyr-342 residues of SARS-CoV nsp15 and the Thr-45/Phe-120 residues of RNase A occupy similar positions in the active-site clefts of the two enzymes (Ulferts and Ziebuhr, 2011). The Ser/Thr residue is conserved in viral and cellular domains, while Tyr-342 is conserved in viral endoU homologs while conservative substitutions (Phe, Trp) are occasionally found in cellular endoU homologs (Renzi et al., 2006; Ricagno et al., 2006). The role of SARS-CoV endoU Ser-293 and Tyr-342 (and equivalent residues in related enzymes) in substrate binding received strong support by molecular modeling and site-directed mutagenesis data, with the conserved Ser/Thr residue being confirmed to have a critical role in the differential cleavage of uridine- and cytidine-containing substrates, respectively (Bhardwaj et al., 2008; Nedialkova et al., 2009; Ricagno et al., 2006).

Similarities in their active-site structures and reaction products containing $2',3'$-cyclic phosphate ends suggest that endoUs and RNase A-like endoribonucleases employ similar catalytic mechanisms (Nedialkova et al., 2009; Ricagno et al., 2006). For RNase A, it has been established that two His residues in the active site act as general base and acid, respectively. The His residue that acts as a general base attracts a hydrogen from the ribose $2'$-hydroxyl group that subsequently attacks the $5'$ P–O bond. The second His donates a hydrogen to the $5'$-O, thus facilitating displacement of this group and subsequent product release (Raines, 1998). In a second step, the $2',3'$-cyclic phosphate is hydrolyzed, resulting in a $3'$-phosphomonoester product and recovery of the enzyme. The latter is essentially a reverse reaction of the transphosphorylation reaction, with the protonated His now acting as an acid and the other His acting as a base. In both reaction steps, Lys interacts with the phosphate to stabilize the pentavalent reaction intermediate.

Although the reaction mechanisms employed by endoUs have not been studied in detail, the enzymes are thought to use an RNase A–like catalytic mechanism. This is supported by several lines of evidence, including (i) the conserved spatial positions in the structure and the critical functional role of two His and one Lys residue(s) (Ricagno et al., 2006), and (ii) the release of $2',3'$-cyclic phosphate-containing reaction products (Bhardwaj et al., 2004; Ivanov et al., 2004a) that are converted to products with $3'$-hydroxyl ends (Nedialkova et al., 2009).

8. SUMMARY AND FUTURE PERSPECTIVES

Since the first in-depth analysis of a CoV replicase in 1989 (Gorbalenya et al., 1989), significant progress has been made in terms of its structural and functional characterization. A multitude of enzymatic functions has been identified and characterized in vitro, although mainly using artificial substrates so far. Protein structures were obtained for most of the subunits from the nsp7–16 region (Neuman et al., 2014b), but unfortunately two prominent remaining "blank spots" on this map concern two key enzymes in CoV RNA synthesis, RdRp and helicase. Filling those gaps would constitute an important step forward, to address basic questions like the priming mechanism employed by the RdRp and the function of the NiRAN domain, and to accelerate targeted drug discovery, for example, in the area of nucleoside inhibitors of CoV RNA synthesis, which has received little attention thus far. Clearly, where CoV mRNA capping is concerned, identification of the elusive GTase remains a research priority (Subissi et al., 2014a). For other nsps, potential functions (nsp9, the N-terminal domain of nsp15) or substrates (nsp15-endoU) remain to be found.

As highlighted by several nsp-wide interaction screening studies (Imbert et al., 2008; Pan et al., 2008; von Brunn et al., 2007) and more specifically by the in vitro data on the interplay between nsp7–8–12–14 (Subissi et al., 2014b) and nsp10–14–16 (Bouvet et al., 2014), CoV nsps need to work together in many ways. The further characterization of the "nsp interactome," now also inside the CoV-infected cell, will undoubtedly provide more clues as to how specific functions are switched on and off or modulated. Likewise, attention should be given to defining the interactions of CoV nsps with the specific RNA signals for genome replication (Madhugiri et al., 2014; Yang and Leibowitz, 2015), discontinuous minus-strand RNA synthesis (attenuation at body TRSs, nascent minus-strand transfer, and reinitiation; Pasternak et al., 2006; Sawicki et al., 2007; Sola et al., 2011), and the transcription, capping, and polyadenylation of sg mRNAs (Fig. 1). These RNA sequences, several of which may be *cis*-acting, could provide starting points for improved biochemical assays, ultimately paving the way for the complete in vitro reconstitution of some of these multi-nsp-driven processes.

The analysis of CoV RTC structure and function in the living infected cell remains an enormous technical challenge, requiring continued toolbox development. It is likely that several functional riddles can only be solved by studying infected cells. For example, the endoribonuclease activity of the nsp15 endoU domain, a potential "suicide enzyme" for an RNA virus, must be controlled tightly in the infected cell. Whereas the enzyme is highly active and displays only very limited substrate specificity in vitro (Ivanov et al., 2004a; Nedialkova et al., 2009), it may be confined to a specific compartment in the infected cell and/or its activity may be modulated by interactions with other nsps or host factors. Such differences between in vitro and in vivo activities will surely emerge for other nsps as well, and they may be better understood following the further characterization (including their lipid composition) of the membranous replication organelles with which the metabolically active CoV RTC presumably is associated (Hagemeijer et al., 2012; Neuman et al., 2014a; van der Hoeven et al., 2016). These studies should also answer the question of how both nsps and viral RNA substrates are targeted to or recruited by the membrane-bound CoV RTC, in particular also during the earlier stages of infection when viral RNA synthesis appears to be taking off in the absence of the prominent membrane rearrangements observed later in infection.

Specific mutations in CoV genomes can now be reverse engineered, but many of the functions encoded by nsp7–nsp12 are so basic that their inactivation will merely result in dead virus mutants that do not provide many deeper insights into nsp function. This in part explains why most progress thus far has been made for some of the functions that can—fortunately— be inactivated without such lethal consequences, like the nsp14 ExoN and nsp16 2′-O-MTase enzymes (Eckerle et al., 2007, 2010; Graham et al., 2012; Menachery et al., 2014; Smith et al., 2013; Züst et al., 2011). For this reason, the field should continue to also employ "traditional" (forward) genetic methods to characterize (and produce more) conditionally defective CoVs, like temperature-sensitive mutants (Sawicki et al., 2005). Thanks to the advent of next-generation sequencing technologies, tracing the evolution of crippled virus mutants and (pseudo) revertants has become much more straightforward than before, and this approach (letting the virus do the work) likely is among the most economical ones in uncovering previously unknown interactions between the protein and RNA players in CoV replication (Züst et al., 2008). Furthermore, it may be possible to develop cell-based assays for the analysis of CoV nsp functions that do not rely on having a replication-competent virus to start with.

Unraveling the molecular mechanisms underlying the presumed mismatch excision function (Bouvet et al., 2012) of the nsp14-ExoN, which is uniquely encoded by RNA virus genomes larger than 20 kb (Nga et al., 2011), connects to the mechanisms driving the evolution of nidoviruses at large (Lauber et al., 2013). Also, replicases from other nidovirus branches will need to be studied to fully understand the basic principles governing the profound divergence and genome expansion of this exceptional order of +RNA viruses. The error rate and genomic plasticity of RNA viruses are among their most fascinating features, and also form the basis for the many problems caused by RNA virus mutation and adaptation, including successful zoonotic transfer. As exemplified by the viable CoV mutants lacking the ExoN or 2′-O-MTase functions (Graham et al., 2012; Habjan et al., 2013; Menachery et al., 2014; Züst et al., 2011), the functional characterization of CoV replicative enzymes can be key to the development of conceptually new live attenuated vaccine prototypes. Likewise, it will contribute to the development of broad-spectrum and highly effective antiviral drugs targeting essential enzyme functions, critical interactions with nsp cofactors, or "nonessential" nsp functions that promote efficient viral replication and/or pathogenesis. As highlighted by the SARS and MERS outbreaks of the past 15 years, having such compounds available would definitely strengthen our first line of defense against CoV infections in humans.

ACKNOWLEDGMENTS

With great respect, we acknowledge the many ground-breaking contributions of our long-term collaborator Dr. Alexander Gorbalenya, founding father of the functional characterization of the replicase of coronaviruses and nidoviruses at large. We are also grateful for the scientific and technical contributions over many years by numerous past and present lab members and colleagues in Leiden (E.J.S.), Marseille (E.D.), and Giessen/Belfast/Würzburg (J.Z.). Specifically, we would like to acknowledge the input of our long-term collaborators Clara Posthuma, Bruno Canard, Bruno Coutard, Isabelle Imbert, Volker Thiel, and Stuart Siddell. We thank Aartjan te Velthuis for his assistance with preparing Fig. 2.

Funding: This work was supported by the Netherlands Organization for Scientific Research (NWO-CW TOP-GO Grant 700.10.352 to E.J.S.), the European Union Seventh Framework Programme (Project SILVER; Grant 260644 to E.J.S. and E.D.), the French National Research Agency (Grants ANR-12 BSV (NidoRNAends) and ANR-14 ASTR-0026 (VMTaseIN) to E.D.), the Deutsche Forschungsgemeinschaft (DFG; SFB1021 and IRTG 1384 to J.Z.), and the German Center for Infection Research (DZIF; to J.Z.).

REFERENCES

Adedeji, A.O., Marchand, B., Te Velthuis, A.J., Snijder, E.J., Weiss, S., Eoff, R.L., et al., 2012a. Mechanism of nucleic acid unwinding by SARS-CoV helicase. PLoS One 7, e36521.

Adedeji, A.O., Singh, K., Calcaterra, N.E., DeDiego, M.L., Enjuanes, L., Weiss, S., et al., 2012b. Severe acute respiratory syndrome coronavirus replication inhibitor that interferes with the nucleic acid unwinding of the viral helicase. Antimicrob. Agents Chemother. 56, 4718–4728.

Adedeji, A.O., Singh, K., Kassim, A., Coleman, C.M., Elliott, R., Weiss, S.R., et al., 2014. Evaluation of SSYA10-001 as a replication inhibitor of severe acute respiratory syndrome, mouse hepatitis, and Middle East respiratory syndrome coronaviruses. Antimicrob. Agents Chemother. 58, 4894–4898.

Ahn, D.G., Choi, J.K., Taylor, D.R., Oh, J.W., 2012. Biochemical characterization of a recombinant SARS coronavirus nsp12 RNA-dependent RNA polymerase capable of copying viral RNA templates. Arch. Virol. 157, 2095–2104.

Ahola, T., Ahlquist, P., 1999. Putative RNA capping activities encoded by brome mosaic virus: methylation and covalent binding of guanylate by replicase protein 1a. J. Virol. 73, 10061–10069.

Almazan, F., Dediego, M.L., Galan, C., Escors, D., Alvarez, E., Ortego, J., et al., 2006. Construction of a severe acute respiratory syndrome coronavirus infectious cDNA clone and a replicon to study coronavirus RNA synthesis. J. Virol. 80, 10900–10906.

Almazan, F., Sola, I., Zuniga, S., Marquez-Jurado, S., Morales, L., Becares, M., et al., 2014. Coronavirus reverse genetic systems: infectious clones and replicons. Virus Res. 189, 262–270.

Anand, K., Palm, G.J., Mesters, J.R., Siddell, S.G., Ziebuhr, J., Hilgenfeld, R., 2002. Structure of coronavirus main proteinase reveals combination of a chymotrypsin fold with an extra alpha-helical domain. EMBO J. 21, 3213–3224.

Anand, K., Ziebuhr, J., Wadhwani, P., Mesters, J.R., Hilgenfeld, R., 2003. Coronavirus main proteinase (3CLpro) structure: basis for design of anti-SARS drugs. Science 300, 1763–1767.

Angelini, M.M., Akhlaghpour, M., Neuman, B.W., Buchmeier, M.J., 2013. Severe acute respiratory syndrome coronavirus nonstructural proteins 3, 4, and 6 induce double-membrane vesicles. mBio 4. e00524–13.

Baez-Santos, Y.M., St John, S.E., Mesecar, A.D., 2015. The SARS-coronavirus papain-like protease: structure, function and inhibition by designed antiviral compounds. Antivir. Res. 115, 21–38.

Barnard, D.L., Day, C.W., Bailey, K., Heiner, M., Montgomery, R., Lauridsen, L., et al., 2006. Enhancement of the infectivity of SARS-CoV in BALB/c mice by IMP dehydrogenase inhibitors, including ribavirin. Antivir. Res. 71, 53–63.

Becares, M., Pascual-Iglesias, A., Nogales, A., Sola, I., Enjuanes, L., Zuniga, S., 2016. Mutagenesis of coronavirus nsp14 reveals its potential role in modulation of the innate immune response. J. Virol. 90, 5399–5414.

Beerens, N., Selisko, B., Ricagno, S., Imbert, I., van der Zanden, L., Snijder, E.J., et al., 2007. De novo initiation of RNA synthesis by the arterivirus RNA-dependent RNA polymerase. J. Virol. 81, 8384–8395.

Beese, L.S., Steitz, T.A., 1991. Structural basis for the 3′-5′ exonuclease activity of Escherichia coli DNA polymerase I: a two metal ion mechanism. EMBO J. 10, 25–33.

Beese, L.S., Derbyshire, V., Steitz, T.A., 1993. Structure of DNA polymerase I Klenow fragment bound to duplex DNA. Science 260, 352–355.

Bernini, A., Spiga, O., Venditti, V., Prischi, F., Bracci, L., Huang, J., et al., 2006. Tertiary structure prediction of SARS coronavirus helicase. Biochem. Biophys. Res. Commun. 343, 1101–1104.

Bhardwaj, K., Guarino, L., Kao, C.C., 2004. The severe acute respiratory syndrome coronavirus Nsp15 protein is an endoribonuclease that prefers manganese as a cofactor. J. Virol. 78, 12218–12224.

Bhardwaj, K., Sun, J., Holzenburg, A., Guarino, L.A., Kao, C.C., 2006. RNA recognition and cleavage by the SARS coronavirus endoribonuclease. J. Mol. Biol. 361, 243–256.

Bhardwaj, K., Palaninathan, S., Alcantara, J.M., Yi, L.L., Guarino, L., Sacchettini, J.C., et al., 2008. Structural and functional analyses of the severe acute respiratory syndrome coronavirus endoribonuclease Nsp15. J. Biol. Chem. 283, 3655–3664.

Bost, A.G., Carnahan, R.H., Lu, X.T., Denison, M.R., 2000. Four proteins processed from the replicase gene polyprotein of mouse hepatitis virus colocalize in the cell periphery and adjacent to sites of virion assembly. J. Virol. 74, 3379–3387.

Boursnell, M.E., Brown, T.D., Foulds, I.J., Green, P.F., Tomley, F.M., Binns, M.M., 1987. Completion of the sequence of the genome of the coronavirus avian infectious bronchitis virus. J. Gen. Virol. 68, 57–77.

Bouvet, M., Debarnot, C., Imbert, I., Selisko, B., Snijder, E.J., Canard, B., et al., 2010. In vitro reconstitution of SARS-coronavirus mRNA cap methylation. PLoS Pathog. 6, e1000863.

Bouvet, M., Imbert, I., Subissi, L., Gluais, L., Canard, B., Decroly, E., 2012. RNA 3′-end mismatch excision by the severe acute respiratory syndrome coronavirus nonstructural protein nsp10/nsp14 exoribonuclease complex. Proc. Natl. Acad. Sci. U.S.A. 109, 9372–9377.

Bouvet, M., Lugari, A., Posthuma, C.C., Zevenhoven, J.C., Bernard, S., Betzi, S., et al., 2014. Coronavirus Nsp10, a critical co-factor for activation of multiple replicative enzymes. J. Biol. Chem. 289, 25783–25796.

Brierley, I., Digard, P., Inglis, S.C., 1989. Characterization of an efficient coronavirus ribosomal frameshifting signal: requirement for an RNA pseudoknot. Cell 57, 537–547.

Brockway, S.M., Clay, C.T., Lu, X.T., Denison, M.R., 2003. Characterization of the expression, intracellular localization, and replication complex association of the putative mouse hepatitis virus RNA-dependent RNA polymerase. J. Virol. 77, 10515–10527.

Brockway, S.M., Lu, X.T., Peters, T.R., Dermody, T.S., Denison, M.R., 2004. Intracellular localization and protein interactions of the gene 1 protein p28 during mouse hepatitis virus replication. J. Virol. 78, 11551–11562.

Cavanagh, D., Horzinek, M.C., 1993. Genus Torovirus assigned to the Coronaviridae. Arch. Virol. 128, 395–396.

Chen, P., Jiang, M., Hu, T., Liu, Q., Chen, X.S., Guo, D., 2007. Biochemical characterization of exoribonuclease encoded by SARS coronavirus. J. Biochem. Mol. Biol. 40, 649–655.

Chen, B., Fang, S., Tam, J.P., Liu, D.X., 2009a. Formation of stable homodimer via the C-terminal alpha-helical domain of coronavirus nonstructural protein 9 is critical for its function in viral replication. Virology 383, 328–337.

Chen, J.Y., Chen, W.N., Poon, K.M., Zheng, B.J., Lin, X., Wang, Y.X., et al., 2009b. Interaction between SARS-CoV helicase and a multifunctional cellular protein

(Ddx5) revealed by yeast and mammalian cell two-hybrid systems. Arch. Virol. 154, 507–512.

Chen, Y., Cai, H., Pan, J., Xiang, N., Tien, P., Ahola, T., et al., 2009c. Functional screen reveals SARS coronavirus nonstructural protein nsp14 as a novel cap N7 methyltransferase. Proc. Natl. Acad. Sci. U.S.A. 106, 3484–3489.

Chen, Y., Su, C., Ke, M., Jin, X., Xu, L., Zhang, Z., et al., 2011. Biochemical and structural insights into the mechanisms of SARS coronavirus RNA ribose 2'-O-methylation by nsp16/nsp10 protein complex. PLoS Pathog. 7, e1002294.

Cheng, A., Zhang, W., Xie, Y., Jiang, W., Arnold, E., Sarafianos, S.G., et al., 2005. Expression, purification, and characterization of SARS coronavirus RNA polymerase. Virology 335, 165–176.

Cheng, Z., Muhlrad, D., Lim, M.K., Parker, R., Song, H., 2007. Structural and functional insights into the human Upf1 helicase core. EMBO J. 26, 253–264.

Chu, C.K., Gadthula, S., Chen, X., Choo, H., Olgen, S., Barnard, D.L., et al., 2006. Antiviral activity of nucleoside analogues against SARS-coronavirus (SARS-coV). Antivir. Chem. Chemother. 17, 285–289.

Clerici, M., Mourao, A., Gutsche, I., Gehring, N.H., Hentze, M.W., Kulozik, A., et al., 2009. Unusual bipartite mode of interaction between the nonsense-mediated decay factors, UPF1 and UPF2. EMBO J. 28, 2293–2306.

Crotty, S., Maag, D., Arnold, J.J., Zhong, W., Lau, J.Y., Hong, Z., et al., 2000. The broadspectrum antiviral ribonucleoside ribavirin is an RNA virus mutagen. Nat. Med. 6, 1375–1379.

Crotty, S., Cameron, C., Andino, R., 2002. Ribavirin's antiviral mechanism of action: lethal mutagenesis? J. Mol. Med. 80, 86–95.

Daffis, S., Szretter, K.J., Schriewer, J., Li, J., Youn, S., Errett, J., et al., 2010. 2'-O methylation of the viral mRNA cap evades host restriction by IFIT family members. Nature 468, 452–456.

de Groot, R.J., Baker, S.C., Baric, R., Enjuanes, L., Gorbalenya, A.E., Holmes, K.V., et al., 2012a. Family *Coronaviridae*. In: King, A.M.Q., Adams, M.J., Carstens, E.B., Lefkowitz, E.J. (Eds.), Virus Taxonomy. Elsevier, Amsterdam, pp. 806–828.

de Groot, R.J., Cowley, J.A., Enjuanes, L., Faaberg, K.S., Perlman, S., Rottier, P.J.M., et al., 2012b. Order *Nidovirales*. In: King, A.M.Q., Adams, M.J., Carstens, E.B., Lefkowitz, E.J. (Eds.), Virus Taxonomy. Elsevier, Amsterdam, pp. 785–795.

Decroly, E., Imbert, I., Coutard, B., Bouvet, M., Selisko, B., Alvarez, K., et al., 2008. Coronavirus nonstructural protein 16 is a cap-0 binding enzyme possessing (nucleoside-2'O)-methyltransferase activity. J. Virol. 82, 8071–8084.

Decroly, E., Debarnot, C., Ferron, F., Bouvet, M., Coutard, B., Imbert, I., et al., 2011. Crystal structure and functional analysis of the SARS-coronavirus RNA cap 2'-O-methyltransferase nsp10/nsp16 complex. PLoS Pathog. 7, e1002059.

Decroly, E., Ferron, F., Lescar, J., Canard, B., 2012. Conventional and unconventional mechanisms for capping viral mRNA. Nat. Rev. Microbiol. 10, 51–65.

Deming, D.J., Graham, R.L., Denison, M.R., Baric, R.S., 2007. Processing of open reading frame 1a replicase proteins nsp7 to nsp10 in murine hepatitis virus strain A59 replication. J. Virol. 81, 10280–10291.

den Boon, J.A., Ahlquist, P., 2010. Organelle-like membrane compartmentalization of positive-strand RNA virus replication factories. Annu. Rev. Microbiol. 64, 241–256.

den Boon, J.A., Snijder, E.J., Chirnside, E.D., de Vries, A.A., Horzinek, M.C., Spaan, W.J., 1991. Equine arteritis virus is not a togavirus but belongs to the coronaviruslike superfamily. J. Virol. 65, 2910–2920.

Deng, Z., Lehmann, K.C., Li, X., Feng, C., Wang, G., Zhang, Q., et al., 2014. Structural basis for the regulatory function of a complex zinc-binding domain in a replicative arterivirus helicase resembling a nonsense-mediated mRNA decay helicase. Nucleic Acids Res. 42, 3464–3477.

Denison, M.R., Spaan, W.J., van der Meer, Y., Gibson, C.A., Sims, A.C., Prentice, E., et al., 1999. The putative helicase of the coronavirus mouse hepatitis virus is processed from the replicase gene polyprotein and localizes in complexes that are active in viral RNA synthesis. J. Virol. 73, 6862–6871.

Derbyshire, V., Grindley, N.D., Joyce, C.M., 1991. The 3′-5′ exonuclease of DNA polymerase I of Escherichia coli: contribution of each amino acid at the active site to the reaction. EMBO J. 10, 17–24.

Devarkar, S.C., Wang, C., Miller, M.T., Ramanathan, A., Jiang, F., Khan, A.G., et al., 2016. Structural basis for m7G recognition and 2′-O-methyl discrimination in capped RNAs by the innate immune receptor RIG-I. Proc. Natl. Acad. Sci. U.S.A. 113, 596–601.

Donaldson, E.F., Sims, A.C., Graham, R.L., Denison, M.R., Baric, R.S., 2007. Murine hepatitis virus replicase protein nsp10 is a critical regulator of viral RNA synthesis. J. Virol. 81, 6356–6368.

Drake, J.W., Holland, J.J., 1999. Mutation rates among RNA viruses. Proc. Natl. Acad. Sci. U.S.A. 96, 13910–13913.

Echols, H., Lu, C., Burgers, P.M., 1983. Mutator strains of Escherichia coli, mutD and dnaQ, with defective exonucleolytic editing by DNA polymerase III holoenzyme. Proc. Natl. Acad. Sci. U.S.A. 80, 2189–2192.

Eckerle, L.D., Lu, X., Sperry, S.M., Choi, L., Denison, M.R., 2007. High fidelity of murine hepatitis virus replication is decreased in nsp14 exoribonuclease mutants. J. Virol. 81, 12135–12144.

Eckerle, L.D., Becker, M.M., Halpin, R.A., Li, K., Venter, E., Lu, X., et al., 2010. Infidelity of SARS-CoV Nsp14-exonuclease mutant virus replication is revealed by complete genome sequencing. PLoS Pathog. 6, e1000896.

Egloff, M.P., Ferron, F., Campanacci, V., Longhi, S., Rancurel, C., Dutartre, H., et al., 2004. The severe acute respiratory syndrome-coronavirus replicative protein nsp9 is a single-stranded RNA-binding subunit unique in the RNA virus world. Proc. Natl. Acad. Sci. U.S.A. 101, 3792–3796.

Falzarano, D., de Wit, E., Rasmussen, A.L., Feldmann, F., Okumura, A., Scott, D.P., et al., 2013. Treatment with interferon-alpha2b and ribavirin improves outcome in MERS-CoV-infected rhesus macaques. Nat. Med. 19, 1313–1317.

Fang, S., Chen, B., Tay, F.P., Ng, B.S., Liu, D.X., 2007. An arginine-to-proline mutation in a domain with undefined functions within the helicase protein (Nsp13) is lethal to the coronavirus infectious bronchitis virus in cultured cells. Virology 358, 136–147.

Fang, S.G., Shen, H., Wang, J., Tay, F.P., Liu, D.X., 2008. Proteolytic processing of polyproteins 1a and 1ab between non-structural proteins 10 and 11/12 of coronavirus infectious bronchitis virus is dispensable for viral replication in cultured cells. Virology 379, 175–180.

Feder, M., Pas, J., Wyrwicz, L.S., Bujnicki, J.M., 2003. Molecular phylogenetics of the RrmJ/fibrillarin superfamily of ribose 2′-O-methyltransferases. Gene 302, 129–138.

Ferron, F., Decroly, E., Selisko, B., Canard, B., 2012. The viral RNA capping machinery as a target for antiviral drugs. Antivir. Res. 96, 21–31.

Filipowicz, W., Furuichi, Y., Sierra, J.M., Muthukrishnan, S., Shatkin, A.J., Ochoa, S., 1976. A protein binding the methylated 5′-terminal sequence, m7GpppN, of eukaryotic messenger RNA. Proc. Natl. Acad. Sci. U.S.A. 73, 1559–1563.

Frieman, M., Yount, B., Agnihothram, S., Page, C., Donaldson, E., Roberts, A., et al., 2012. Molecular determinants of severe acute respiratory syndrome coronavirus pathogenesis and virulence in young and aged mouse models of human disease. J. Virol. 86, 884–897.

Ge, X.Y., Li, J.L., Yang, X.L., Chmura, A.A., Zhu, G., Epstein, J.H., et al., 2013. Isolation and characterization of a bat SARS-like coronavirus that uses the ACE2 receptor. Nature 503, 535–538.

Gioia, U., Laneve, P., Dlakic, M., Arceci, M., Bozzoni, I., Caffarelli, E., 2005. Functional characterization of XendoU, the endoribonuclease involved in small nucleolar RNA biosynthesis. J. Biol. Chem. 280, 18996–19002.

Gorbalenya, A.E., 2001. Big nidovirus genome. When count and order of domains matter. Adv. Exp. Med. Biol. 494, 1–17.

Gorbalenya, A.E., Koonin, E.V., Donchenko, A.P., Blinov, V.M., 1989. Coronavirus genome: prediction of putative functional domains in the non-structural polyprotein by comparative amino acid sequence analysis. Nucleic Acids Res. 17, 4847–4861.

Gorbalenya, A.E., Pringle, F.M., Zeddam, J.L., Luke, B.T., Cameron, C.E., Kalmakoff, J., et al., 2002. The palm subdomain-based active site is internally permuted in viral RNA-dependent RNA polymerases of an ancient lineage. J. Mol. Biol. 324, 47–62.

Gorbalenya, A.E., Enjuanes, L., Ziebuhr, J., Snijder, E.J., 2006. Nidovirales: evolving the largest RNA virus genome. Virus Res. 117, 17–37.

Gosert, R., Kanjanahaluethai, A., Egger, D., Bienz, K., Baker, S.C., 2002. RNA replication of mouse hepatitis virus takes place at double-membrane vesicles. J. Virol. 76, 3697–3708.

Graham, R.L., Baric, R.S., 2010. Recombination, reservoirs, and the modular spike: mechanisms of coronavirus cross-species transmission. J. Virol. 84, 3134–3146.

Graham, R.L., Becker, M.M., Eckerle, L.D., Bolles, M., Denison, M.R., Baric, R.S., 2012. A live, impaired-fidelity coronavirus vaccine protects in an aged, immunocompromised mouse model of lethal disease. Nat. Med. 18, 1820–1826.

Graham, R.L., Donaldson, E.F., Baric, R.S., 2013. A decade after SARS: strategies for controlling emerging coronaviruses. Nat. Rev. Microbiol. 11, 836–848.

Guarino, L.A., Bhardwaj, K., Dong, W., Sun, J., Holzenburg, A., Kao, C., 2005. Mutational analysis of the SARS virus Nsp15 endoribonuclease: identification of residues affecting hexamer formation. J. Mol. Biol. 353, 1106–1117.

Habjan, M., Hubel, P., Lacerda, L., Benda, C., Holze, C., Eberl, C.H., et al., 2013. Sequestration by IFIT1 impairs translation of 2′O-unmethylated capped RNA. PLoS Pathog. 9, e1003663.

Hagemeijer, M.C., Rottier, P.J., de Haan, C.A., 2012. Biogenesis and dynamics of the coronavirus replicative structures. Viruses 4, 3245–3269.

Hagemeijer, M.C., Monastyrska, I., Griffith, J., van der Sluijs, P., Voortman, J., van Bergen en Henegouwen, P.M., et al., 2014. Membrane rearrangements mediated by coronavirus nonstructural proteins 3 and 4. Virology 458–459, 125–135.

Hamdan, S., Carr, P.D., Brown, S.E., Ollis, D.L., Dixon, N.E., 2002. Structural basis for proofreading during replication of the Escherichia coli chromosome. Structure 10, 535–546.

He, R., Adonov, A., Traykova-Adonova, M., Cao, J., Cutts, T., Grudesky, E., et al., 2004. Potent and selective inhibition of SARS coronavirus replication by aurintricarboxylic acid. Biochem. Biophys. Res. Commun. 320, 1199–1203.

Hilgenfeld, R., 2014. From SARS to MERS: crystallographic studies on coronaviral proteases enable antiviral drug design. FEBS J. 281, 4085–4096.

Hoffmann, M., Eitner, K., von Grotthuss, M., Rychlewski, L., Banachowicz, E., Grabarkiewicz, T., et al., 2006. Three dimensional model of severe acute respiratory syndrome coronavirus helicase ATPase catalytic domain and molecular design of severe acute respiratory syndrome coronavirus helicase inhibitors. J. Comput. Aided Mol. Des. 20, 305–319.

Hyde, J.L., Diamond, M.S., 2015. Innate immune restriction and antagonism of viral RNA lacking 2-O methylation. Virology 479–480, 66–74.

Hyde, J.L., Gardner, C.L., Kimura, T., White, J.P., Liu, G., Trobaugh, D.W., et al., 2014. A viral RNA structural element alters host recognition of nonself RNA. Science 343, 783–787.

Ikejiri, M., Saijo, M., Morikawa, S., Fukushi, S., Mizutani, T., Kurane, I., et al., 2007. Synthesis and biological evaluation of nucleoside analogues having 6-chloropurine as anti-SARS-CoV agents. Bioorg. Med. Chem. Lett. 17, 2470–2473.

Imbert, I., Guillemot, J.C., Bourhis, J.M., Bussetta, C., Coutard, B., Egloff, M.P., et al., 2006. A second, non-canonical RNA-dependent RNA polymerase in SARS coronavirus. EMBO J. 25, 4933–4942.

Imbert, I., Snijder, E.J., Dimitrova, M., Guillemot, J.C., Lecine, P., Canard, B., 2008. The SARS-Coronavirus PLnc domain of nsp3 as a replication/transcription scaffolding protein. Virus Res. 133, 136–148.

Irigoyen, N., Firth, A.E., Jones, J.D., Chung, B.Y., Siddell, S.G., Brierley, I., 2016. High-resolution analysis of coronavirus gene expression by RNA sequencing and ribosome profiling. PLoS Pathog. 12, e1005473.

Ivanov, K.A., Ziebuhr, J., 2004. Human coronavirus 229E nonstructural protein 13: characterization of duplex-unwinding, nucleoside triphosphatase, and RNA 5'-triphosphatase activities. J. Virol. 78, 7833–7838.

Ivanov, K.A., Hertzig, T., Rozanov, M., Bayer, S., Thiel, V., Gorbalenya, A.E., et al., 2004a. Major genetic marker of nidoviruses encodes a replicative endoribonuclease. Proc. Natl. Acad. Sci. U.S.A. 101, 12694–12699.

Ivanov, K.A., Thiel, V., Dobbe, J.C., van der Meer, Y., Snijder, E.J., Ziebuhr, J., 2004b. Multiple enzymatic activities associated with severe acute respiratory syndrome coronavirus helicase. J. Virol. 78, 5619–5632.

Jin, X., Chen, Y., Sun, Y., Zeng, C., Wang, Y., Tao, J., et al., 2013. Characterization of the guanine-N7 methyltransferase activity of coronavirus nsp14 on nucleotide GTP. Virus Res. 176, 45–52.

Johnson, M.A., Jaudzems, K., Wuthrich, K., 2010. NMR structure of the SARS-CoV nonstructural protein 7 in solution at pH 6.5. J. Mol. Biol. 402, 619–628.

Joseph, J.S., Saikatendu, K.S., Subramanian, V., Neuman, B.W., Buchmeier, M.J., Stevens, R.C., et al., 2007. Crystal structure of a monomeric form of severe acute respiratory syndrome coronavirus endonuclease nsp15 suggests a role for hexamerization as an allosteric switch. J. Virol. 81, 6700–6708.

Kadare, G., Haenni, A.L., 1997. Virus-encoded RNA helicases. J. Virol. 71, 2583–2590.

Kang, H., Bhardwaj, K., Li, Y., Palaninathan, S., Sacchettini, J., Guarino, L., et al., 2007. Biochemical and genetic analyses of murine hepatitis virus Nsp15 endoribonuclease. J. Virol. 81, 13587–13597.

Ke, M., Chen, Y., Wu, A., Sun, Y., Su, C., Wu, H., et al., 2012. Short peptides derived from the interaction domain of SARS coronavirus nonstructural protein nsp10 can suppress the 2'-O-methyltransferase activity of nsp10/nsp16 complex. Virus Res. 167, 322–328.

Kesel, A.J., 2005. Synthesis of novel test compounds for antiviral chemotherapy of severe acute respiratory syndrome (SARS). Curr. Med. Chem. 12, 2095–2162.

Kim, M.K., Yu, M.S., Park, H.R., Kim, K.B., Lee, C., Cho, S.Y., et al., 2011. 2,6-Bis-arylmethyloxy-5-hydroxychromones with antiviral activity against both hepatitis C virus (HCV) and SARS-associated coronavirus (SCV). Eur. J. Med. Chem. 46, 5698–5704.

Knoops, K., Kikkert, M., Worm, S.H., Zevenhoven-Dobbe, J.C., van der Meer, Y., Koster, A.J., et al., 2008. SARS-coronavirus replication is supported by a reticulovesicular network of modified endoplasmic reticulum. PLoS Biol. 6, e226.

Kwong, A.D., Rao, B.G., Jeang, K.T., 2005. Viral and cellular RNA helicases as antiviral targets. Nat. Rev. Drug Discov. 4, 845–853.

Lai, M.M., Cavanagh, D., 1997. The molecular biology of coronaviruses. Adv. Virus Res. 48, 1–100.

Lai, M.M., Stohlman, S.A., 1981. Comparative analysis of RNA genomes of mouse hepatitis viruses. J. Virol. 38, 661–670.

Laneve, P., Altieri, F., Fiori, M.E., Scaloni, A., Bozzoni, I., Caffarelli, E., 2003. Purification, cloning, and characterization of XendoU, a novel endoribonuclease involved in processing of intron-encoded small nucleolar RNAs in Xenopus laevis. J. Biol. Chem. 278, 13026–13032.

Laneve, P., Gioia, U., Ragno, R., Altieri, F., Di Franco, C., Santini, T., et al., 2008. The tumor marker human placental protein 11 is an endoribonuclease. J. Biol. Chem. 283, 34712–34719.

Lauber, C., Ziebuhr, J., Junglen, S., Drosten, C., Zirkel, F., Nga, P.T., et al., 2012. Mesoniviridae: a proposed new family in the order Nidovirales formed by a single species of mosquito-borne viruses. Arch. Virol. 157, 1623–1628.

Lauber, C., Goeman, J.J., Parquet Mdel, C., Nga, P.T., Snijder, E.J., Morita, K., et al., 2013. The footprint of genome architecture in the largest genome expansion in RNA viruses. PLoS Pathog. 9, e1003500.

Lehmann, K.C., Gulyaeva, A., Zevenhoven-Dobbe, J.C., Janssen, G.M., Ruben, M., Overkleeft, H.S., et al., 2015a. Discovery of an essential nucleotidylating activity associated with a newly delineated conserved domain in the RNA polymerase-containing protein of all nidoviruses. Nucleic Acids Res. 43, 8416–8434.

Lehmann, K.C., Hooghiemstra, L., Gulyaeva, A., Samborskiy, D.V., Zevenhoven-Dobbe, J.C., Snijder, E.J., et al., 2015b. Arterivirus nsp12 versus the coronavirus nsp16 2′-O-methyltransferase: comparison of the C-terminal cleavage products of two nidovirus pp1ab polyproteins. J. Gen. Virol. 96, 2643–2655.

Lehmann, K.C., Snijder, E.J., Posthuma, C.C., Gorbalenya, A.E., 2015c. What we know but do not understand about nidovirus helicases. Virus Res. 202, 12–32.

Lei, Y., Moore, C.B., Liesman, R.M., O'Connor, B.P., Bergstralh, D.T., Chen, Z.J., et al., 2009. MAVS-mediated apoptosis and its inhibition by viral proteins. PLoS One 4, e5466.

Liu, H., Kiledjian, M., 2006. Decapping the message: a beginning or an end. Biochem. Soc. Trans. 34, 35–38.

Liu, W.J., Sedlak, P.L., Kondratieva, N., Khromykh, A.A., 2002. Complementation analysis of the flavivirus Kunjin NS3 and NS5 proteins defines the minimal regions essential for formation of a replication complex and shows a requirement of NS3 in cis for virus assembly. J. Virol. 76, 10766–10775.

Liu, D.X., Fung, T.S., Chong, K.K., Shukla, A., Hilgenfeld, R., 2014. Accessory proteins of SARS-CoV and other coronaviruses. Antivir. Res. 109, 97–109.

Lugari, A., Betzi, S., Decroly, E., Bonnaud, E., Hermant, A., Guillemot, J.C., et al., 2010. Molecular mapping of the RNA Cap 2′-O-methyltransferase activation interface between severe acute respiratory syndrome coronavirus nsp10 and nsp16. J. Biol. Chem. 285, 33230–33241.

Ma, Y., Yates, J., Liang, Y., Lemon, S.M., Yi, M., 2008. NS3 helicase domains involved in infectious intracellular hepatitis C virus particle assembly. J. Virol. 82, 7624–7639.

Ma, Y., Wu, L., Shaw, N., Gao, Y., Wang, J., Sun, Y., et al., 2015. Structural basis and functional analysis of the SARS coronavirus nsp14-nsp10 complex. Proc. Natl. Acad. Sci. U.S.A. 112, 9436–9441.

Madhugiri, R., Fricke, M., Marz, M., Ziebuhr, J., 2014. RNA structure analysis of alphacoronavirus terminal genome regions. Virus Res. 194, 76–89.

Maier, H.J., Hawes, P.C., Cottam, E.M., Mantell, J., Verkade, P., Monaghan, P., et al., 2013. Infectious bronchitis virus generates spherules from zippered endoplasmic reticulum membranes. mBio 4, e00801–e00813.

Matthes, N., Mesters, J.R., Coutard, B., Canard, B., Snijder, E.J., Moll, R., et al., 2006. The non-structural protein Nsp10 of mouse hepatitis virus binds zinc ions and nucleic acids. FEBS Lett. 580, 4143–4149.

Menachery, V.D., Baric, R.S., 2013. Bugs in the system. Immunol. Rev. 255, 256–274.

Menachery, V.D., Yount Jr., B.L., Josset, L., Gralinski, L.E., Scobey, T., Agnihothram, S., et al., 2014. Attenuation and restoration of severe acute respiratory syndrome coronavirus mutant lacking 2'-O-methyltransferase activity. J. Virol. 88, 4251–4264.

Menachery, V.D., Yount Jr., B.L., Debbink, K., Agnihothram, S., Gralinski, L.E., Plante, J.A., et al., 2015. A SARS-like cluster of circulating bat coronaviruses shows potential for human emergence. Nat. Med. 21, 1508–1513.

Mielech, A.M., Chen, Y., Mesecar, A.D., Baker, S.C., 2014. Nidovirus papain-like proteases: multifunctional enzymes with protease, deubiquitinating and deISGylating activities. Virus Res. 194, 184–190.

Miknis, Z.J., Donaldson, E.F., Umland, T.C., Rimmer, R.A., Baric, R.S., Schultz, L.W., 2009. Severe acute respiratory syndrome coronavirus nsp9 dimerization is essential for efficient viral growth. J. Virol. 83, 3007–3018.

Minskaia, E., Hertzig, T., Gorbalenya, A.E., Campanacci, V., Cambillau, C., Canard, B., et al., 2006. Discovery of an RNA virus $3' \rightarrow 5'$ exoribonuclease that is critically involved in coronavirus RNA synthesis. Proc. Natl. Acad. Sci. U.S.A. 103, 5108–5113.

Mukherjee, C., Patil, D.P., Kennedy, B.A., Bakthavachalu, B., Bundschuh, R., Schoenberg, D.R., 2012. Identification of cytoplasmic capping targets reveals a role for cap homeostasis in translation and mRNA stability. Cell Rep. 2, 674–684.

Narayanan, K., Huang, C., Makino, S., 2008. SARS coronavirus accessory proteins. Virus Res. 133, 113–121.

Nedialkova, D.D., Ulferts, R., van den Born, E., Lauber, C., Gorbalenya, A.E., Ziebuhr, J., et al., 2009. Biochemical characterization of arterivirus nonstructural protein 11 reveals the nidovirus-wide conservation of a replicative endoribonuclease. J. Virol. 83, 5671–5682.

Nedialkova, D.D., Gorbalenya, A.E., Snijder, E.J., 2010. Arterivirus Nsp1 modulates the accumulation of minus-strand templates to control the relative abundance of viral mRNAs. PLoS Pathog. 6, e1000772.

Neuman, B.W., Angelini, M.M., Buchmeier, M.J., 2014a. Does form meet function in the coronavirus replicative organelle? Trends Microbiol. 22, 642–647.

Neuman, B.W., Chamberlain, P., Bowden, F., Joseph, J., 2014b. Atlas of coronavirus replicase structure. Virus Res. 194, 49–66.

Ng, L.F., Xu, H.Y., Liu, D.X., 2001. Further identification and characterization of products processed from the coronavirus avian infectious bronchitis virus (IBV) 1a polyprotein by the 3C-like proteinase. Adv. Exp. Med. Biol. 494, 291–298.

Ng, K.K., Arnold, J.J., Cameron, C.E., 2008. Structure-function relationships among RNA-dependent RNA polymerases. Curr. Top. Microbiol. Immunol. 320, 137–156.

Nga, P.T., Parquet Mdel, C., Lauber, C., Parida, M., Nabeshima, T., Yu, F., et al., 2011. Discovery of the first insect nidovirus, a missing evolutionary link in the emergence of the largest RNA virus genomes. PLoS Pathog. 7, e1002215.

Ohlmann, T., Rau, M., Pain, V.M., Morley, S.J., 1996. The C-terminal domain of eukaryotic protein synthesis initiation factor (eIF) 4G is sufficient to support cap-independent translation in the absence of eIF4E. EMBO J. 15, 1371–1382.

Pan, J., Peng, X., Gao, Y., Li, Z., Lu, X., Chen, Y., et al., 2008. Genome-wide analysis of protein-protein interactions and involvement of viral proteins in SARS-CoV replication. PLoS One 3, e3299.

Pasternak, A.O., Spaan, W.J., Snijder, E.J., 2006. Nidovirus transcription: how to make sense…? J. Gen. Virol. 87, 1403–1421.

Paul, A.V., Rieder, E., Kim, D.W., van Boom, J.H., Wimmer, E., 2000. Identification of an RNA hairpin in poliovirus RNA that serves as the primary template in the in vitro uridylylation of VPg. J. Virol. 74, 10359–10370.

Peters, H.L., Jochmans, D., de Wilde, A.H., Posthuma, C.C., Snijder, E.J., Neyts, J., et al., 2015. Design, synthesis and evaluation of a series of acyclic fleximer nucleoside analogues with anti-coronavirus activity. Bioorg. Med. Chem. Lett. 25, 2923–2926.

Peti, W., Johnson, M.A., Herrmann, T., Neuman, B.W., Buchmeier, M.J., Nelson, M., et al., 2005. Structural genomics of the severe acute respiratory syndrome coronavirus: nuclear magnetic resonance structure of the protein nsP7. J. Virol. 79, 12905–12913.

Pettersen, E.F., Goddard, T.D., Huang, C.C., Couch, G.S., Greenblatt, D.M., Meng, E.C., et al., 2004. UCSF Chimera—a visualization system for exploratory research and analysis. J. Comput. Chem. 25, 1605–1612.

Pichlmair, A., Lassnig, C., Eberle, C.A., Gorna, M.W., Baumann, C.L., Burkard, T.R., et al., 2011. IFIT1 is an antiviral protein that recognizes 5′-triphosphate RNA. Nat. Immunol. 12, 624–630.

Ponnusamy, R., Moll, R., Weimar, T., Mesters, J.R., Hilgenfeld, R., 2008. Variable oligomerization modes in coronavirus non-structural protein 9. J. Mol. Biol. 383, 1081–1096.

Posthuma, C.C., Nedialkova, D.D., Zevenhoven-Dobbe, J.C., Blokhuis, J.H., Gorbalenya, A.E., Snijder, E.J., 2006. Site-directed mutagenesis of the Nidovirus replicative endoribonuclease NendoU exerts pleiotropic effects on the arterivirus life cycle. J. Virol. 80, 1653–1661.

Pyrc, K., Bosch, B.J., Berkhout, B., Jebbink, M.F., Dijkman, R., Rottier, P., et al., 2006. Inhibition of human coronavirus NL63 infection at early stages of the replication cycle. Antimicrob. Agents Chemother. 50, 2000–2008.

Pyrc, K., Berkhout, B., van der Hoek, L., 2007. The novel human coronaviruses NL63 and HKU1. J. Virol. 81, 3051–3057.

Raines, R.T., 1998. Ribonuclease A. Chem. Rev. 98, 1045–1066.

Ratia, K., Saikatendu, K.S., Santarsiero, B.D., Barretto, N., Baker, S.C., Stevens, R.C., et al., 2006. Severe acute respiratory syndrome coronavirus papain-like protease: structure of a viral deubiquitinating enzyme. Proc. Natl. Acad. Sci. U.S.A. 103, 5717–5722.

Renzi, F., Caffarelli, E., Laneve, P., Bozzoni, I., Brunori, M., Vallone, B., 2006. The structure of the endoribonuclease XendoU: from small nucleolar RNA processing to severe acute respiratory syndrome coronavirus replication. Proc. Natl. Acad. Sci. U.S.A. 103, 12365–12370.

Reusken, C.B., Haagmans, B.L., Muller, M.A., Gutierrez, C., Godeke, G.J., Meyer, B., et al., 2013. Middle East respiratory syndrome coronavirus neutralising serum antibodies in dromedary camels: a comparative serological study. Lancet Infect. Dis. 13, 859–866.

Ricagno, S., Egloff, M.P., Ulferts, R., Coutard, B., Nurizzo, D., Campanacci, V., et al., 2006. Crystal structure and mechanistic determinants of SARS coronavirus nonstructural protein 15 define an endoribonuclease family. Proc. Natl. Acad. Sci. U.S.A. 103, 11892–11897.

Robert, X., Gouet, P., 2014. Deciphering key features in protein structures with the new ENDscript server. Nucleic Acids Res. 42, W320–W324.

Romero-Brey, I., Bartenschlager, R., 2016. Endoplasmic reticulum: the favorite intracellular niche for viral replication and assembly. Viruses 8, 160.

Saif, L.J., 2004. Animal coronavirus vaccines: lessons for SARS. Dev. Biol. 119, 129–140.

Sanjuan, R., Nebot, M.R., Chirico, N., Mansky, L.M., Belshaw, R., 2010. Viral mutation rates. J. Virol. 84, 9733–9748.

Sawicki, S.G., Sawicki, D.L., 1995. Coronaviruses use discontinuous extension for synthesis of subgenome-length negative strands. Adv. Exp. Med. Biol. 380, 499–506.

Sawicki, S.G., Sawicki, D.L., Younker, D., Meyer, Y., Thiel, V., Stokes, H., et al., 2005. Functional and genetic analysis of coronavirus replicase-transcriptase proteins. PLoS Pathog. 1, e39.

Sawicki, S.G., Sawicki, D.L., Siddell, S.G., 2007. A contemporary view of coronavirus transcription. J. Virol. 81, 20–29.

Scheuermann, R., Tam, S., Burgers, P.M., Lu, C., Echols, H., 1983. Identification of the epsilon-subunit of Escherichia coli DNA polymerase III holoenzyme as the dnaQ gene product: a fidelity subunit for DNA replication. Proc. Natl. Acad. Sci. U.S.A. 80, 7085–7089.

Schoenberg, D.R., Maquat, L.E., 2009. Re-capping the message. Trends Biochem. Sci. 34, 435–442.

Schuberth-Wagner, C., Ludwig, J., Bruder, A.K., Herzner, A.M., Zillinger, T., Goldeck, M., et al., 2015. A conserved histidine in the RNA sensor RIG-I controls immune tolerance to N1-2'O-methylated self RNA. Immunity 43, 41–51.

Sethna, P.B., Hung, S.L., Brian, D.A., 1989. Coronavirus subgenomic minus-strand RNAs and the potential for mRNA replicons. Proc. Natl. Acad. Sci. U.S.A. 86, 5626–5630.

Sevajol, M., Subissi, L., Decroly, E., Canard, B., Imbert, I., 2014. Insights into RNA synthesis, capping, and proofreading mechanisms of SARS-coronavirus. Virus Res. 194, 90–99.

Sexton, N.R., Smith, E.C., Blanc, H., Vignuzzi, M., Peersen, O.B., Denison, M.R., 2016. Homology-based identification of a mutation in the coronavirus RNA-dependent RNA polymerase that confers resistance to multiple mutagens. J. Virol. 90, 7415–7428.

Seybert, A., Hegyi, A., Siddell, S.G., Ziebuhr, J., 2000a. The human coronavirus 229E superfamily 1 helicase has RNA and DNA duplex-unwinding activities with 5'-to-3' polarity. RNA 6, 1056–1068.

Seybert, A., van Dinten, L.C., Snijder, E.J., Ziebuhr, J., 2000b. Biochemical characterization of the equine arteritis virus helicase suggests a close functional relationship between arterivirus and coronavirus helicases. J. Virol. 74, 9586–9593.

Seybert, A., Posthuma, C.C., van Dinten, L.C., Snijder, E.J., Gorbalenya, A.E., Ziebuhr, J., 2005. A complex zinc finger controls the enzymatic activities of nidovirus helicases. J. Virol. 79, 696–704.

Shatkin, A.J., 1976. Capping of eucaryotic mRNAs. Cell 9, 645–653.

Sievers, F., Wilm, A., Dineen, D., Gibson, T.J., Karplus, K., Li, W., et al., 2011. Fast, scalable generation of high-quality protein multiple sequence alignments using clustal omega. Mol. Syst. Biol. 7, 539.

Singleton, M.R., Dillingham, M.S., Wigley, D.B., 2007. Structure and mechanism of helicases and nucleic acid translocases. Annu. Rev. Biochem. 76, 23–50.

Smith, E.C., Blanc, H., Surdel, M.C., Vignuzzi, M., Denison, M.R., 2013. Coronaviruses lacking exoribonuclease activity are susceptible to lethal mutagenesis: evidence for proofreading and potential therapeutics. PLoS Pathog. 9, e1003565.

Smith, E.C., Sexton, N.R., Denison, M.R., 2014. Thinking outside the triangle: replication fidelity of the largest RNA viruses. Annu. Rev. Virol. 1, 111–132.

Snijder, E.J., den Boon, J.A., Bredenbeek, P.J., Horzinek, M.C., Rijnbrand, R., Spaan, W.J., 1990. The carboxyl-terminal part of the putative Berne virus polymerase is expressed by ribosomal frameshifting and contains sequence motifs which indicate that toro- and coronaviruses are evolutionarily related. Nucleic Acids Res. 18, 4535–4542.

Snijder, E.J., Horzinek, M.C., Spaan, W.J., 1993. The coronaviruslike superfamily. Adv. Exp. Med. Biol. 342, 235–244.

Snijder, E.J., Bredenbeek, P.J., Dobbe, J.C., Thiel, V., Ziebuhr, J., Poon, L.L., et al., 2003. Unique and conserved features of genome and proteome of SARS-coronavirus, an early split-off from the coronavirus group 2 lineage. J. Mol. Biol. 331, 991–1004.

Sola, I., Mateos-Gomez, P.A., Almazan, F., Zuniga, S., Enjuanes, L., 2011. RNA-RNA and RNA-protein interactions in coronavirus replication and transcription. RNA Biol. 8, 237–248.

Song, H.D., Tu, C.C., Zhang, G.W., Wang, S.Y., Zheng, K., Lei, L.C., et al., 2005. Cross-host evolution of severe acute respiratory syndrome coronavirus in palm civet and human. Proc. Natl. Acad. Sci. U.S.A. 102, 2430–2435.

Steitz, T.A., Steitz, J.A., 1993. A general two-metal-ion mechanism for catalytic RNA. Proc. Natl. Acad. Sci. U.S.A. 90, 6498–6502.

Steuber, H., Hilgenfeld, R., 2010. Recent advances in targeting viral proteases for the discovery of novel antivirals. Curr. Top. Med. Chem. 10, 323–345.

Su, D., Lou, Z., Sun, F., Zhai, Y., Yang, H., Zhang, R., et al., 2006. Dodecamer structure of severe acute respiratory syndrome coronavirus nonstructural protein nsp10. J. Virol. 80, 7902–7908.

Subissi, L., Imbert, I., Ferron, F., Collet, A., Coutard, B., Decroly, E., et al., 2014a. SARS-CoV ORF1b-encoded nonstructural proteins 12–16: replicative enzymes as antiviral targets. Antivir. Res. 101, 122–130.

Subissi, L., Posthuma, C.C., Collet, A., Zevenhoven-Dobbe, J.C., Gorbalenya, A.E., Decroly, E., et al., 2014b. One severe acute respiratory syndrome coronavirus protein complex integrates processive RNA polymerase and exonuclease activities. Proc. Natl. Acad. Sci. U.S.A. 111, E3900–E3909.

Sun, Y., Wang, Z., Tao, J., Wang, Y., Wu, A., Yang, Z., et al., 2014. Yeast-based assays for the high-throughput screening of inhibitors of coronavirus RNA cap guanine-N7-methyltransferase. Antivir. Res. 104, 156–164.

Sutton, G., Fry, E., Carter, L., Sainsbury, S., Walter, T., Nettleship, J., et al., 2004. The nsp9 replicase protein of SARS-coronavirus, structure and functional insights. Structure 12, 341–353.

Tanner, J.A., Watt, R.M., Chai, Y.B., Lu, L.Y., Lin, M.C., Peiris, J.S., et al., 2003. The severe acute respiratory syndrome (SARS) coronavirus NTPase/helicase belongs to a distinct class of 5′ to 3′ viral helicases. J. Biol. Chem. 278, 39578–39582.

Tanner, J.A., Zheng, B.J., Zhou, J., Watt, R.M., Jiang, J.Q., Wong, K.L., et al., 2005. The adamantane-derived bananins are potent inhibitors of the helicase activities and replication of SARS coronavirus. Chem. Biol. 12, 303–311.

te Velthuis, A.J., 2014. Common and unique features of viral RNA-dependent polymerases. Cell. Mol. Life Sci. 71, 4403–4420.

te Velthuis, A.J., Arnold, J.J., Cameron, C.E., van den Worm, S.H., Snijder, E.J., 2010. The RNA polymerase activity of SARS-coronavirus nsp12 is primer dependent. Nucleic Acids Res. 38, 203–214.

te Velthuis, A.J., van den Worm, S.H., Snijder, E.J., 2012. The SARS-coronavirus nsp7 +nsp8 complex is a unique multimeric RNA polymerase capable of both de novo initiation and primer extension. Nucleic Acids Res. 40, 1737–1747.

Ulferts, R., Ziebuhr, J., 2011. Nidovirus ribonucleases: structures and functions in viral replication. RNA Biol. 8, 295–304.

van der Hoeven, B., Oudshoorn, D., Koster, A.J., Snijder, E.J., Kikkert, M., Barcena, M., 2016. Biogenesis and architecture of arterivirus replication organelles. Virus Res. 220, 70–90.

van Dijk, A.A., Makeyev, E.V., Bamford, D.H., 2004. Initiation of viral RNA-dependent RNA polymerization. J. Gen. Virol. 85, 1077–1093.

van Dinten, L.C., den Boon, J.A., Wassenaar, A.L., Spaan, W.J., Snijder, E.J., 1997. An infectious arterivirus cDNA clone: identification of a replicase point mutation that abolishes discontinuous mRNA transcription. Proc. Natl. Acad. Sci. U.S.A. 94, 991–996.

van Dinten, L.C., van Tol, H., Gorbalenya, A.E., Snijder, E.J., 2000. The predicted metal-binding region of the arterivirus helicase protein is involved in subgenomic mRNA synthesis, genome replication, and virion biogenesis. J. Virol. 74, 5213–5223.

van Hemert, M.J., de Wilde, A.H., Gorbalenya, A.E., Snijder, E.J., 2008. The in vitro RNA synthesizing activity of the isolated arterivirus replication/transcription complex is dependent on a host factor. J. Biol. Chem. 283, 16525–16536.

van Vliet, A.L., Smits, S.L., Rottier, P.J., de Groot, R.J., 2002. Discontinuous and non-discontinuous subgenomic RNA transcription in a nidovirus. EMBO J. 21, 6571–6580.

Velankar, S.S., Soultanas, P., Dillingham, M.S., Subramanya, H.S., Wigley, D.B., 1999. Crystal structures of complexes of PcrA DNA helicase with a DNA substrate indicate an inchworm mechanism. Cell 97, 75–84.

V'Kovski, P., Al-Mulla, H., Thiel, V., Neuman, B.W., 2015. New insights on the role of paired membrane structures in coronavirus replication. Virus Res. 202, 33–40.

von Brunn, A., Teepe, C., Simpson, J.C., Pepperkok, R., Friedel, C.C., Zimmer, R., et al., 2007. Analysis of intraviral protein-protein interactions of the SARS coronavirus ORFeome. PLoS One 2, e459.

von Grotthuss, M., Wyrwicz, L.S., Rychlewski, L., 2003. mRNA cap-1 methyltransferase in the SARS genome. Cell 113, 701–702.

Walker, J.E., Saraste, M., Runswick, M.J., Gay, N.J., 1982. Distantly related sequences in the alpha- and beta-subunits of ATP synthase, myosin, kinases and other ATP-requiring enzymes and a common nucleotide binding fold. EMBO J. 1, 945–951.

Wang, Y., Sun, Y., Wu, A., Xu, S., Pan, R., Zeng, C., et al., 2015. Coronavirus nsp10/nsp16 methyltransferase can be targeted by nsp10-derived peptide in vitro and in vivo to reduce replication and pathogenesis. J. Virol. 89, 8416–8427.

Warren, T.K., Wells, J., Panchal, R.G., Stuthman, K.S., Garza, N.L., Van Tongeren, S.A., et al., 2014. Protection against filovirus diseases by a novel broad-spectrum nucleoside analogue BCX4430. Nature 508, 402–405.

Xiao, Y., Ma, Q., Restle, T., Shang, W., Svergun, D.I., Ponnusamy, R., et al., 2012. Non-structural proteins 7 and 8 of feline coronavirus form a 2:1 heterotrimer that exhibits primer-independent RNA polymerase activity. J. Virol. 86, 4444–4454.

Xu, K., Nagy, P.D., 2014. Expanding use of multi-origin subcellular membranes by positive-strand RNA viruses during replication. Curr. Opin. Virol. 9, 119–126.

Xu, X., Liu, Y., Weiss, S., Arnold, E., Sarafianos, S.G., Ding, J., 2003. Molecular model of SARS coronavirus polymerase: implications for biochemical functions and drug design. Nucleic Acids Res. 31, 7117–7130.

Xu, X., Zhai, Y., Sun, F., Lou, Z., Su, D., Xu, Y., et al., 2006. New antiviral target revealed by the hexameric structure of mouse hepatitis virus nonstructural protein nsp15. J. Virol. 80, 7909–7917.

Xu, L.H., Huang, M., Fang, S.G., Liu, D.X., 2011. Coronavirus infection induces DNA replication stress partly through interaction of its nonstructural protein 13 with the p125 subunit of DNA polymerase delta. J. Biol. Chem. 286, 39546–39559.

Yang, D., Leibowitz, J.L., 2015. The structure and functions of coronavirus genomic 3′ and 5′ ends. Virus Res. 206, 120–133.

Yang, N., Tanner, J.A., Wang, Z., Huang, J.D., Zheng, B.J., Zhu, N., et al., 2007. Inhibition of SARS coronavirus helicase by bismuth complexes. Chem. Commun., 4413–4415.

Yarranton, G.T., Gefter, M.L., 1979. Enzyme-catalyzed DNA unwinding: studies on Escherichia coli rep protein. Proc. Natl. Acad. Sci. U.S.A. 76, 1658–1662.

Yu, M.S., Lee, J., Lee, J.M., Kim, Y., Chin, Y.W., Jee, J.G., et al., 2012. Identification of myricetin and scutellarein as novel chemical inhibitors of the SARS coronavirus helicase, nsp13. Bioorg. Med. Chem. Lett. 22, 4049–4054.

Zeng, C., Wu, A., Wang, Y., Xu, S., Tang, Y., Jin, X., et al., 2016. Identification and characterization of a ribose 2′-O-methyltransferase encoded by the ronivirus branch of Nidovirales. J. Virol. 90, 6675–6685.

Zhai, Y., Sun, F., Li, X., Pang, H., Xu, X., Bartlam, M., et al., 2005. Insights into SARS-CoV transcription and replication from the structure of the nsp7-nsp8 hexadecamer. Nat. Struct. Mol. Biol. 12, 980–986.

Ziebuhr, J., Siddell, S.G., 1999. Processing of the human coronavirus 229E replicase polyproteins by the virus-encoded 3C-like proteinase: identification of proteolytic products and cleavage sites common to pp1a and pp1ab. J. Virol. 73, 177–185.

Ziebuhr, J., Snijder, E.J., Gorbalenya, A.E., 2000. Virus-encoded proteinases and proteolytic processing in the Nidovirales. J. Gen. Virol. 81, 853–879.

Zirkel, F., Kurth, A., Quan, P.L., Briese, T., Ellerbrok, H., Pauli, G., et al., 2011. An insect nidovirus emerging from a primary tropical rainforest. mBio 2. e00077–00011.

Zirkel, F., Roth, H., Kurth, A., Drosten, C., Ziebuhr, J., Junglen, S., 2013. Identification and characterization of genetically divergent members of the newly established family Mesoniviridae. J. Virol. 87, 6346–6358.

Zumla, A., Chan, J.F., Azhar, E.I., Hui, D.S., Yuen, K.Y., 2016. Coronaviruses—drug discovery and therapeutic options. Nat. Rev. Drug Discov. 15, 327–347.

Zuo, Y., Deutscher, M.P., 2001. Exoribonuclease superfamilies: structural analysis and phylogenetic distribution. Nucleic Acids Res. 29, 1017–1026.

Züst, R., Miller, T.B., Goebel, S.J., Thiel, V., Masters, P.S., 2008. Genetic interactions between an essential 3′ cis-acting RNA pseudoknot, replicase gene products, and the extreme 3′ end of the mouse coronavirus genome. J. Virol. 82, 1214–1228.

Züst, R., Cervantes-Barragan, L., Habjan, M., Maier, R., Neuman, B.W., Ziebuhr, J., et al., 2011. Ribose 2′-O-methylation provides a molecular signature for the distinction of self and non-self mRNA dependent on the RNA sensor Mda5. Nat. Immunol. 12, 137–143.

CHAPTER FOUR

Coronavirus *cis*-Acting RNA Elements

R. Madhugiri*, M. Fricke†, M. Marz†,‡, J. Ziebuhr*,1

*Institute of Medical Virology, Justus Liebig University Giessen, Giessen, Germany
†Faculty of Mathematics and Computer Science, Friedrich Schiller University Jena, Jena, Germany
‡FLI Leibniz Institute for Age Research, Jena, Germany
[1]Corresponding author: e-mail address: john.ziebuhr@viro.med.uni-giessen.de

Contents

Abstract

Coronaviruses have exceptionally large RNA genomes of approximately 30 kilobases. Genome replication and transcription is mediated by a multisubunit protein complex comprised of more than a dozen virus-encoded proteins. The protein complex is thought to bind specific *cis*-acting RNA elements primarily located in the 5′- and 3′-terminal genome regions and upstream of the open reading frames located in the 3′-proximal one-third of the genome. Here, we review our current understanding of coronavirus *cis*-acting RNA elements, focusing on elements required for genome replication and packaging. Recent bioinformatic, biochemical, and genetic studies suggest a previously unknown level of conservation of *cis*-acting RNA structures among different coronavirus genera and, in some cases, even beyond genus boundaries. Also, there is increasing evidence to suggest that individual *cis*-acting elements may be part of higher-order RNA structures involving long-range and dynamic RNA–RNA interactions between RNA structural elements separated by thousands of nucleotides in the viral genome. We discuss the structural and functional features of these *cis*-acting RNA elements and their specific functions in coronavirus RNA synthesis.

Advances in Virus Research, Volume 96
ISSN 0065-3527
http://dx.doi.org/10.1016/bs.aivir.2016.08.007

1. INTRODUCTION

Coronaviruses are enveloped, positive-strand RNA viruses. They have been united in the subfamily *Coronavirinae* within the family *Coronaviridae* (de Groot et al., 2012a; Masters and Perlman, 2013). Together with three other families (*Arteriviridae*, *Roniviridae*, and *Mesoniviridae*), the *Coronaviridae* form the order *Nidovirales* (de Groot et al., 2012b). According to the current classification, the family *Coronaviridae* comprises four genera called *Alpha-*, *Beta-*, *Gamma-*, and *Deltacoronavirus*. In some cases, these genera have been further subdivided into lineages. Coronaviruses infect a wide range of mammals and birds and include pathogens of major medical, veterinary, and economic interest (de Groot et al., 2012a; Fehr and Perlman, 2015; Masters and Perlman, 2013), with severe acute respiratory syndrome (SARS) coronavirus (SARS-CoV), and Middle East respiratory syndrome (MERS) coronavirus (MERS-CoV) providing two prominent examples of zoonotic coronaviruses causing severe respiratory disease in humans (Drosten et al., 2003; Ksiazek et al., 2003; Vijay and Perlman, 2016; Zaki et al., 2012; Zumla et al., 2015).

Among plus-strand RNA viruses, coronaviruses and related nidoviruses stick out by their large genome size of about 30 kilobases (kb), the synthesis of numerous subgenomic mRNAs, and the large number of nonstructural proteins (nsps) involved in viral RNA synthesis and interactions with host cell functions (reviewed in Masters and Perlman, 2013; Ziebuhr, 2008). Most of the nsps are encoded by the viral replicase gene that occupies the 5′-terminal two-thirds of the genome and is comprised of two large open reading frames, ORF1a and ORF1b. Translation of ORF1a yields polyprotein (pp) 1a (~450 kDa). Translation of ORF1b requires a programmed ribosomal frameshift event (Brierley et al., 1987, 1989) that occurs just upstream of the ORF1a stop codon and results in pp1ab (~750 kDa). Co- and posttranslational cleavage of pp1a/1ab by two types of virus-encoded proteases associated with nsp3 and nsp5 (Mielech et al., 2014; Ziebuhr et al., 2000) gives rise to a total of 15–16 mature proteins that form the viral replication–transcription complex (RTC) which is thought to also involve the nucleocapsid protein and several cellular proteins (Almazan et al., 2004; Schelle et al., 2005; Ziebuhr, 2008; Ziebuhr et al., 2000). This multiprotein complex replicates the viral genome and produces an extensive set of 3′-coterminal subgenomic messenger RNAs (sg mRNAs), the latter representing a hallmark of corona- and other nidoviruses (Pasternak et al.,

2006; Sawicki et al., 2007; Ziebuhr and Snijder, 2007). The sg mRNAs are used to express the genes located downstream of the replicase gene, involving the viral structural proteins (nucleocapsid (N), membrane (M), spike (S), and envelope (E) protein) and several accessory proteins that, in many cases, have been implicated in functions that interfere with antiviral host responses (Liu et al., 2014; Masters and Perlman, 2013; Narayanan et al., 2008b).

In this chapter, we will briefly summarize coronavirus RNA synthesis and then discuss the structural and functional features of currently known *cis*-acting RNA elements located in the 5′- and 3′-terminal untranslated regions (UTR) and neighboring coding regions. Also, we will review the current knowledge of signals required for packaging and of cellular proteins presumed to be involved in viral RNA synthesis.

2. CORONAVIRUS GENOME REPLICATION AND TRANSCRIPTION

Following receptor-mediated entry into the host cell, the viral genome RNA, which is 5′-capped and 3′-polyadenylated, is released from the nucleocapsid and used for translation of the 5′-terminal ORFs 1a and 1b to produce the key components of the viral RTC. The complex is anchored by membrane-spanning domains (residing in nsp3, 4, and 6) to virus-induced membranous structures that provide a scaffold for the protein machinery involved in viral RNA synthesis (den Boon and Ahlquist, 2010; Gosert et al., 2002; Kanjanahaluethai et al., 2007; Knoops et al., 2008; Oostra et al., 2007, 2008; Snijder et al., 2006; van Hemert et al., 2008). Over the past years, a wealth of information has been obtained on enzymatic and other functions, three-dimensional structures and interactions of individual nsps produced from pp1a and pp1ab (reviewed in Imbert et al., 2010; Masters, 2006; Ulferts et al., 2010; Ziebuhr, 2008). The studies show that, in addition to common enzymes conserved in most +RNA viruses, such as RNA-dependent RNA polymerase (RdRp) (te Velthuis et al., 2010), helicase/NTPase (Seybert et al., 2000), proteases (Baker et al., 1989; Ziebuhr et al., 1995), 5′ cap-specific methylases (Chen et al., 2009b; Decroly et al., 2008, 2011), coronaviruses encode an extra set of proteins in their replicase genes. These additional (sometimes even unique) enzymatic functions include a 3′–5′ exoribonuclease (Minskaia et al., 2006; Snijder et al., 2003) that is thought to be involved in mechanisms required for high-fidelity replication of nidovirus (including coronavirus) genomes of more than 20 kb (Eckerle et al., 2010; Minskaia et al., 2006; Smith et al., 2013, 2014) and a uridylate-specific endoribonuclease of currently unknown function that was found to be conserved in all vertebrate nidoviruses (Ivanov et al., 2004; Nga et al., 2011; Ulferts and Ziebuhr, 2011). In some cases, the replicase gene-encoded enzymes could be linked to specific steps of viral RNA synthesis and/or RNA processing or were shown to interfere with cellular functions (reviewed in Fehr and Perlman, 2015; Masters and Perlman, 2013; Ziebuhr, 2008). Interactions between different nsps have been predicted and characterized for a large number of proteins and the structural basis and possible functional implications of these interactions has been a major topic of research. For example, it has been shown that the exoribonuclease and ribose $2'$-O-methyltransferase activities associated with nsp14 and nsp16, respectively, are stimulated by nsp10 and the interacting surfaces have been identified by mutagenesis and structural studies (Bouvet et al., 2014; Decroly et al., 2011; Ma et al., 2015). Also, there is evidence that a hexadecameric complex formed by eight molecules of nsp7 and eight molecules of nsp8 assists the RdRp by acting as a processivity factor (Subissi et al., 2014; Zhai et al., 2005). Additional interactions between individual subunits of the RTC have been suggested on the basis of two-hybrid screening data (Pan et al., 2008; von Brunn et al., 2007) and there is evidence that a large number of coronavirus nsps assemble to form homo- or heterooligomeric complexes (Anand et al., 2002, 2003; Bouvet et al., 2014; Chen et al., 2011; Ma et al., 2015; Ricagno et al., 2006; Su et al., 2006; Xiao et al., 2012; Zhai et al., 2005).

Coronaviruses produce a set of $5'$- and $3'$-coterminal sg mRNAs that contain a common $5'$-leader sequence of about 60–95 nt (Spaan et al., 1983). The sequence of this leader is identical to the $5'$-terminal sequence of the viral genome RNA. Synthesis of coronavirus sg mRNAs is thought to involve a "discontinuous" step during negative-strand RNA synthesis (Sawicki and Sawicki, 1995). Specific proteins of the RTC that are required for (or involved in) this discontinuous extension step remain to be identified while important *cis*-acting RNA elements, called "transcription-regulating sequences" (TRSs), that are required for this step have been characterized for a number of coronaviruses (reviewed in Sola et al., 2011b, 2015). TRSs are located downstream of the $5'$-leader on the genome ("leader-TRS," TRS-L) and upstream of each of the major ORFs present in the $3'$-proximal genome region ("body-TRSs," TRS-B). They play a vital role in supporting the transfer of the nascent minus strand from a distant position in the $3'$-proximal genome region to the TRS-L located near the $5'$-end of the genome following attenuation of minus-strand RNA synthesis at one of the TRS-B. Coronavirus TRSs contain an AU–rich motif of about 10 nucleotides that is involved in base-pairing interactions between the TRS-L and the complement of a body-TRS (Sawicki and Sawicki, 1995, 1998; Sawicki et al., 2007; Sethna et al., 1991). Following transfer of the nascent minus strand from its downstream position on the template (at the TRS-B) to the TRS-L close to the 5' end of the genome, negative-strand RNA synthesis is resumed and completed by copying the 5' leader sequence. The resulting set of 3' antileader-containing sg minus-strand RNAs is subsequently used as templates for the production of the characteristic nested set of 5' leader-containing mRNAs in coronavirus-infected cells (Lai et al., 1983; Sawicki and Sawicki, 1995; Sawicki et al., 2001; Sethna et al., 1989; Spaan et al., 1983). Sg minus-strand RNAs contain a U-stretch at their 5' end, providing a possible template for 3' polyadenylation of sg mRNAs (Hofmann and Brian, 1991; Wu et al., 2013).

As mentioned earlier, the *cis*-acting RNA elements required for coronavirus replication (and transcription) are located in the 5'- and 3'-terminal genome regions and largely (but not exclusively) encompass noncoding regions (Chang et al., 1994; Dalton et al., 2001; Izeta et al., 1999; Kim et al., 1993; Liao and Lai, 1994; Lin et al., 1994, 1996; Zhang et al., 1994). Additional *cis*-acting elements are located at internal positions and include the TRS elements involved in transcription as well as specific RNA signals required for genome packaging (Chen et al., 2007; Escors et al., 2003; Makino et al., 1990; Morales et al., 2013; Penzes et al., 1994). Another important RNA structural element is located in the ORF1a–ORF1b overlap region. This complex pseudoknot structure mediates a (−1) ribosomal frameshift event and thus controls the expression of the second large ORF on the coronavirus genome RNA (ORF1b) (Brierley et al., 1987, 1989; de Haan et al., 2002; Namy et al., 2006).

3. CORONAVIRUS *cis*-ACTING RNA ELEMENTS

Historically, *cis*-acting RNA elements essential for coronavirus RNA synthesis have mainly been characterized using naturally occurring and genetically engineered defective interfering RNAs (DI RNAs) (reviewed in Brian and Baric, 2005; Brian and Spaan, 1997; Masters, 2007; Sola et al., 2011b). DI RNAs are relatively short RNAs that are derived from viral genome RNA but lack large (internal) sequence parts. DI RNAs are replicated in cells provided that a suitable (i.e., closely related) helper virus provides functional replicase complexes *in trans* (Levis et al., 1986; Weiss et al., 1983) and that the DI RNA contains all the *cis*-acting RNA signals required for replication. In general, DI RNAs contain the entire 5′- and 3′-untranslated genome regions and, in most cases, also small parts of neighboring (or internal) coding regions (Lin and Lai, 1993). Coronavirus DI RNAs were first reported and most extensively studied for the betacoronaviruses MHV and BCoV (Chang et al., 1994; de Groot et al., 1992; Hofmann et al., 1990; Luytjes et al., 1996; Makino et al., 1984, 1985, 1988a,b). Subsequently, DI RNAs were also identified and characterized in alpha- and gammacoronaviruses (Izeta et al., 1999; Mendez et al., 1996; Penzes et al., 1994, 1996).

Identification and characterization of DI RNAs in various coronaviruses have been instrumental in mapping the minimal RNA sequences and structures required for replication and packaging. A major problem in studies using DI RNAs for defining elements required for replication was the high-frequency homologous recombination between the RNA replicon and the helper virus genome. For example, BCoV-derived artificial DI RNAs containing base substitutions within 5′ leader sequences rapidly acquired the leader sequence of the helper virus (Chang et al., 1994, 1996; Makino and Lai, 1990). This "leader switching" was regularly observed in serial passaging experiments aimed to rescue (or amplify) DI RNAs for further phenotypic characterization. With the development of a range of coronavirus reverse genetic systems, the manipulation of full-length coronavirus cDNA copies for functional characterization of *cis*-acting RNA elements at the genome level (including long-range RNA–RNA interactions) has now become an attractive alternative to overcome some of the limitations of the DI RNA-based systems used previously (Almazan et al., 2000; Casais et al., 2001; Scobey et al., 2013; Tekes et al., 2008; Thiel et al., 2001; van den Worm et al., 2012; Yount et al., 2000, 2003).

3.1 5′-Terminal *cis*-Acting RNA Elements

DI RNA-based studies performed with representative betacoronaviruses (MHV and BCoV) revealed that approximately 500 nt from the genomic 5′ end (467 nt in MHV and 498 nt in BCoV) are required for replication (Chang et al., 1994; Kim et al., 1993; Luytjes et al., 1996). Similar 5′-terminal sequence requirements were established in subsequent studies for the alphacoronavirus TGEV (649 nt) (Escors et al., 2003) and the gammacoronavirus IBV (544 nt) (Dalton et al., 2001). These DI RNAs contained the entire 5′ UTR, ranging in size from 210 nt (MHV, BCoV, and HCoV-OC43) to 314 nt (TGEV), and a part of the replicase gene (from the nsp1-coding region) (see later). In contrast to alpha- and betacoronaviruses, the gammacoronavirus IBV features a larger 5′ UTR (528 nt) (Boursnell et al., 1987) and lacks an equivalent of nsp1 (Ziebuhr et al., 2001). In this case, the 5′ UTR alone appears to contain all the signals required for genome replication.

3.1.1 Structural Features of Coronavirus 5′-Terminal cis-Acting Elements

The majority of the 5′-proximal RNA structures and sequences essential for coronavirus genome replication have first been characterized for BCoV using DI RNA-based systems (Brown et al., 2007; Chang et al., 1994, 1996; Gustin et al., 2009; Raman and Brian, 2005; Raman et al., 2003). The 5′-proximal 215 nts of the BCoV genome were predicted to harbor four stem-loops (SLs) that, in the older literature, were termed SL I (comprised of Ia and Ib), II, III, and IV. The structures were identified by in vitro structure probing analysis of appropriate DI RNAs and their *cis*-acting functions were investigated by DI RNA replication studies and mutation analysis. More recently, two additional SLs called SL-V and SL-VI were identified in the BCoV nsp1-coding region, with SL-VI being essential for DI RNA replication (Brown et al., 2007).

Unlike BCoV, MHV is predicted to contain three conserved SLs, SL1, SL2, and SL4, in this 5′-terminal genome region (Fig. 1). Using 5′-terminal genome sequences of about 140 nts of nine coronaviruses, including five betacoronaviruses (BCoV, human coronavirus (HCoV) OC43, HCoV-HKU1, SARS-CoV, and MHV-A59), three alphacoronaviruses (HCoV-NL63, HCoV-229E, and TGEV), and one gammacoronavirus (IBV), the Leibowitz and Giedroc laboratories proposed a consensus 5′-terminal RNA secondary structure model (Kang et al., 2006; Liu et al., 2007) that includes three highly conserved hairpin structures, SL1, SL2, and SL4. This model was confirmed and extended by genus-wide alignment-based secondary structure predictions using LocARNA (Madhugiri et al., 2014; Smith et al., 2010; Will et al., 2007, 2012) in which, despite profound sequence diversity in this genome region, three highly conserved SLs SL1, SL2, and SL4 were identified in the 5′-terminal 150-nt betacoronavirus genome regions (Madhugiri et al., 2014) (Fig. 1).

Interestingly, the BCoV and SARS-CoV genome RNAs were predicted to accommodate an additional SL (called SL3) in the region between SL2

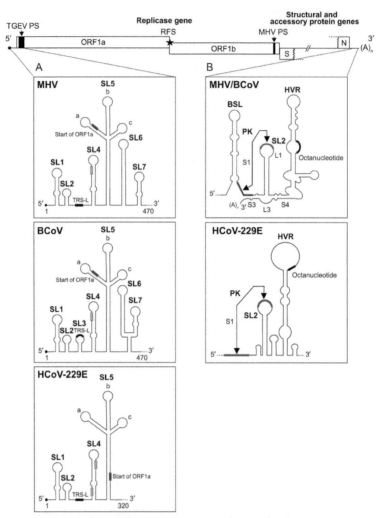

Fig. 1 Conserved *cis*-acting RNA elements in the 5′- and 3′-proximal genome regions of coronaviruses. Shown is the coronavirus genome organization with the two large 5′ ORFs, 1a and 1b, that together constitute the replicase gene, while details of structural and accessory protein ORFs are not shown. *Black circles* at the RNA 5′ ends indicate the 5′ cap structure, while (A)$_n$ indicates the 3′ poly(A) tail. The −1 ribosomal frameshift signal (RFS) at the ORF1a/1b junction site is indicated by an asterisk. *S*, S gene; *N*, N gene. Approximate positions of the packaging signals (PS) determined for MHV and TGEV are indicated by *arrows*. (A) Schematic representation of RNA structural elements in the 5′-terminal genome regions of MHV, BCoV, and HCoV-229E. *Filled boxes* indicate the leader-TRS (TRS-L). *Boxes* in *light gray* indicate the start codons of the uORF(s) located upstream of ORF1a. *Boxes* in *dark gray* indicate the position of the ORF1a start codon. (B) Schematic representation of RNA structural elements in the 3′-terminal genome regions of MHV, BCoV, and HCoV-229E. Major conserved RNA structural elements are shown, together with base-pairing interactions required to form a pseudoknot (PK) structure. Also shown is the position of a highly conserved octanucleotide sequence that is located in a single-stranded region. *BSL*, bulged stem-loop; *L*, loop; *S*, stem; *SL*, stem-loop structure; *HVR*, hypervariable region; *PK*, pseudoknot.

and SL4. SL3 is predicted to adopt a stable hairpin structure containing the TRS-L (Fig. 1). The formation of an equivalent SL3 structure can also be forced for MHV and several other betacoronaviruses (Chen and Olsthoorn, 2010; Madhugiri et al., 2014), although this structural element would only contain two conserved base pairs and was predicted to be unstable at 37°C (Liu et al., 2007). In a recent study, we extended these studies and used multiple alignments calculated with LocARNA (Madhugiri et al., 2014; Smith et al., 2010; Will et al., 2007, 2012) to identify conserved RNA structural elements conserved in the 5′-proximal genome regions of alphacoronaviruses (Madhugiri et al., 2014). The predicted structures were verified and refined by RNA structure probing analyses (Ehresmann et al., 1987; Qu et al., 1983) using in vitro-transcribed RNAs with sequences corresponding to the 5′-terminal genome regions of HCoV-229E and HCoV-NL63, respectively. The combined structural and phylogenetic analyses performed in different laboratories produce a rather coherent picture, with SL1, SL2, and SL4 representing *cis*-acting RNA elements that are highly conserved across different coronavirus genera despite pronounced sequence diversity in the respective 5′-terminal genome regions (Chen and Olsthoorn, 2010; Kang et al., 2006; Liu et al., 2007; Madhugiri et al., 2014).

To further confirm the previously identified conserved betacoronavirus 5′-proximal RNA secondary structures, a recent study used a selective 2′-hydroxyl acylation and primer extension (SHAPE) methodology to determine the secondary structure of the 5′-terminal 474 nts region of the MHV-A59 genome RNA in the virus (in virio), after gentle extraction and deproteinization (ex virio) and an in vitro-transcribed RNA (Yang et al., 2015). With very few exceptions, the RNA secondary structures determined in this study essentially confirmed the previously characterized or predicted SL1, SL2, and SL4 structures (Fig. 1) (Li et al., 2008; Liu et al., 2007, 2009; Yang et al., 2011). The SHAPE analyses also confirmed that the (weak) TRS-L-containing SL3 hairpin predicted previously by phylogenetic algorithms (Chen and Olsthoorn, 2010) is part of a single-stranded region, consistent with previous predictions that this region is weekly paired or unpaired (Liu et al., 2007; Madhugiri et al., 2014). Also several other RNA secondary structures identified by SHAPE analysis corresponded very well to the previous models of MHV-A59 RNA secondary structures proposed by Brian and coworkers (Guan et al., 2011, 2012; Yang et al., 2015). Furthermore, the study provides biochemical support for the presence of additional hairpin structures in the MHV 5′-terminal genome region, including SL5a (designated earlier as SL-IV), SL5b, SL5c, SL6, and SL7. An *Alphacoronavirus* genus-wide bioinformatics study revealed a very well conserved higher-order RNA structure (comprising 5a, 5b, and 5c) in an equivalent genome region (Madhugiri et al., 2014). The predicted SL5a, b, and c structures were confirmed and refined by in vitro RNA structure probing information obtained for the 5′-terminal 600 nts of HCoV-229E and HCoV-NL63 (Madhugiri et al., 2014; unpublished data). Also, the study identified significant constraints in the alphacoronavirus SL5 as judged by the large number of covariant base pairs, suggesting an important function in alphacoronavirus RNA synthesis, possibly related to that described for the betacoronavirus MHV-A59 SL-IV (=SL5a) in supporting efficient viral replication. Furthermore, SL5 was suggested to be involved in long-range RNA–RNA interactions (Guan et al., 2012), which was found to be in good agreement with the SHAPE analysis data (Yang et al., 2015).

Downstream of SL5, additional SL structures (SL6, 7, and 8) were identified. The available evidence suggests that these structures are less well conserved among MHV, BCoV, and SARS-CoV and probably play a less important role in viral replication (Brockway and Denison, 2005; Yang et al., 2015).

Taken together, the available information suggests a model in which the 5′-terminal ∼320-nt genome regions of both alpha- and betacoronaviruses contain four major RNA structural elements called SL1, SL2, SL4, and SL5 (Chen and Olsthoorn, 2010; Kang et al., 2006; Liu et al., 2007; Madhugiri et al., 2014; Yang et al., 2015) (see Fig. 1). The conservation of the SL1, SL2, SL4, and SL5abc RNA structural elements (despite pronounced nucleotide sequence divergency) suggests important functions for these structures in the coronavirus life cycle. Functional features of individual structural elements will be discussed later in more detail.

3.1.2 Functional Roles of Coronavirus 5′-Terminal cis-Acting Elements

In contrast to the growing body of information on structures and their conservation in the coronavirus 5′-terminal genome region across all genera of coronaviruses, the functional significance of the individual SL structures has almost exclusively been studied for two (closely related) betacoronaviruses, MHV and BCoV. The structural and functional conservation inferred from these studies for 5′-terminal betacoronavirus *cis*-acting elements was substantiated by reverse genetic data demonstrating that SARS-CoV SL1, SL2, and SL4 can functionally replace their counterparts in the MHV genome when introduced individually (Kang et al., 2006). Unlike the individual hairpin structure substitution, replacement of the entire MHV 5′ UTR with that of SARS-CoV did not yield a viable MHV mutant, possibly indicating a requirement for stable or transient long-range RNA–RNA interactions of the 5′ UTR with other genome regions. Evidence to support this hypothesis was obtained in subsequent studies. For example, the energetically unstable lower part of MHV SL1 was found to be involved in long-range RNA interactions with the 3′ UTR (Li et al., 2008) (see later). Similar to the SARS-CoV data mentioned earlier, each of the four BCoV 5′-terminal SLs, SL1, SL2, SL4, and SL5a, was shown to functionally replace its MHV counterpart, yielding chimeric viruses with near-wild-type replication kinetics. Furthermore, using MHV/BCoV chimera, a region downstream of SL5 was revealed to be engaged in long-range interactions with the nsp1-coding region, possibly forming an extensive higher-order RNA structure (Guan et al., 2012). Furthermore, a mutagenesis study using BCoV DI RNA (Su et al., 2014) indicated that this multipartite RNA structure may involve several SL substructures identified in earlier studies (Gustin et al., 2009; Raman and Brian, 2005) but require refolding of other RNA structures suggested earlier to be essential for DI RNA replication (Brown et al., 2007). A recent study (Su et al., 2014) provided evidence that a short oligopeptide from the N-terminal domain of nsp1 may be an essential *cis*-acting protein factor involved in betacoronavirus replication, thus adding to the multiple other functions of this protein (Brockway and Denison, 2005; Huang et al., 2011a,b; Kamitani et al., 2006, 2009; Lei et al., 2013; Lokugamage et al., 2012; Narayanan et al., 2008a; Tanaka et al., 2012; Tohya et al., 2009; Wathelet et al., 2007; Züst et al., 2007).

3.1.2.1 Stem-Loops 1 and 2

The 5′-proximal SL1 and SL2 are predicted to be conserved across all genera of the *Coronavirinae* (Chen and Olsthoorn, 2010; Liu et al., 2007). Nuclear magnetic resonance spectroscopy studies of MHV and HCoV-OC43 SL1 RNAs revealed a functionally and structurally bipartite structure for this SL (Li et al., 2008). SL1 was proposed to exist in an equilibrium with higher-energy (partially unfolded) conformers. Characterization of MHV mutants containing specific replacements in SL1 and sequence analysis of second-site revertants support a "dynamic SL1" model in which the lower part of SL1 is required to have an optimally balanced stability/lability. The structural destabilization of the upper part of SL1 by disrupting specific base-pair interactions proved to be lethal or resulted in viruses with replication defects, while compensatory mutations that restored the base pairing in the upper part of SL1 restored viral replication to near–wild-type levels, suggesting that efficient virus replication requires this part of SL1 to be base-paired. In contrast, disruption of the basal part of SL1 was largely tolerated while compensatory mutations that restored base pairing in the lower part proved to be lethal, suggesting a prominent role for RNA sequence rather than structure conservation in the lower part of SL1. Interestingly, the study also identified a possible link between SL1 and minus-strand subgenomic RNA synthesis (Li et al., 2008). The combined data presented in this study suggest that SL1 requires an optimized stability suitable to establish or fine-tune transient long-range (RNA- or protein-mediated) interactions between the 5′ and 3′ UTRs that may be required for sgRNA transcription and genome replication. This hypothesis is also supported by deletion mutagenesis studies in which viable second-site (pseudo)revertants acquired other destabilizing mutations, most likely, to keep the stability of this structure below a certain threshold. Finally, several viable viruses were revealed to contain mutations in the 3′-UTR, providing genetic evidence for interactions between the 5′- and 3′-UTRs.

SL2 is the most conserved structure in the coronavirus 5′ UTR (Chen and Olsthoorn, 2010; Liu et al., 2007). It is comprised of a 5-bp stem and a highly conserved loop sequence, 5′-CUUGY-3′, that was shown to adopt a 5′-uYNMG(U)a- or 5′-uCUYG(U)a-like tetraloop structure (Lee et al., 2011; Liu et al., 2009). Reverse genetics data confirmed that SL2 is required for MHV replication and, possibly, sg mRNA synthesis. Within certain structural constraints, nucleotide replacements were found to be tolerated or could be rescued by increasing the stem stability, suggesting a limited plasticity of this conserved *cis*-acting RNA element (Liu et al., 2009).

3.1.2.2 Stem-Loop 3

As mentioned earlier, SL3 (named SL-II in previous BCoV DI RNA studies) appears to be conserved in a small subset of beta- and gammacoronaviruses (Chen and Olsthoorn, 2010). For BCoV and SARS-CoV, the TRS-L core sequence (CS) has been predicted to be part of this SL3 hairpin loop, a structure similar to the TRS-L hairpin structure reported for the related arterivirus equine arteritis virus (Chang et al., 1996; van den Born et al., 2004, 2005). In contrast to the situation in BCoV and SARS-CoV, the structure probing data obtained for MHV, HCoV-229E, and HCoV-NL63 suggest that the TRS-L CS and flanking regions are located in single-stranded regions (Chen and

Olsthoorn, 2010; Madhugiri et al., 2014; Stirrups et al., 2000; Wang and Zhang, 2000; Yang et al., 2015).

3.1.2.3 Stem-Loop 4

SL4 is a long hairpin structure located downstream of the TRS-L CS and has been suggested to be conserved across all coronavirus genera (Chen and Olsthoorn, 2010; Raman and Brian, 2005; Raman et al., 2003). Using a BCoV DI RNA system, a SL structure that was designated SLIII was mapped between nts 97 and 116 in the 5′-terminal genome region. The *cis*-acting function of SLIII was corroborated by studying effects of destabilizing mutations in this structural element (Raman et al., 2003). Subsequent studies by other laboratories confirmed these findings (Kang et al., 2006; Liu et al., 2007). Genus-wide bioinformatics analyses revealed that SL4 is conserved in alpha- and betacoronaviruses (Madhugiri et al., 2014). It is predicted to form a bipartite SL structure, comprised of 4a and 4b, the latter substructures being separated by a bulge (Kang et al., 2006; Liu et al., 2007; Madhugiri et al., 2014). SL4b identified by various groups corresponds to the SLIII identified by Brian and coworkers (see earlier). Furthermore, SL4 was shown to contain a short ORF comprised of just a few codons. Because of its position in the genome, upstream of the large ORF1a, it is generally referred to as the uORF. Recent reverse genetics work in the MHV system (Wu et al., 2014; Yang et al., 2011) showed that disruption of the uORF yields viable mutants that, however, evolve other uORFs upon serial passaging in cell culture. In vitro, uORF-disrupted RNAs showed enhanced translation of the downstream ORF, suggesting that the uORF represses ORF1a/1b translation and has a beneficial but nonessential role in coronavirus replication in cell culture.

Even though the 5′-terminal SL4 is conserved across the *Coronavirinae*, this hairpin structure tolerates extensive mutations. For example, it was shown for MHV that base pairing in SL4a is not required for replication and also separate deletions of SL4a and SL4b were tolerated. By contrast, deletion of the entire SL4 and a 3-nt deletion immediately downstream of SL4 abolished or profoundly impaired viral RNA synthesis. Analysis of second-site mutations and experiments using a viable MHV mutant in which SL4 was replaced with a shorter SL with a heterologous sequence led to a model in which SL4 acts as a spacer element that controls the proper orientation of SL1, SL2, and TRS-L required for subgenomic RNA synthesis (Yang et al., 2011). The SL4 sequence overlaps with the "hotspot" of the 5′-proximal genomic acceptor required for BCoV discontinuous transcription (Wu et al., 2006), thus further supporting a role of the region immediately downstream of TRS-L in subgenomic RNA synthesis. Based on these observations, it is reasonable to think that the structural flexibility of SL4 may be required to establish transient long-range RNA–RNA interactions. In line with this idea, a previous TGEV reverse genetic study showed that mutants permitting additional base-pairing interactions of the copy TRS-B upstream of a reporter sgRNA with the 5′-GAAA-3′ sequence immediately downstream of the TGEV TRS-L CS (5′-ACUAAAC-3′) enhance the production of this particular reporter sgRNA (Zúñiga et al., 2004). Based on the available functional data and structural analyses of alphacoronavirus 5′-terminal genome regions, it was proposed that the basal part of SL4 exists in a flexible state, thereby possibly facilitating strand transfer during sg minus-strand RNA synthesis (Zúñiga et al., 2004). In addition to the inherent SL4 structural flexibility, proteins known to bind to this region may additionally modulate the stability of the SL4 structure, a hypothesis that remains to be investigated in further experiments. Of particular interest in this context, heterogeneous nuclear ribonucleoprotein (hnRNP) family members and the viral N protein have been shown to bind to this region and there is evidence that the N protein has chaperone functions and TRS-L/TRS-B unwinding activities (Galan et al., 2009; Grossoehme et al., 2009; Huang and Lai, 1999; Keane et al., 2012; Li et al., 1997, 1999; Shi and Lai, 2005; Sola et al., 2011a,b; Zúñiga et al., 2007, 2010). It is therefore tempting to speculate that cellular and/or viral proteins bind and unwind the energetically labile SL4 substructure to facilitate the strand transfer during sg minus-strand RNA synthesis.

3.1.2.4 Stem-Loop 5

A 5′-terminal SL designated earlier as SL-IV that extends into the nsp1 coding sequence was described as an RNA element required for optimal MHV replication (Guan et al., 2011). The SHAPE analysis mentioned earlier suggests that SL5 contains three hairpin substructures, SL5a (previously designated as SL-IV), 5b, and 5c (Yang et al., 2015). Genus-wide analyses of 5′-terminal genome regions suggest a similar SL5 structure to be conserved in alphacoronaviruses, which includes three substructures called SL5a, 5b, and 5c (Chen and Olsthoorn, 2010; Madhugiri et al., 2014). In both alpha- and betacoronaviruses, SL5 extends into ORF1a. Depending on the lineage studied, conserved loop sequences could be identified in the hairpin substructures of SL5. This sequence conservation was more pronounced in alpha- than in betacoronaviruses. In alphacoronaviruses, each of the three hairpins (SL5a, 5b, and 5c) was found to contain a 5′-UUCCGU-3′ loop sequence (Madhugiri et al., 2014). Equivalent structures in betacoronaviruses were only partly conserved, with significant lineage-specific variations being detectable in the substructural hairpins and their terminal loop sequences. A possible SL5 equivalent in gammacoronaviruses was predicted to adopt a rod-like structure that lacks conserved loop sequences (Chen and Olsthoorn, 2010).

As outlined earlier, possible betacoronavirus SL5 substructures located within (or extending into) the nsp1-coding region (previously termed SLs IV, V, VI, and VII) have been characterized structurally and functionally using BCoV DI RNA and MHV reverse genetics systems (Brown et al., 2007; Guan et al., 2011, 2012; Raman and Brian, 2005). In a BCoV-based DI RNA system, SL5A (previously designated as SL-IV) was revealed to be a *cis*-acting element essential for DI RNA replication (Brown et al., 2007). In a recent MHV reverse genetic study, nucleotide substitutions that disrupt SL5C while preserving the N-terminal nsp1 amino acid sequence resulted in the recovery of viable mutant viruses with only moderate impairment of virus replication compared to wild-type virus, implying that SL5C is dispensable for viral replication (Yang et al., 2015) while earlier studies suggested this region to be required for accumulation and replication of a BCoV-based DI RNA (Brown et al., 2007). The reasons for these contradictory results are not clear but may be linked to limitations of DI-based replication assays in which even small functional defects may result in a complete loss of DI RNA replication. Similar observations were made in other cases. For example, DI RNAs and recombinant viruses containing identical mutations in the 5′- and 3′-UTRs led to quite different phenotypes in some cases (Johnson et al., 2005; Yang et al., 2011), illustrating that reverse genetics systems based on full-length genomes are powerful and, in some cases, essential tools in functional studies of *cis*-acting elements.

3.2 3′-Terminal *cis*-Acting RNA Elements

The first studies of 3′ *cis*-acting elements required for RNA replication were based on betacoronavirus DI RNA systems (Kim et al., 1993; Lin and Lai, 1993; Luytjes et al., 1996). Coronavirus 3′ UTRs range in size from ∼300 to ∼500 nts (excluding the 3′ poly(A) tail). Using MHV DI RNAs, the minimal length of 3′-terminal sequence required for replication was determined to involve 436 nts, including the entire 301-nt 3′ UTR, part of the N protein-coding sequence and the poly(A) tail (Lin et al., 1996; Luytjes et al., 1996). In subsequent studies, the minimal 3′-terminal sequences required for TGEV (492 nts) and IBV (338 nts) DI RNA replication were determined (Dalton et al., 2001; Mendez et al., 1996). In both viruses, the *cis*-acting signal required for RNA synthesis could be mapped to the 3′-UTR (only), while N protein-coding sequences were not required. Similar observations were made for betacoronaviruses using recombinant MHV mutants. These studies demonstrated that the structural protein genes (including the N protein-coding region) tolerate substantial alterations including combinations of single-site mutations and rearrangements of entire genes, suggesting that the 3′-proximal coding regions are not part of the 3′ *cis*-acting elements (de Haan et al., 2002; Goebel et al., 2004b; Lorenz et al., 2011). Furthermore, studies by Enjuanes and coworkers suggested that the N gene was dispensable for replication of *Alphacoronavirus 1* using both TGEV and FCoV (Izeta et al., 1999). Also, deletions of the FCoV accessory protein genes 7a and 7b were shown to be tolerated, demonstrating that the 3′ *cis*-acting replication signals of this virus involve only 283 nts plus poly(A) tail (Haijema et al., 2004). For MHV, the minimal 3′-terminal *cis*-acting signal required for negative-strand (but not plus-strand) RNA synthesis was mapped to no more than 55 nts using a DI RNA-based system (Lin et al., 1994). Furthermore, a short poly(A) tract of at least 5–10 nts was shown to be an essential *cis*-acting signal to support BCoV DI RNA replication (Spagnolo and Hogue, 2000).

3.2.1 Structural Features of Coronavirus 3′ cis-Acting Elements

Also in this case, our knowledge of coronavirus 3′ *cis*-acting elements is largely based on studies using betacoronaviruses, such as MHV. A combination of bioinformatics, biochemical analyses, and functional studies was used to identify and characterize *cis*-acting RNA elements in the 3′ UTR (Goebel et al., 2004a, 2007; Hsue and Masters, 1997; Hsue et al., 2000; Liu et al., 2001, 2013; Stammler et al., 2011; Williams et al., 1999; Züst et al., 2008). More recently, these studies were extended to alphacoronaviruses using genus-wide bioinformatics analyses. A combination of sequence and structural alignments of all currently recognized alphacoronavirus species was used to identify conserved RNA structures in the 3′-terminal genome region and the predicted structures were then confirmed and refined using structure probing data obtained for HCoV-229E and HCoV-NL63 (Madhugiri et al., 2014). Fig. 1 provides a simplistic representation of the 3′-proximal RNA structures identified in beta- and alphacoronaviruses.

The 5'-most RNA structure in this region is a bulged stem–loop (BSL) of 68 nts. It is located immediately downstream of the N gene stop codon and was shown to be required for MHV DI RNA replication (Hsue and Masters, 1997; Hsue et al., 2000). Despite limited sequence similarity in this genome region, the BSL structure is predicted to be conserved in betacoronaviruses (Goebel et al., 2004a; Hsue and Masters, 1997). A possible BSL equivalent was also identified in IBV and other gammacoronaviruses and its functional importance was supported using IBV DI RNA constructs (Dalton et al., 2001). The nearly perfect SL structure in IBV comprises 42 nts and is located at the upstream end of region II, a conserved region in the gamma-coronavirus 3' UTR. Recent structural and bioinformatics analyses suggest that alphacoronavirus 3' UTRs do not contain a structural equivalent of the betacoronavirus BSL (Madhugiri et al., 2014).

The second essential RNA structure positioned 3' to the BSL is a classical hairpin-type pseudoknot (PK) structure, which was first identified in BCoV. This 54-nt RNA element was identified as a *cis*-acting element required for BCoV DI RNA replication (Williams et al., 1999). Also the 3'-terminal genome regions of other betacoronaviruses, such as HCoV-HKU1 (Woo et al., 2005) and SARS-CoV, were found to contain this PK structure (Goebel et al., 2004b). Other studies suggested that this PK structure was conserved in beta- and alphacoronaviruses while gammacoronaviruses retained only some of the PK features or lacked this structure entirely (Williams et al., 1999). An interesting structural property of the BSL and the PK is that the elements overlap by five nucleotides in the primary structure. This implies that they cannot exist simultaneously, at least not completely, which led to a model in which the BSL and PK are part of a "molecular switch" that regulates viral RNA synthesis. Evidence to support this model was obtained in an extensive MHV mutagenesis study (Goebel et al., 2004a).

A recent bioinformatics study revisited the conservation of RNA structural elements in the betacoronavirus 3' UTR, including the BSL and the two SL structures that form the PK. The predictions were in excellent agreement with previous studies (Goebel et al., 2004a) and confirmed that, in all established betacoronavirus species, the formation of the PK requires structural rearrangements at the base of the BSL to permit the base-pairing interactions required to form PK stem 1, the latter involving the loop sequence of the PK-SL2 element and the BSL 3'-terminal sequence (Madhugiri et al., 2014). Interestingly, this study also revealed another conserved structural element, a short hairpin, immediately upstream of the PK-SL2 and suggested that the formation of this hairpin may compete with base-pairing interactions required to form the basal part of the BSL and the PK stem 1, respectively. Furthermore, this hairpin overlaps partly with the PK loop 1 region that, in a previous study, was suggested to interact with the extreme 3′ end of the MHV genome (Züst et al., 2008). The conservation of both structure and sequence of this hairpin supports a biological function of this element. In this context, it may be worth mentioning that the hairpin structure is predicted to be disrupted by the 6-nt insertion in loop 1 that, previously, was reported to cause a poorly replicating and unstable phenotype in MHV (Goebel et al., 2004a). It remains to be seen if the small hairpin represents yet another element in the intricate network of base-pairing interactions between the BSL, the PK, and the 3′ end that together constitute the complex molecular switch proposed by the Masters laboratory (Goebel et al., 2004a).

Our recent study using representative viruses from all currently recognized alphacoronavirus species identified a number of conserved RNA structural elements in the alphacoronavirus 3′ UTR (Madhugiri et al., 2014). As described earlier, a counterpart of the betacoronavirus BSL structure (Goebel et al., 2004a; Hsue and Masters, 1997) could not be identified in the alphacoronavirus 3′ UTR, while structural elements required to form a PK structure were identified in all alphacoronaviruses (Madhugiri et al., 2014). Intriguingly, despite the absence of an upstream BSL in alphacoronaviruses, the formation of this putative PK structure was predicted to require the disruption of a short hairpin immediately upstream of PK-SL2, a scenario that is similar to (but less complex than) that described for betacoronaviruses. Further studies are required to answer the question of whether or not alphacoronaviruses employ a molecular switch mechanism similar to that employed by betacoronaviruses (Goebel et al., 2004a). Furthermore, our structure probing analyses supported the predicted PK-SL2 structure for both HCoV-229E and HCoV-NL63 (Madhugiri et al., 2014). They also supported base-pairing interactions upstream of the HCoV-NL63 SL2, thus supporting the formation of the predicted small hairpin in this region, while we failed to obtain experimental support for this hairpin in HCoV-229E. Also, the structure probing data did not support the formation of a stable PK structure, possibly reflecting a low thermodynamic stability as previously reported for the equivalent PK in betacoronaviruses (Stammler et al., 2011). Further experiments including reverse genetics studies are required to confirm the existence and biological significance of the predicted alphacoronavirus PK structure.

The 3′-most RNA secondary structure, a long multibranched SL structure downstream of the pseudoknot was predicted and further confirmed by biochemical probing (Liu et al., 2001). For MHV, several of the stems in this region were reported to be required for efficient DI RNA replication. Using an MHV reverse genetic approach, Masters and coworkers demonstrated that the long hypervariable BSL structure is dispensable for viral replication (Goebel et al., 2007). The study by Madhugiri et al. (2014) revealed the conservation of this RNA structural element downstream of PK-SL2 in all betacoronaviruses and, as expected, confirmed the conservation of the octanucleotide sequence, 5′-GGAAGAGC-3′, that has been identified previously in the 3′ UTR of most coronaviruses (Goebel et al., 2007). The octanucleotide sequence was confirmed to be part of a single-stranded region. As pointed out earlier, the role of this conserved element is currently unclear as both the HVR and the octanucleotide sequence appear to be dispensable for MHV replication in vitro (Goebel et al., 2007; Liu et al., 2001).

With respect to the HVR downstream of PK-SL2, an extensive SL structure was predicted in bioinformatics analyses of alphacoronavirus 3′ UTRs (Madhugiri et al., 2014). The structure is supported by a large number of covariant base pairs and contains the conserved octanucleotide sequence in a single-stranded region, which could be corroborated by structure probing data obtained for HCoV-229E and HCoV-NL63. Of note, the cell culture-adapted HCoV-NL63 isolate used in our study for structure probing analysis contained a short deletion (apparently acquired during serial passaging in cell culture) that resulted in a smaller loop but retained the octanucleotide sequence (with one G-to-A replacement) in a position identical to that predicted for HCoV-229E (Madhugiri et al., 2014). This serendipitous deletion shows that the distal part of the extended SL structure is dispensable for HCoV-NL63 replication in cell culture. The data also suggested that, despite the deletion, the octanucleotide sequence retains a position in the loop region of the SL structure and tolerates minimal changes, the latter being consistent with MHV reverse genetics data obtained for the HVR/octanucleotide region (Goebel et al., 2007).

3.2.2 Functional Roles of Coronavirus 3′-Terminal cis-Acting Elements

Possible functions of RNA elements residing in the 3′-proximal genome regions have been studied most extensively in betacoronaviruses. Although the betacoronavirus 3′ UTRs have minimal sequence identity, the RNA structures conserved across different betacoronavirus lineages appear to be functionally equivalent as demonstrated in studies using viable chimeric viruses. Intriguingly, this functional conservation of 3'-proximal RNA structures does not extend to alpha- and gammacoronaviruses because replacements of the MHV 3' UTR with that of TGEV and IBV, respectively, did not give rise to viable MHV mutants (Goebel et al., 2004b). The available evidence suggests that coronaviruses evolved several genus-specific *cis*-acting RNA elements. For example, the presence of a BSL followed by a PK structure is limited to betacoronaviruses, while other genera appear to contain only one of these elements, with the PK being conserved in alphacoronaviruses and the BSL in gammacoronaviruses (Dalton et al., 2001; Hsue and Masters, 1997; Williams et al., 1999).

3.2.2.1 BSL and Pseudoknot

The structures and several potentially important substructures of both the BSL and PK have been characterized in significant detail for BCoV and MHV (Goebel et al., 2004a; Hsue et al., 2000; Williams et al., 1999). As indicated earlier, the BSL and PK regions overlap by several nucleotides. Formation of the first stem of the PK structure requires base-pairing interactions with the downstream segment F of the BSL, thereby disrupting the basal part of this structure. In a comprehensive MHV mutagenesis study, the functional significance of both structures was demonstrated conclusively. Because the two structures cannot exist simultaneously and, yet, each of them is essential for viral replication, it was proposed that the two elements may adopt substructures that act as a "molecular switch" that controls the transition between different steps of the viral replication cycle (Goebel et al., 2004a). In a subsequent study, the proposed "molecular switch" was characterized in more detail and evidence was obtained to suggest a direct interaction between loop 1 of the PK with the extreme 3' end of the MHV genome (Züst et al., 2008). The characterization of second-site revertants arising from MHV mutants with genetically engineered insertions in loop 1 revealed distinct replacements at the extreme 3' end, thereby retaining specific base-pairing interactions with the loop 1 region and thus precluding the formation of stem 1 of the PK. Other mutants were found to contain second-site replacements indicative of RNA:protein interactions between the PK region and nsp8 and nsp9. Based on these data, a model was proposed in which the formation and disruption of the PK by differential base-pairing interactions with the BSL and 3'-terminal genome sequences, respectively, may lead to alternate substructures that govern different steps of the initiation and continuation of negative-strand RNA synthesis (Züst et al., 2008). Further evidence to support this model was obtained in a subsequent MHV reverse genetics study by Liu et al. (2013). Thermodynamic investigations revealed a limited stability of the PK structure (Stammler et al., 2011). This structural flexibility is consistent with the proposed role as a "molecular switch."

3.2.2.2 Hypervariable Region

The region downstream of the PK is less conserved among betacoronaviruses. It is generally referred to as the "hypervariable region (HVR)" and is not identical to the "HVR" identified at the 5′ end of the 3′ UTR in IBV (Dalton et al., 2001; Williams et al., 1993). The betacoronavirus HVR was predicted to contain a complex and functionally relevant RNA structure based on enzymatic probing and MHV DI RNA mutagenesis data (Liu et al., 2001). By contrast, more recent studies showed that large parts or even the entire HVR region can be deleted without causing major defects in MHV replication, arguing against an essential role of this genome region in viral replication (Goebel et al., 2007; Züst et al., 2008). However, some of the MHV HVR mutants proved to be highly attenuated in vivo, suggesting a possible role in pathogenesis (Goebel et al., 2007).

The conserved octanucleotide sequence mentioned earlier, 5′-GGAAGAGC-3′, was identified in early coronavirus sequence analyses performed in the late 1980s (Boursnell et al., 1985; Lapps et al., 1987; Schreiber et al., 1989). Subsequent studies confirmed its universal conservation across all coronavirus genera, with only very few viruses containing single replacements in this sequence (Goebel et al., 2007). Obviously, this strict conservation suggests an important functional role which, however, could not be confirmed to date. As mentioned earlier, the entire HVR including the octanucleotide sequence can be deleted from the MHV genome without causing major defects in viral replication in vitro (Goebel et al., 2007). In line with this, replacements of single nucleotides within the octanucleotide motif were tolerated although most of these mutants exhibited small-plaque phenotypes and/or delayed single-step growth kinetics. In both high- and low-multiplicity-of-infection experiments, octanucleotide and HVR deletion mutants lagged behind the wild-type virus but reached near-wild-type titers at later time points and had no detectable defect in viral RNA synthesis (Goebel et al., 2007).

3.2.2.3 3′-Terminal Poly(A) Tail

MHV and BCoV DI RNA studies showed that the poly(A) tail at the 3′ end of coronavirus genomes is essential, with a minimum of 5–10 adenylate residues being required for DI RNA replication (Spagnolo and Hogue, 2000). This requirement corresponds well to the minimal binding site of the poly(A)-binding protein (PABP) on DI RNAs poly(A) sequences (Spagnolo and Hogue, 2000). Recent studies further suggest that 3′ poly(A) tail lengths may vary between 30 and 65 nt in the course of viral replication in vitro (Wu et al., 2013) as was shown for both beta- and gammacoronavirus infections and in a range of cell types, both in vitro and in vivo (Shien et al., 2014). The biological significance of these observations is currently unclear.

4. RNA ELEMENTS INVOLVED IN CORONAVIRUS GENOME PACKAGING

To selectively package their genome RNA (rather than other viral and cellular RNAs), viruses employ distinct *cis*-acting sequences in the viral genome RNA and *trans*-acting viral factors (Annamalai and Rao, 2006; D'Souza and Summers, 2005; Nugent et al., 1999). Even though coronaviruses produce large amounts of subgenomic mRNAs in infected cells, these RNAs are not (or extremely inefficiently) incorporated into virus particles (Escors et al., 2003), suggesting that coronaviruses have evolved specific mechanisms to efficiently package their genome RNA into progeny virus particles.

Like for other coronavirus *cis*-acting RNA elements, the genomic packaging signal (PS) was first discovered by DI RNA studies using MHV (Makino et al., 1990; van der Most et al., 1991). PSs of alpha- and gammacoronaviruses were first identified for TGEV and IBV (Escors et al., 2003; Penzes et al., 1994). MHV DI RNA studies revealed a 69-nt SL structure that was (i) located in the 3′ region of ORF1b, (ii) confirmed to be required for DI RNA packaging, and (iii) shown to interact with the viral N protein (Fosmire et al., 1992; Molenkamp and Spaan, 1997; Woo et al., 1997). Subsequent studies indicated that a larger PS element and, possibly, additional factors are required for optimal packaging efficiency (Bos et al., 1997; Cologna and Hogue, 2000; Narayanan and Makino, 2001). More recently, a PS that is conserved in lineage A betacoronaviruses and a novel 95-nt BSL were predicted and supported by chemical and enzymatic probing experiments (Chen et al., 2007). The conservation of this PS among lineage A coronaviruses is consistent with earlier observations that the BCoV PS is functionally replaceable with its MHV counterpart (Cologna and Hogue, 2000). Remarkably, this structurally and functionally conserved PS of lineage A betacoronaviruses is not conserved in other lineages of betacoronaviruses and other coronavirus genera (Kuo and Masters, 2013), suggesting differential requirements for genome packaging among closely related coronaviruses.

For TGEV, the PS was identified using genetically engineered DI RNAs (Izeta et al., 1999). Deletion analyses revealed a minimal TGEV PS required for efficient packaging. This PS contained nts 100–649 from the 5′-proximal genome region (Escors et al., 2003). Also for IBV, a DI RNA that was efficiently packaged has been isolated and characterized (Penzes et al., 1994), even though, in this case, the mapping of a possible PS produced inconclusive data (Dalton et al., 2001). The TGEV and IBV studies support the notion above that coronavirus PSs are found in different genome regions (Escors et al., 2003; Penzes et al., 1994) which, to some extent, is reminiscent of the situation described for picornavirus *cre* elements (Steil and Barton, 2009).

Further insight into the role of PS in genome RNA packaging into virions was obtained in a recent study using MHV (Kuo and Masters, 2013). The study provides conclusive evidence that (i) the PS supports selective packaging of the viral genome RNA into virions and (ii) remains functional when transposed to an ectopic genomic site. Surprisingly, this study also revealed that the PS is not essential for MHV viability and viral growth in cell culture, suggesting that the principal role of the MHV PS is to ensure selective packaging of viral genome RNA into virions. Further insight into conserved and distinct properties of coronaviruses PSs can be expected from future studies using viruses representing all established coronavirus genera.

5. POSSIBLE ROLES OF CELLULAR PROTEINS IN CORONAVIRUS REPLICATION

A number of studies have addressed possible roles of cellular proteins in coronavirus (mainly MHV) RNA synthesis. In these studies, several members of the hnRNP family (PTB or hnRNP A1, SYNCRIP) were identified based on their ability to bind to viral RNA fragments containing TRS (TRS-L as well as TRS-B) in vitro and, in some cases, to affect MHV replication (Choi et al., 2004; Furuya and Lai, 1993; Li et al., 1997; Zhang and Lai, 1995). Deletion analysis and site-directed mutagenesis of the binding regions of PTB or hnRNP A1 further demonstrated significant inhibition of RNA transcription (Li et al., 1999). Furthermore, the functional importance of hnRNP in coronavirus RNA replication was demonstrated by overexpressing wild-type hnRNP A1 or a dominant-negative form of hnRNP A1 in cells (Shi et al., 2000, 2003). It was also shown that hnRNP A1 interacts with the viral N protein (both in vitro and in vivo), suggesting that this protein may become part of the RTC (Shi et al., 2000; Wang and Zhang, 1999). Members of the hnRNP family (hnRNP A1 and PTB) were shown to bind to 5' UTR and 3' UTR and were suggested to mediate a cross talk between 5'- and 3'-terminal genome regions (Huang and Lai, 1999, 2001). Furthermore, it was reported that interactions of hnRNP A1 and PTB modulate viral RNA synthesis and SYNCRIP silencing leads to reduced virus production (Choi et al., 2004). Similar observations were made for TGEV. It was shown that PTB binds to the TGEV TRS-L sequence while other hnRNP family members were found to bind to the 3' end of the genome (hnRNP Q, hnRNP A2B1, and hnRNP A0) (Galan et al., 2009; Sola et al., 2011b). Furthermore, silencing of hnRNP Q expression showed a significant reduction in TGEV RNA synthesis and virus production, supporting biologically relevant functions of hnRNP family members in coronavirus RNA synthesis (Galan et al., 2009). Host factors that interact with specific 5' *cis*-acting structures have only been described for BCoV (Raman and Brian, 2005). These host factors were revealed to bind to SL 5a (previously designated as SL-IV).

Proteins that specifically interact with coronavirus 3' UTRs have mainly been identified by UV-crosslinking experiments using MHV, BCoV, and TGEV terminal genome sequences (Sola et al., 2011b). Members of the hnRNP family and several other proteins were shown to interact with the 3' UTR and have been suggested to have a role in negative-strand as well as positive-strand (genomic and subgenomic) RNA synthesis (Sola et al., 2011b). For BCoV, immunoprecipitation experiments revealed several proteins, including PABP, to interact with the poly(A) tail, another important *cis*-acting element for coronavirus replication (Spagnolo and Hogue, 2000). As mentioned earlier, several proteins, including PABP, could be enriched by RNA affinity chromatography using the TGEV 3' UTR (Galan et al., 2009). Silencing of PABP, hnRNP Q, and glutamyl-prolyl-tRNA synthetase expression led to a two- to threefold reduction in viral RNA synthesis, suggesting that host factors that specifically interact with viral *cis*-acting elements may affect (or even be essential for) viral RNA replication. Clearly, the possible functions of these cellular factors deserve further investigation.

In addition to their interactions with *cis*-acting RNA elements, cellular proteins were found to interact with specific coronavirus nsps. For example, the purification and characterization of enzymatically active SARS-CoV RTCs showed that cellular factors may enhance viral RdRp activity (van Hemert et al., 2008). Also, cellular DEAD-box-family helicases, such as DDX5 and DDX1, have been implicated in coronavirus RNA synthesis. Specific interactions of the DDX5 protein with the SARS-CoV helicase, nsp13, were confirmed in yeast and mammalian two-hybrid and co-immunoprecipitation experiments. Silencing of DDX5 expression led to reduced viral RNA replication and virus titers, supporting the biological significance of this interaction (Chen et al., 2009a). Similarly, in IBV and SARS-CoV, interactions between DDX1 and nsp14 were identified by yeast two-hybrid and coimmunoprecipitation assays (Xu et al., 2010) and validated by showing that knockdown of DDX1 expression affects coronavirus RNA replication and transcription. Similar conclusions were drawn from TGEV TRS interaction studies. Also in this context, the DDX1 helicase was suggested to have a role in coronavirus replication (Sola et al., 2011b).

6. CONCLUSIONS AND OUTLOOK

Over the past years, a large number of studies using structural, biochemical, and reverse genetics approaches have provided important new insight into *cis*-acting elements that drive and control coronavirus RNA replication (reviewed in Masters, 2007; Sola et al., 2011b). In many cases, these studies used betacoronaviruses, while alpha- and gammacoronaviruses were studied to a lesser extent and there is essentially no information on deltacoronaviruses. This work also identified a growing number of cellular and viral proteins that bind to these structures and may have functions in genomic and/or subgenomic RNA synthesis, genome packaging, genome expression, or intracellular targeting of factors/structures engaged in viral RNA synthesis (reviewed in Narayanan and Makino, 2007; Sola et al., 2011b).

Recent bioinformatic studies suggest that the RNA secondary structure elements identified to date for only a small number of coronaviruses may be significantly more conserved than previously thought, both within and across the four coronavirus genera (Chen and Olsthoorn, 2010; Madhugiri et al., 2014; Yang et al., 2015). These studies also provide evidence to suggest a coevolution of RNA structures in the terminal genome regions with the viral replication machinery. Consistent with this hypothesis, the level of conservation of 5'- and 3'-terminal *cis*-active RNA elements among different coronavirus genera and lineages was found to be largely consistent with the replicase gene-based classification of the *Coronavirinae* (de Groot et al., 2012a; Madhugiri et al., 2014). The most conserved elements identified to date include SL 1, 2, and 4 (possibly, also SL 5) in the 5' genome region and a putative PK in the 3'-UTR. The precise roles of these structures and the viral and cellular proteins that bind these structures to perform specific steps in viral RNA synthesis remain to be investigated in more detail. Another interesting aspect to be explored in future studies should address a possible role of the coronavirus 3'-UTR in specific virus–host interactions and/or pathogenesis (Goebel et al., 2007).

ACKNOWLEDGMENTS

The work of R.M. and J.Z. is supported by grants from the Deutsche Forschungsgemeinschaft (SFB 1021, A01, and B01).

REFERENCES

Almazan, F., Gonzalez, J.M., Penzes, Z., Izeta, A., Calvo, E., Plana-Duran, J., et al., 2000. Engineering the largest RNA virus genome as an infectious bacterial artificial chromosome. Proc. Natl. Acad. Sci. U.S.A. 97, 5516–5521.

Almazan, F., Galan, C., Enjuanes, L., 2004. The nucleoprotein is required for efficient coronavirus genome replication. J. Virol. 78, 12683–12688.

Anand, K., Palm, G.J., Mesters, J.R., Siddell, S.G., Ziebuhr, J., Hilgenfeld, R., 2002. Structure of coronavirus main proteinase reveals combination of a chymotrypsin fold with an extra alpha-helical domain. EMBO J. 21, 3213–3224.

Anand, K., Ziebuhr, J., Wadhwani, P., Mesters, J.R., Hilgenfeld, R., 2003. Coronavirus main proteinase (3CLpro) structure: basis for design of anti-SARS drugs. Science 300, 1763–1767.

Annamalai, P., Rao, A.L., 2006. Packaging of brome mosaic virus subgenomic RNA is functionally coupled to replication-dependent transcription and translation of coat protein. J. Virol. 80, 10096–10108.

Baker, S.C., Shieh, C.K., Soe, L.H., Chang, M.F., Vannier, D.M., Lai, M.M., 1989. Identification of a domain required for autoproteolytic cleavage of murine coronavirus gene A polyprotein. J. Virol. 63, 3693–3699.

Bos, E.C., Dobbe, J.C., Luytjes, W., Spaan, W.J., 1997. A subgenomic mRNA transcript of the coronavirus mouse hepatitis virus strain A59 defective interfering (DI) RNA is packaged when it contains the DI packaging signal. J. Virol. 71, 5684–5687.

Boursnell, M.E., Binns, M.M., Foulds, I.J., Brown, T.D., 1985. Sequences of the nucleocapsid genes from two strains of avian infectious bronchitis virus. J. Gen. Virol. 66, 573–580.

Boursnell, M.E., Brown, T.D., Foulds, I.J., Green, P.F., Tomley, F.M., Binns, M.M., 1987. Completion of the sequence of the genome of the coronavirus avian infectious bronchitis virus. J. Gen. Virol. 68, 57–77.

Bouvet, M., Lugari, A., Posthuma, C.C., Zevenhoven, J.C., Bernard, S., Betzi, S., et al., 2014. Coronavirus Nsp10: a critical co-factor for activation of multiple replicative enzymes. J. Biol. Chem. 289, 25783–25796.

Brian, D.A., Baric, R.S., 2005. Coronavirus genome structure and replication. Curr. Top. Microbiol. Immunol. 287, 1–30.

Brian, D.A., Spaan, W.J.M., 1997. Recombination and coronavirus defective interfering RNAs. Semin. Virol. 8, 101–111.

Brierley, I., Boursnell, M.E., Binns, M.M., Bilimoria, B., Blok, V.C., Brown, T.D., et al., 1987. An efficient ribosomal frame-shifting signal in the polymerase-encoding region of the coronavirus IBV. EMBO J. 6, 3779–3785.

Brierley, I., Digard, P., Inglis, S.C., 1989. Characterization of an efficient coronavirus ribosomal frameshifting signal: requirement for an RNA pseudoknot. Cell 57, 537–547.

Brockway, S.M., Denison, M.R., 2005. Mutagenesis of the murine hepatitis virus nsp1-coding region identifies residues important for protein processing, viral RNA synthesis, and viral replication. Virology 340, 209–223.

Brown, C.G., Nixon, K.S., Senanayake, S.D., Brian, D.A., 2007. An RNA stem-loop within the bovine coronavirus nsp1 coding region is a cis-acting element in defective interfering RNA replication. J. Virol. 81, 7716–7724.

Casais, R., Thiel, V., Siddell, S.G., Cavanagh, D., Britton, P., 2001. Reverse genetics system for the avian coronavirus infectious bronchitis virus. J. Virol. 75, 12359–12369.

Chang, R.Y., Hofmann, M.A., Sethna, P.B., Brian, D.A., 1994. A cis-acting function for the coronavirus leader in defective interfering RNA replication. J. Virol. 68, 8223–8231.

Chang, R.Y., Krishnan, R., Brian, D.A., 1996. The UCUAAAC promoter motif is not required for high-frequency leader recombination in bovine coronavirus defective interfering RNA. J. Virol. 70, 2720–2729.

Chen, S.C., Olsthoorn, R.C., 2010. Group-specific structural features of the 5′-proximal sequences of coronavirus genomic RNAs. Virology 401, 29–41.

Chen, S.C., van den Born, E., van den Worm, S.H., Pleij, C.W., Snijder, E.J., Olsthoorn, R.C., 2007. New structure model for the packaging signal in the genome of group IIa coronaviruses. J. Virol. 81, 6771–6774.

Chen, J.Y., Chen, W.N., Poon, K.M., Zheng, B.J., Lin, X., Wang, Y.X., et al., 2009a. Interaction between SARS-CoV helicase and a multifunctional cellular protein (Ddx5) revealed by yeast and mammalian cell two-hybrid systems. Arch. Virol. 154, 507–512.

Chen, Y., Cai, H., Pan, J., Xiang, N., Tien, P., Ahola, T., et al., 2009b. Functional screen reveals SARS coronavirus nonstructural protein nsp14 as a novel cap N7 methyltransferase. Proc. Natl. Acad. Sci. U.S.A. 106, 3484–3489.

Chen, Y., Su, C., Ke, M., Jin, X., Xu, L., Zhang, Z., et al., 2011. Biochemical and structural insights into the mechanisms of SARS coronavirus RNA ribose 2′-O-methylation by nsp16/nsp10 protein complex. PLoS Pathog. 7, e1002294.

Choi, K.S., Mizutani, A., Lai, M.M., 2004. SYNCRIP, a member of the heterogeneous nuclear ribonucleoprotein family, is involved in mouse hepatitis virus RNA synthesis. J. Virol. 78, 13153–13162.

Cologna, R., Hogue, B.G., 2000. Identification of a bovine coronavirus packaging signal. J. Virol. 74, 580–583.

Dalton, K., Casais, R., Shaw, K., Stirrups, K., Evans, S., Britton, P., et al., 2001. cis-acting sequences required for coronavirus infectious bronchitis virus defective-RNA replication and packaging. J. Virol. 75, 125–133.

de Groot, R.J., van der Most, R.G., Spaan, W.J., 1992. The fitness of defective interfering murine coronavirus DI-a and its derivatives is decreased by nonsense and frameshift mutations. J. Virol. 66, 5898–5905.

de Groot, R.J., Baker, S.C., Baric, R., Enjuanes, L., Gorbalenya, A.E., Holmes, K.V., et al., 2012a. Family Coronaviridae. In: King, A.M.Q., Adams, M.J., Carstens, E.B., Lefkowitz, E.J. (Eds.), Virus Taxonomy. Elsevier, Amsterdam, pp. 806–828.

de Groot, R.J., Cowley, J.A., Enjuanes, L., Faaberg, K.S., Perlman, S., Rottier, P.J.M., et al., 2012b. Order Nidovirales. In: King, A.M.Q., Adams, M.J., Carstens, E.B., Lefkowitz, E.J. (Eds.), Virus Taxonomy. Elsevier, Amsterdam, pp. 785–795.

de Haan, C.A., Volders, H., Koetzner, C.A., Masters, P.S., Rottier, P.J., 2002. Coronaviruses maintain viability despite dramatic rearrangements of the strictly conserved genome organization. J. Virol. 76, 12491–12502.

Decroly, E., Imbert, I., Coutard, B., Bouvet, M., Selisko, B., Alvarez, K., et al., 2008. Coronavirus nonstructural protein 16 is a cap-0 binding enzyme possessing (nucleoside-2'O)-methyltransferase activity. J. Virol. 82, 8071–8084.

Decroly, E., Debarnot, C., Ferron, F., Bouvet, M., Coutard, B., Imbert, I., et al., 2011. Crystal structure and functional analysis of the SARS-coronavirus RNA cap 2'-O-methyltransferase nsp10/nsp16 complex. PLoS Pathog. 7, e1002059.

den Boon, J.A., Ahlquist, P., 2010. Organelle-like membrane compartmentalization of positive-strand RNA virus replication factories. Annu. Rev. Microbiol. 64, 241–256.

Drosten, C., Günther, S., Preiser, W., van der Werf, S., Brodt, H.R., Becker, S., et al., 2003. Identification of a novel coronavirus in patients with severe acute respiratory syndrome. N. Engl. J. Med. 348, 1967–1976.

D'Souza, V., Summers, M.F., 2005. How retroviruses select their genomes. Nat. Rev. Microbiol. 3, 643–655.

Eckerle, L.D., Becker, M.M., Halpin, R.A., Li, K., Venter, E., Lu, X., et al., 2010. Infidelity of SARS-CoV Nsp14-exonuclease mutant virus replication is revealed by complete genome sequencing. PLoS Pathog. 6, e1000896.

Ehresmann, C., Baudin, F., Mougel, M., Romby, P., Ebel, J.P., Ehresmann, B., 1987. Probing the structure of RNAs in solution. Nucleic Acids Res. 15, 9109–9128.

Escors, D., Izeta, A., Capiscol, C., Enjuanes, L., 2003. Transmissible gastroenteritis coronavirus packaging signal is located at the 5' end of the virus genome. J. Virol. 77, 7890–7902.

Fehr, A.R., Perlman, S., 2015. Coronaviruses: an overview of their replication and pathogenesis. Methods Mol. Biol. 1282, 1–23.

Fosmire, J.A., Hwang, K., Makino, S., 1992. Identification and characterization of a coronavirus packaging signal. J. Virol. 66, 3522–3530.

Furuya, T., Lai, M.M., 1993. Three different cellular proteins bind to complementary sites on the 5'-end-positive and 3'-end-negative strands of mouse hepatitis virus RNA. J. Virol. 67, 7215–7222.

Galan, C., Sola, I., Nogales, A., Thomas, B., Akoulitchev, A., Enjuanes, L., et al., 2009. Host cell proteins interacting with the 3' end of TGEV coronavirus genome influence virus replication. Virology 391, 304–314.

Goebel, S.J., Hsue, B., Dombrowski, T.F., Masters, P.S., 2004a. Characterization of the RNA components of a putative molecular switch in the 3' untranslated region of the murine coronavirus genome. J. Virol. 78, 669–682.

Goebel, S.J., Taylor, J., Masters, P.S., 2004b. The 3' cis-acting genomic replication element of the severe acute respiratory syndrome coronavirus can function in the murine coronavirus genome. J. Virol. 78, 7846–7851.

Goebel, S.J., Miller, T.B., Bennett, C.J., Bernard, K.A., Masters, P.S., 2007. A hypervariable region within the 3' cis-acting element of the murine coronavirus genome is nonessential for RNA synthesis but affects pathogenesis. J. Virol. 81, 1274–1287.

Gosert, R., Kanjanahaluethai, A., Egger, D., Bienz, K., Baker, S.C., 2002. RNA replication of mouse hepatitis virus takes place at double-membrane vesicles. J. Virol. 76, 3697–3708.

Grossoehme, N.E., Li, L., Keane, S.C., Liu, P., Dann 3rd, C.E., Leibowitz, J.L., et al., 2009. Coronavirus N protein N-terminal domain (NTD) specifically binds the transcriptional regulatory sequence (TRS) and melts TRS-cTRS RNA duplexes. J. Mol. Biol. 394, 544–557.

Guan, B.J., Wu, H.Y., Brian, D.A., 2011. An optimal cis-replication stem-loop IV in the 5′ untranslated region of the mouse coronavirus genome extends 16 nucleotides into open reading frame 1. J. Virol. 85, 5593–5605.

Guan, B.J., Su, Y.P., Wu, H.Y., Brian, D.A., 2012. Genetic evidence of a long-range RNA-RNA interaction between the genomic 5′ untranslated region and the nonstructural protein 1 coding region in murine and bovine coronaviruses. J. Virol. 86, 4631–4643.

Gustin, K.M., Guan, B.J., Dziduszko, A., Brian, D.A., 2009. Bovine coronavirus nonstructural protein 1 (p28) is an RNA binding protein that binds terminal genomic cis-replication elements. J. Virol. 83, 6087–6097.

Haijema, B.J., Volders, H., Rottier, P.J., 2004. Live, attenuated coronavirus vaccines through the directed deletion of group-specific genes provide protection against feline infectious peritonitis. J. Virol. 78, 3863–3871.

Hofmann, M.A., Brian, D.A., 1991. The 5′ end of coronavirus minus-strand RNAs contains a short poly(U) tract. J. Virol. 65, 6331–6333.

Hofmann, M.A., Sethna, P.B., Brian, D.A., 1990. Bovine coronavirus mRNA replication continues throughout persistent infection in cell culture. J. Virol. 64, 4108–4114.

Hsue, B., Masters, P.S., 1997. A bulged stem-loop structure in the 3′ untranslated region of the genome of the coronavirus mouse hepatitis virus is essential for replication. J. Virol. 71, 7567–7578.

Hsue, B., Hartshorne, T., Masters, P.S., 2000. Characterization of an essential RNA secondary structure in the 3′ untranslated region of the murine coronavirus genome. J. Virol. 74, 6911–6921.

Huang, P., Lai, M.M., 1999. Polypyrimidine tract-binding protein binds to the complementary strand of the mouse hepatitis virus 3′ untranslated region, thereby altering RNA conformation. J. Virol. 73, 9110–9116.

Huang, P., Lai, M.M., 2001. Heterogeneous nuclear ribonucleoprotein a1 binds to the 3′-untranslated region and mediates potential 5′-3′-end cross talks of mouse hepatitis virus RNA. J. Virol. 75, 5009–5017.

Huang, C., Lokugamage, K.G., Rozovics, J.M., Narayanan, K., Semler, B.L., Makino, S., 2011a. Alphacoronavirus transmissible gastroenteritis virus nsp1 protein suppresses protein translation in mammalian cells and in cell-free HeLa cell extracts but not in rabbit reticulocyte lysate. J. Virol. 85, 638–643.

Huang, C., Lokugamage, K.G., Rozovics, J.M., Narayanan, K., Semler, B.L., Makino, S., 2011b. SARS coronavirus nsp1 protein induces template-dependent endonucleolytic cleavage of mRNAs: viral mRNAs are resistant to nsp1-induced RNA cleavage. PLoS Pathog. 7, e1002433.

Imbert, I., Ulferts, R., Ziebuhr, J., Canard, B., 2010. SARS coronavirus replicative enzymes: structures and mechanisms. In: Lal, S.K. (Ed.), Molecular Biology of the SARS-Coronavirus. Springer, Berlin and Heidelberg, pp. 99–114.

Ivanov, K.A., Hertzig, T., Rozanov, M., Bayer, S., Thiel, V., Gorbalenya, A.E., et al., 2004. Major genetic marker of nidoviruses encodes a replicative endoribonuclease. Proc. Natl. Acad. Sci. U.S.A. 101, 12694–12699.

Izeta, A., Smerdou, C., Alonso, S., Penzes, Z., Mendez, A., Plana-Duran, J., et al., 1999. Replication and packaging of transmissible gastroenteritis coronavirus-derived synthetic minigenomes. J. Virol. 73, 1535–1545.

Johnson, R.F., Feng, M., Liu, P., Millership, J.J., Yount, B., Baric, R.S., et al., 2005. Effect of mutations in the mouse hepatitis virus 3′(+)42 protein binding element on RNA replication. J. Virol. 79, 14570–14585.

Kamitani, W., Narayanan, K., Huang, C., Lokugamage, K., Ikegami, T., Ito, N., et al., 2006. Severe acute respiratory syndrome coronavirus nsp1 protein suppresses host gene expression by promoting host mRNA degradation. Proc. Natl. Acad. Sci. U.S.A. 103, 12885–12890.

Kamitani, W., Huang, C., Narayanan, K., Lokugamage, K.G., Makino, S., 2009. A two-pronged strategy to suppress host protein synthesis by SARS coronavirus Nsp1 protein. Nat. Struct. Mol. Biol. 16, 1134–1140.

Kang, H., Feng, M., Schroeder, M.E., Giedroc, D.P., Leibowitz, J.L., 2006. Putative cis-acting stem-loops in the 5′ untranslated region of the severe acute respiratory syndrome coronavirus can substitute for their mouse hepatitis virus counterparts. J. Virol. 80, 10600–10614.

Kanjanahaluethai, A., Chen, Z., Jukneliene, D., Baker, S.C., 2007. Membrane topology of murine coronavirus replicase nonstructural protein 3. Virology 361, 391–401.

Keane, S.C., Liu, P., Leibowitz, J.L., Giedroc, D.P., 2012. Functional transcriptional regulatory sequence (TRS) RNA binding and helix destabilizing determinants of murine hepatitis virus (MHV) nucleocapsid (N) protein. J. Biol. Chem. 287, 7063–7073.

Kim, Y.N., Jeong, Y.S., Makino, S., 1993. Analysis of cis-acting sequences essential for coronavirus defective interfering RNA replication. Virology 197, 53–63.

Knoops, K., Kikkert, M., Worm, S.H., Zevenhoven-Dobbe, J.C., van der Meer, Y., Koster, A.J., et al., 2008. SARS-coronavirus replication is supported by a reticulovesicular network of modified endoplasmic reticulum. PLoS Biol. 6, e226.

Ksiazek, T.G., Erdman, D., Goldsmith, C.S., Zaki, S.R., Peret, T., Emery, S., et al., 2003. A novel coronavirus associated with severe acute respiratory syndrome. N. Engl. J. Med. 348, 1953–1966.

Kuo, L., Masters, P.S., 2013. Functional analysis of the murine coronavirus genomic RNA packaging signal. J. Virol. 87, 5182–5192.

Lai, M.M., Patton, C.D., Baric, R.S., Stohlman, S.A., 1983. Presence of leader sequences in the mRNA of mouse hepatitis virus. J. Virol. 46, 1027–1033.

Lapps, W., Hogue, B.G., Brian, D.A., 1987. Sequence analysis of the bovine coronavirus nucleocapsid and matrix protein genes. Virology 157, 47–57.

Lee, C.W., Li, L., Giedroc, D.P., 2011. The solution structure of coronaviral stem-loop 2 (SL2) reveals a canonical CUYG tetraloop fold. FEBS Lett. 585, 1049–1053.

Lei, L., Ying, S., Baojun, L., Yi, Y., Xiang, H., Wenli, S., et al., 2013. Attenuation of mouse hepatitis virus by deletion of the LLRKxGxKG region of Nsp1. PLoS One 8, e61166.

Levis, R., Weiss, B.G., Tsiang, M., Huang, H., Schlesinger, S., 1986. Deletion mapping of Sindbis virus DI RNAs derived from cDNAs defines the sequences essential for replication and packaging. Cell 44, 137–145.

Li, H.P., Zhang, X., Duncan, R., Comai, L., Lai, M.M., 1997. Heterogeneous nuclear ribonucleoprotein A1 binds to the transcription-regulatory region of mouse hepatitis virus RNA. Proc. Natl. Acad. Sci. U.S.A. 94, 9544–9549.

Li, H.P., Huang, P., Park, S., Lai, M.M., 1999. Polypyrimidine tract-binding protein binds to the leader RNA of mouse hepatitis virus and serves as a regulator of viral transcription. J. Virol. 73, 772–777.

Li, L., Kang, H., Liu, P., Makkinje, N., Williamson, S.T., Leibowitz, J.L., et al., 2008. Structural lability in stem-loop 1 drives a 5′ UTR-3′ UTR interaction in coronavirus replication. J. Mol. Biol. 377, 790–803.

Liao, C.L., Lai, M.M., 1994. Requirement of the 5′-end genomic sequence as an upstream cis-acting element for coronavirus subgenomic mRNA transcription. J. Virol. 68, 4727–4737.

Lin, Y.J., Lai, M.M., 1993. Deletion mapping of a mouse hepatitis virus defective interfering RNA reveals the requirement of an internal and discontiguous sequence for replication. J. Virol. 67, 6110–6118.

Lin, Y.J., Liao, C.L., Lai, M.M., 1994. Identification of the cis-acting signal for minus-strand RNA synthesis of a murine coronavirus: implications for the role of minus-strand RNA in RNA replication and transcription. J. Virol. 68, 8131–8140.

Lin, Y.J., Zhang, X., Wu, R.C., Lai, M.M., 1996. The 3′ untranslated region of coronavirus RNA is required for subgenomic mRNA transcription from a defective interfering RNA. J. Virol. 70, 7236–7240.

Liu, Q., Johnson, R.F., Leibowitz, J.L., 2001. Secondary structural elements within the 3′ untranslated region of mouse hepatitis virus strain JHM genomic RNA. J. Virol. 75, 12105–12113.

Liu, P., Li, L., Millership, J.J., Kang, H., Leibowitz, J.L., Giedroc, D.P., 2007. A U-turn motif-containing stem-loop in the coronavirus 5′ untranslated region plays a functional role in replication. RNA 13, 763–780.

Liu, P., Li, L., Keane, S.C., Yang, D., Leibowitz, J.L., Giedroc, D.P., 2009. Mouse hepatitis virus stem-loop 2 adopts a uYNMG(U)a-like tetraloop structure that is highly functionally tolerant of base substitutions. J. Virol. 83, 12084–12093.

Liu, P., Yang, D., Carter, K., Masud, F., Leibowitz, J.L., 2013. Functional analysis of the stem loop S3 and S4 structures in the coronavirus 3′UTR. Virology 443, 40–47.

Liu, D.X., Fung, T.S., Chong, K.K., Shukla, A., Hilgenfeld, R., 2014. Accessory proteins of SARS-CoV and other coronaviruses. Antiviral Res. 109, 97–109.

Lokugamage, K.G., Narayanan, K., Huang, C., Makino, S., 2012. Severe acute respiratory syndrome coronavirus protein nsp1 is a novel eukaryotic translation inhibitor that represses multiple steps of translation initiation. J. Virol. 86, 13598–13608.

Lorenz, R., Bernhart, S.H., Honer Zu Siederdissen, C., Tafer, H., Flamm, C., Stadler, P.F., et al., 2011. ViennaRNA package 2.0. Algorithms Mol. Biol. 6, 26.

Luytjes, W., Gerritsma, H., Spaan, W.J., 1996. Replication of synthetic defective interfering RNAs derived from coronavirus mouse hepatitis virus-A59. Virology 216, 174–183.

Ma, Y., Wu, L., Shaw, N., Gao, Y., Wang, J., Sun, Y., et al., 2015. Structural basis and functional analysis of the SARS coronavirus nsp14-nsp10 complex. Proc. Natl. Acad. Sci. U.S.A. 112, 9436–9441.

Madhugiri, R., Fricke, M., Marz, M., Ziebuhr, J., 2014. RNA structure analysis of alphacoronavirus terminal genome regions. Virus Res. 194, 76–89.

Makino, S., Lai, M.M., 1990. Studies of coronavirus DI RNA replication using in vitro constructed DI cDNA clones. Adv. Exp. Med. Biol. 276, 341–347.

Makino, S., Taguchi, F., Fujiwara, K., 1984. Defective interfering particles of mouse hepatitis virus. Virology 133, 9–17.

Makino, S., Fujioka, N., Fujiwara, K., 1985. Structure of the intracellular defective viral RNAs of defective interfering particles of mouse hepatitis virus. J. Virol. 54, 329–336.

Makino, S., Shieh, C.K., Keck, J.G., Lai, M.M., 1988a. Defective-interfering particles of murine coronavirus: mechanism of synthesis of defective viral RNAs. Virology 163, 104–111.

Makino, S., Shieh, C.K., Soe, L.H., Baker, S.C., Lai, M.M., 1988b. Primary structure and translation of a defective interfering RNA of murine coronavirus. Virology 166, 550–560.

Makino, S., Yokomori, K., Lai, M.M., 1990. Analysis of efficiently packaged defective interfering RNAs of murine coronavirus: localization of a possible RNA-packaging signal. J. Virol. 64, 6045–6053.

Masters, P.S., 2006. The molecular biology of coronaviruses. Adv. Virus Res. 66, 193–292.

Masters, P.S., 2007. Genomic cis-acting elements in coronavirus RNA replication. In: Thiel, V. (Ed.), Coronaviruses—Molecular and Cellular Biology. Caister Academic Press, Norfolk, UK, pp. 65–80.

Masters, P.S., Perlman, S., 2013. Coronaviridae. In: Knipe, D.M., Howley, P.M. (Eds.), sixth ed. Fields Virology, vol. 1. Lippincott Williams & Wilkins, Philadelphia, PA, pp. 825–858.

Mendez, A., Smerdou, C., Izeta, A., Gebauer, F., Enjuanes, L., 1996. Molecular characterization of transmissible gastroenteritis coronavirus defective interfering genomes: packaging and heterogeneity. Virology 217, 495–507.

Mielech, A.M., Chen, Y., Mesecar, A.D., Baker, S.C., 2014. Nidovirus papain-like proteases: multifunctional enzymes with protease, deubiquitinating and deISGylating activities. Virus Res. 194, 184–190.

Minskaia, E., Hertzig, T., Gorbalenya, A.E., Campanacci, V., Cambillau, C., Canard, B., et al., 2006. Discovery of an RNA virus $3'\rightarrow5'$ exoribonuclease that is critically involved in coronavirus RNA synthesis. Proc. Natl. Acad. Sci. U.S.A. 103, 5108–5113.

Molenkamp, R., Spaan, W.J., 1997. Identification of a specific interaction between the coronavirus mouse hepatitis virus A59 nucleocapsid protein and packaging signal. Virology 239, 78–86.

Morales, L., Mateos-Gomez, P.A., Capiscol, C., del Palacio, L., Enjuanes, L., Sola, I., 2013. Transmissible gastroenteritis coronavirus genome packaging signal is located at the 5′ end of the genome and promotes viral RNA incorporation into virions in a replication-independent process. J. Virol. 87, 11579–11590.

Namy, O., Moran, S.J., Stuart, D.I., Gilbert, R.J., Brierley, I., 2006. A mechanical explanation of RNA pseudoknot function in programmed ribosomal frameshifting. Nature 441, 244–247.

Narayanan, K., Makino, S., 2001. Cooperation of an RNA packaging signal and a viral envelope protein in coronavirus RNA packaging. J. Virol. 75, 9059–9067.

Narayanan, K., Makino, S., 2007. Coronavirus genome packaging. In: Thiel, V. (Ed.), Coronaviruses—Molecular and Cellular Biology. Caister Academic Press, Norfolk, UK, pp. 131–142.

Narayanan, K., Huang, C., Lokugamage, K., Kamitani, W., Ikegami, T., Tseng, C.T., et al., 2008a. Severe acute respiratory syndrome coronavirus nsp1 suppresses host gene expression, including that of type I interferon, in infected cells. J. Virol. 82, 4471–4479.

Narayanan, K., Huang, C., Makino, S., 2008b. Coronavirus accessory proteins. In: - Perlman, S., Gallagher, T., Snijder, E.J. (Eds.), Nidoviruses. ASM Press, Washington, DC, pp. 235–244.

Nga, P.T., Parquet Mdel, C., Lauber, C., Parida, M., Nabeshima, T., Yu, F., et al., 2011. Discovery of the first insect nidovirus, a missing evolutionary link in the emergence of the largest RNA virus genomes. PLoS Pathog. 7, e1002215.

Nugent, C.I., Johnson, K.L., Sarnow, P., Kirkegaard, K., 1999. Functional coupling between replication and packaging of poliovirus replicon RNA. J. Virol. 73, 427–435.

Oostra, M., te Lintelo, E.G., Deijs, M., Verheije, M.H., Rottier, P.J., de Haan, C.A., 2007. Localization and membrane topology of coronavirus nonstructural protein 4: involvement of the early secretory pathway in replication. J. Virol. 81, 12323–12336.

Oostra, M., Hagemeijer, M.C., van Gent, M., Bekker, C.P., te Lintelo, E.G., Rottier, P.J., et al., 2008. Topology and membrane anchoring of the coronavirus replication complex: not all hydrophobic domains of nsp3 and nsp6 are membrane spanning. J. Virol. 82, 12392–12405.

Pan, J., Peng, X., Gao, Y., Li, Z., Lu, X., Chen, Y., et al., 2008. Genome-wide analysis of protein-protein interactions and involvement of viral proteins in SARS-CoV replication. PLoS One 3, e3299.

Pasternak, A.O., Spaan, W.J.M., Snijder, E.J., 2006. Nidovirus transcription: how to make sense…? J. Gen. Virol. 87, 1403–1421.

Penzes, Z., Tibbles, K., Shaw, K., Britton, P., Brown, T.D., Cavanagh, D., 1994. Characterization of a replicating and packaged defective RNA of avian coronavirus infectious bronchitis virus. Virology 203, 286–293.

Penzes, Z., Wroe, C., Brown, T.D., Britton, P., Cavanagh, D., 1996. Replication and packaging of coronavirus infectious bronchitis virus defective RNAs lacking a long open reading frame. J. Virol. 70, 8660–8668.

Qu, H.L., Michot, B., Bachellerie, J.P., 1983. Improved methods for structure probing in large RNAs: a rapid 'heterologous' sequencing approach is coupled to the direct mapping of nuclease accessible sites. Application to the 5′ terminal domain of eukaryotic 28S rRNA. Nucleic Acids Res. 11, 5903–5920.

Raman, S., Brian, D.A., 2005. Stem-loop IV in the 5′ untranslated region is a cis-acting element in bovine coronavirus defective interfering RNA replication. J. Virol. 79, 12434–12446.

Raman, S., Bouma, P., Williams, G.D., Brian, D.A., 2003. Stem-loop III in the 5′ untranslated region is a cis-acting element in bovine coronavirus defective interfering RNA replication. J. Virol. 77, 6720–6730.

Ricagno, S., Egloff, M.P., Ulferts, R., Coutard, B., Nurizzo, D., Campanacci, V., et al., 2006. Crystal structure and mechanistic determinants of SARS coronavirus nonstructural protein 15 define an endoribonuclease family. Proc. Natl. Acad. Sci. U.S.A. 103, 11892–11897.

Sawicki, S.G., Sawicki, D.L., 1995. Coronaviruses use discontinuous extension for synthesis of subgenome-length negative strands. Adv. Exp. Med. Biol. 380, 499–506.

Sawicki, S.G., Sawicki, D.L., 1998. A new model for coronavirus transcription. Adv. Exp. Med. Biol. 440, 215–219.

Sawicki, D., Wang, T., Sawicki, S., 2001. The RNA structures engaged in replication and transcription of the A59 strain of mouse hepatitis virus. J. Gen. Virol. 82, 385–396.

Sawicki, S.G., Sawicki, D.L., Siddell, S.G., 2007. A contemporary view of coronavirus transcription. J. Virol. 81, 20–29.

Schelle, B., Karl, N., Ludewig, B., Siddell, S.G., Thiel, V., 2005. Selective replication of coronavirus genomes that express nucleocapsid protein. J. Virol. 79, 6620–6630.

Schreiber, S.S., Kamahora, T., Lai, M.M., 1989. Sequence analysis of the nucleocapsid protein gene of human coronavirus 229E. Virology 169, 142–151.

Scobey, T., Yount, B.L., Sims, A.C., Donaldson, E.F., Agnihothram, S.S., Menachery, V.D., et al., 2013. Reverse genetics with a full-length infectious cDNA of the Middle East respiratory syndrome coronavirus. Proc. Natl. Acad. Sci. U.S.A. 110, 16157–16162.

Sethna, P.B., Hung, S.L., Brian, D.A., 1989. Coronavirus subgenomic minus-strand RNAs and the potential for mRNA replicons. Proc. Natl. Acad. Sci. U.S.A. 86, 5626–5630.

Sethna, P.B., Hofmann, M.A., Brian, D.A., 1991. Minus-strand copies of replicating coronavirus mRNAs contain antileaders. J. Virol. 65, 320–325.

Seybert, A., Hegyi, A., Siddell, S.G., Ziebuhr, J., 2000. The human coronavirus 229E superfamily 1 helicase has RNA and DNA duplex-unwinding activities with 5′-to-3′ polarity. RNA 6, 1056–1068.

Shi, S.T., Lai, M.M., 2005. Viral and cellular proteins involved in coronavirus replication. Curr. Top. Microbiol. Immunol. 287, 95–131.

Shi, S.T., Huang, P., Li, H.P., Lai, M.M., 2000. Heterogeneous nuclear ribonucleoprotein A1 regulates RNA synthesis of a cytoplasmic virus. EMBO J. 19, 4701–4711.

Shi, S.T., Yu, G.Y., Lai, M.M., 2003. Multiple type A/B heterogeneous nuclear ribonucleoproteins (hnRNPs) can replace hnRNP A1 in mouse hepatitis virus RNA synthesis. J. Virol. 77, 10584–10593.

Shien, J.H., Su, Y.D., Wu, H.Y., 2014. Regulation of coronaviral poly(A) tail length during infection is not coronavirus species- or host cell-specific. Virus Genes 49, 383–392.

Smith, C., Heyne, S., Richter, A.S., Will, S., Backofen, R., 2010. Freiburg RNA Tools: a web server integrating INTARNA, EXPARNA and LOCARNA. Nucleic Acids Res. 38, W373–W377.

Smith, E.C., Blanc, H., Surdel, M.C., Vignuzzi, M., Denison, M.R., 2013. Coronaviruses lacking exoribonuclease activity are susceptible to lethal mutagenesis: evidence for proof-reading and potential therapeutics. PLoS Pathog. 9, e1003565.

Smith, E.C., Sexton, N.R., Denison, M.R., 2014. Thinking outside the triangle: replication fidelity of the largest RNA viruses. Annu. Rev. Virol. 1, 111–132.

Snijder, E.J., Bredenbeek, P.J., Dobbe, J.C., Thiel, V., Ziebuhr, J., Poon, L.L., et al., 2003. Unique and conserved features of genome and proteome of SARS-coronavirus, an early split-off from the coronavirus group 2 lineage. J. Mol. Biol. 331, 991–1004.

Snijder, E.J., van der Meer, Y., Zevenhoven-Dobbe, J., Onderwater, J.J., van der Meulen, J., Koerten, H.K., et al., 2006. Ultrastructure and origin of membrane vesicles associated with the severe acute respiratory syndrome coronavirus replication complex. J. Virol. 80, 5927–5940.

Sola, I., Galán, C., Mateos-Gómez, P.A., Palacio, L., Zúñiga, S., Cruz, J.L., et al., 2011a. The polypyrimidine tract-binding protein affects coronavirus RNA accumulation levels and relocalizes viral RNAs to novel cytoplasmic domains different from replication-transcription sites. J. Virol. 85, 5136–5149.

Sola, I., Mateos-Gomez, P.A., Almazan, F., Zuniga, S., Enjuanes, L., 2011b. RNA-RNA and RNA-protein interactions in coronavirus replication and transcription. RNA Biol. 8, 237–248.

Sola, I., Almazan, F., Zuniga, S., Enjuanes, L., 2015. Continuous and discontinuous RNA synthesis in coronaviruses. Annu. Rev. Virol. 2, 265–288.

Spaan, W., Delius, H., Skinner, M., Armstrong, J., Rottier, P., Smeekens, S., et al., 1983. Coronavirus mRNA synthesis involves fusion of non-contiguous sequences. EMBO J. 2, 1839–1844.

Spagnolo, J.F., Hogue, B.G., 2000. Host protein interactions with the 3′ end of bovine coronavirus RNA and the requirement of the poly(A) tail for coronavirus defective genome replication. J. Virol. 74, 5053–5065.

Stammler, S.N., Cao, S., Chen, S.J., Giedroc, D.P., 2011. A conserved RNA pseudoknot in a putative molecular switch domain of the 3′-untranslated region of coronaviruses is only marginally stable. RNA 17, 1747–1759.

Steil, B.P., Barton, D.J., 2009. Cis-active RNA elements (CREs) and picornavirus RNA replication. Virus Res. 139, 240–252.

Stirrups, K., Shaw, K., Evans, S., Dalton, K., Cavanagh, D., Britton, P., 2000. Leader switching occurs during the rescue of defective RNAs by heterologous strains of the coronavirus infectious bronchitis virus. J. Gen. Virol. 81, 791–801.

Su, D., Lou, Z., Sun, F., Zhai, Y., Yang, H., Zhang, R., et al., 2006. Dodecamer structure of severe acute respiratory syndrome coronavirus nonstructural protein nsp10. J. Virol. 80, 7902–7908.

Su, Y.P., Fan, Y.H., Brian, D.A., 2014. Dependence of coronavirus RNA replication on an NH2-terminal partial nonstructural protein 1 in cis. J. Virol. 88, 8868–8882.

Subissi, L., Posthuma, C.C., Collet, A., Zevenhoven-Dobbe, J.C., Gorbalenya, A.E., Decroly, E., et al., 2014. One severe acute respiratory syndrome coronavirus protein complex integrates processive RNA polymerase and exonuclease activities. Proc. Natl. Acad. Sci. U.S.A. 111, E3900–E3909.

Tanaka, T., Kamitani, W., DeDiego, M.L., Enjuanes, L., Matsuura, Y., 2012. Severe acute respiratory syndrome coronavirus nsp1 facilitates efficient propagation in cells through a specific translational shutoff of host mRNA. J. Virol. 86, 11128–11137.

te Velthuis, A.J., Arnold, J.J., Cameron, C.E., van den Worm, S.H., Snijder, E.J., 2010. The RNA polymerase activity of SARS-coronavirus nsp12 is primer dependent. Nucleic Acids Res. 38, 203–214.

Tekes, G., Hofmann-Lehmann, R., Stallkamp, I., Thiel, V., Thiel, H.J., 2008. Genome organization and reverse genetic analysis of a type I feline coronavirus. J. Virol. 82, 1851–1859.

Thiel, V., Herold, J., Schelle, B., Siddell, S.G., 2001. Infectious RNA transcribed in vitro from a cDNA copy of the human coronavirus genome cloned in vaccinia virus. J. Gen. Virol. 82, 1273–1281.

Tohya, Y., Narayanan, K., Kamitani, W., Huang, C., Lokugamage, K., Makino, S., 2009. Suppression of host gene expression by nsp1 proteins of group 2 bat coronaviruses. J. Virol. 83, 5282–5288.

Ulferts, R., Ziebuhr, J., 2011. Nidovirus ribonucleases: structures and functions in viral replication. RNA Biol. 8, 295–304.

Ulferts, R., Imbert, I., Canard, B., Ziebuhr, J., 2010. Expression and functions of SARS coronavirus replicative proteins. In: Lal, S.K. (Ed.), Molecular Biology of the SARS-Coronavirus. Springer, Berlin and Heidelberg, pp. 75–98.

van den Born, E., Gultyaev, A.P., Snijder, E.J., 2004. Secondary structure and function of the 5'-proximal region of the equine arteritis virus RNA genome. RNA 10, 424–437.

van den Born, E., Posthuma, C.C., Gultyaev, A.P., Snijder, E.J., 2005. Discontinuous subgenomic RNA synthesis in arteriviruses is guided by an RNA hairpin structure located in the genomic leader region. J. Virol. 79, 6312–6324.

van den Worm, S.H., Eriksson, K.K., Zevenhoven, J.C., Weber, F., Zust, R., Kuri, T., et al., 2012. Reverse genetics of SARS-related coronavirus using vaccinia virus-based recombination. PLoS One 7, e32857.

van der Most, R.G., Bredenbeek, P.J., Spaan, W.J., 1991. A domain at the 3' end of the polymerase gene is essential for encapsidation of coronavirus defective interfering RNAs. J. Virol. 65, 3219–3226.

van Hemert, M.J., van den Worm, S.H., Knoops, K., Mommaas, A.M., Gorbalenya, A.E., Snijder, E.J., 2008. SARS-coronavirus replication/transcription complexes are membrane-protected and need a host factor for activity in vitro. PLoS Pathog. 4, e1000054.

Vijay, R., Perlman, S., 2016. Middle East respiratory syndrome and severe acute respiratory syndrome. Curr. Opin. Virol. 16, 70–76.

von Brunn, A., Teepe, C., Simpson, J.C., Pepperkok, R., Friedel, C.C., Zimmer, R., et al., 2007. Analysis of intraviral protein-protein interactions of the SARS coronavirus ORFeome. PLoS One 2, e459.

Wang, Y., Zhang, X., 1999. The nucleocapsid protein of coronavirus mouse hepatitis virus interacts with the cellular heterogeneous nuclear ribonucleoprotein A1 in vitro and in vivo. Virology 265, 96–109.

Wang, Y., Zhang, X., 2000. The leader RNA of coronavirus mouse hepatitis virus contains an enhancer-like element for subgenomic mRNA transcription. J. Virol. 74, 10571–10580.

Wathelet, M.G., Orr, M., Frieman, M.B., Baric, R.S., 2007. Severe acute respiratory syndrome coronavirus evades antiviral signaling: role of nsp1 and rational design of an attenuated strain. J. Virol. 81, 11620–11633.

Weiss, B., Levis, R., Schlesinger, S., 1983. Evolution of virus and defective-interfering RNAs in BHK cells persistently infected with Sindbis virus. J. Virol. 48, 676–684.

Will, S., Reiche, K., Hofacker, I.L., Stadler, P.F., Backofen, R., 2007. Inferring noncoding RNA families and classes by means of genome-scale structure-based clustering. PLoS Comput. Biol. 3, e65.

Will, S., Joshi, T., Hofacker, I.L., Stadler, P.F., Backofen, R., 2012. LocARNA-P: accurate boundary prediction and improved detection of structural RNAs. RNA 18, 900–914.

Williams, A.K., Wang, L., Sneed, L.W., Collisson, E.W., 1993. Analysis of a hypervariable region in the 3' non-coding end of the infectious bronchitis virus genome. Virus Res. 28, 19–27.

Williams, G.D., Chang, R.Y., Brian, D.A., 1999. A phylogenetically conserved hairpin-type 3' untranslated region pseudoknot functions in coronavirus RNA replication. J. Virol. 73, 8349–8355.

Woo, K., Joo, M., Narayanan, K., Kim, K.H., Makino, S., 1997. Murine coronavirus packaging signal confers packaging to nonviral RNA. J. Virol. 71, 824–827.

Woo, P.C., Lau, S.K., Chu, C.M., Chan, K.H., Tsoi, H.W., Huang, Y., et al., 2005. Characterization and complete genome sequence of a novel coronavirus, coronavirus HKU1, from patients with pneumonia. J. Virol. 79, 884–895.

Wu, H.Y., Ozdarendeli, A., Brian, D.A., 2006. Bovine coronavirus 5'-proximal genomic acceptor hotspot for discontinuous transcription is 65 nucleotides wide. J. Virol. 80, 2183–2193.

Wu, H.Y., Ke, T.Y., Liao, W.Y., Chang, N.Y., 2013. Regulation of coronaviral poly(A) tail length during infection. PLoS One 8, e70548.

Wu, H.Y., Guan, B.J., Su, Y.P., Fan, Y.H., Brian, D.A., 2014. Reselection of a genomic upstream open reading frame in mouse hepatitis coronavirus 5'-untranslated-region mutants. J. Virol. 88, 846–858.

Xiao, Y., Ma, Q., Restle, T., Shang, W., Svergun, D.I., Ponnusamy, R., et al., 2012. Nonstructural proteins 7 and 8 of feline coronavirus form a 2:1 heterotrimer that exhibits primer-independent RNA polymerase activity. J. Virol. 86, 4444–4454.

Xu, L., Khadijah, S., Fang, S., Wang, L., Tay, F.P., Liu, D.X., 2010. The cellular RNA helicase DDX1 interacts with coronavirus nonstructural protein 14 and enhances viral replication. J. Virol. 84, 8571–8583.

Yang, D., Liu, P., Giedroc, D.P., Leibowitz, J., 2011. Mouse hepatitis virus stem-loop 4 functions as a spacer element required to drive subgenomic RNA synthesis. J. Virol. 85, 9199–9209.

Yang, D., Liu, P., Wudeck, E.V., Giedroc, D.P., Leibowitz, J.L., 2015. SHAPE analysis of the RNA secondary structure of the Mouse Hepatitis Virus 5' untranslated region and N-terminal nsp1 coding sequences. Virology 475, 15–27.

Yount, B., Curtis, K.M., Baric, R.S., 2000. Strategy for systematic assembly of large RNA and DNA genomes: transmissible gastroenteritis virus model. J. Virol. 74, 10600–10611.

Yount, B., Curtis, K.M., Fritz, E.A., Hensley, L.E., Jahrling, P.B., Prentice, E., et al., 2003. Reverse genetics with a full-length infectious cDNA of severe acute respiratory syndrome coronavirus. Proc. Natl. Acad. Sci. U.S.A. 100, 12995–13000.

Zaki, A.M., van Boheemen, S., Bestebroer, T.M., Osterhaus, A.D., Fouchier, R.A., 2012. Isolation of a novel coronavirus from a man with pneumonia in Saudi Arabia. N. Engl. J. Med. 367, 1814–1820.

Zhai, Y., Sun, F., Li, X., Pang, H., Xu, X., Bartlam, M., et al., 2005. Insights into SARS-CoV transcription and replication from the structure of the nsp7-nsp8 hexadecamer. Nat. Struct. Mol. Biol. 12, 980–986.

Zhang, X., Lai, M.M., 1995. Interactions between the cytoplasmic proteins and the intergenic (promoter) sequence of mouse hepatitis virus RNA: correlation with the amounts of subgenomic mRNA transcribed. J. Virol. 69, 1637–1644.

Zhang, X., Liao, C.L., Lai, M.M., 1994. Coronavirus leader RNA regulates and initiates subgenomic mRNA transcription both in trans and in cis. J. Virol. 68, 4738–4746.

Ziebuhr, J., 2008. Coronavirus replicative proteins. In: Perlman, S., Gallagher, T., Snijder, E.J. (Eds.), Nidoviruses. ASM Press, Washington, DC, pp. 65–81.

Ziebuhr, J., Snijder, E.J., 2007. The coronavirus replicase gene: special enzymes for special viruses. In: Thiel, V. (Ed.), Coronaviruses—Molecular and Cellular Biology. Caister Academic Press, Norfolk, UK, pp. 33–63.

Ziebuhr, J., Herold, J., Siddell, S.G., 1995. Characterization of a human coronavirus (strain 229E) 3C-like proteinase activity. J. Virol. 69, 4331–4338.

Ziebuhr, J., Snijder, E.J., Gorbalenya, A.E., 2000. Virus-encoded proteinases and proteolytic processing in the Nidovirales. J. Gen. Virol. 81, 853–879.

Ziebuhr, J., Thiel, V., Gorbalenya, A.E., 2001. The autocatalytic release of a putative RNA virus transcription factor from its polyprotein precursor involves two paralogous papain-like proteases that cleave the same peptide bond. J. Biol. Chem. 276, 33220–33232.

Zumla, A., Hui, D.S., Perlman, S., 2015. Middle East respiratory syndrome. Lancet 386, 995–1007.

Zúñiga, S., Sola, I., Alonso, S., Enjuanes, L., 2004. Sequence motifs involved in the regulation of discontinuous coronavirus subgenomic RNA synthesis. J. Virol. 78, 980–994.

Zúñiga, S., Sola, I., Moreno, J.L., Sabella, P., Plana-Duran, J., Enjuanes, L., 2007. Coronavirus nucleocapsid protein is an RNA chaperone. Virology 357, 215–227.

Zúñiga, S., Cruz, J.L., Sola, I., Mateos-Gomez, P.A., Palacio, L., Enjuanes, L., 2010. Coronavirus nucleocapsid protein facilitates template switching and is required for efficient transcription. J. Virol. 84, 2169–2175.

Züst, R., Cervantes-Barragan, L., Kuri, T., Blakqori, G., Weber, F., Ludewig, B., et al., 2007. Coronavirus non-structural protein 1 is a major pathogenicity factor: implications for the rational design of coronavirus vaccines. PLoS Pathog. 3, e109.

Züst, R., Miller, T.B., Goebel, S.J., Thiel, V., Masters, P.S., 2008. Genetic interactions between an essential 3′ cis-acting RNA pseudoknot, replicase gene products, and the extreme 3′ end of the mouse coronavirus genome. J. Virol. 82, 1214–1228.

CHAPTER FIVE

Viral and Cellular mRNA Translation in Coronavirus-Infected Cells

K. Nakagawa*,2, K.G. Lokugamage*,2, S. Makino*,†,‡,§,¶,1

*The University of Texas Medical Branch, Galveston, TX, United States
†Center for Biodefense and Emerging Infectious Diseases, The University of Texas Medical Branch, Galveston, TX, United States
‡UTMB Center for Tropical Diseases, The University of Texas Medical Branch, Galveston, TX, United States
§Sealy Center for Vaccine Development, The University of Texas Medical Branch, Galveston, TX, United States
¶Institute for Human Infections and Immunity, The University of Texas Medical Branch, Galveston, TX, United States
1Corresponding author: e-mail address: shmakino@utmb.edu

Contents

---

2 These authors contributed equally to this work.

Advances in Virus Research, Volume 96
ISSN 0065-3527
http://dx.doi.org/10.1016/bs.aivir.2016.08.001

Abstract

Coronaviruses have large positive-strand RNA genomes that are 5′ capped and 3′ polyadenylated. The 5′-terminal two-thirds of the genome contain two open reading frames (ORFs), 1a and 1b, that together make up the viral replicase gene and encode two large polyproteins that are processed by viral proteases into 15–16 nonstructural proteins, most of them being involved in viral RNA synthesis. ORFs located in the 3′-terminal one-third of the genome encode structural and accessory proteins and are expressed from a set of 5′ leader-containing subgenomic mRNAs that are synthesized by a process called discontinuous transcription. Coronavirus protein synthesis not only involves cap-dependent translation mechanisms but also employs regulatory mechanisms, such as ribosomal frameshifting. Coronavirus replication is known to affect cellular translation, involving activation of stress-induced signaling pathways, and employing viral proteins that affect cellular mRNA translation and RNA stability. This chapter describes our current understanding of the mechanisms involved in coronavirus mRNA translation and changes in host mRNA translation observed in coronavirus-infected cells.

1. INTRODUCTION

1.1 Overview of Translation Mechanism in Animal Cells

Being obligate intracellular parasites, viruses heavily depend on host cell structures and functions to complete their life cycle, and they also use the translational apparatus of the infected cell to express their proteins. In several cases, viruses have been shown to affect and/or modulate the status of the host translational machinery to achieve efficient viral protein synthesis and replication, while cellular mRNA translation is inhibited (Hilton et al., 1986; Narayanan et al., 2008a; Siddell et al., 1980, 1981a,b). In eukaryotic cells, translation occurs in the cytoplasm and essentially involves four steps: initiation, elongation, termination, and recycling (Kapp and Lorsch, 2004). The translational initiation step includes the recognition of an mRNA by the host translational machinery and assembly of the 80S complex, in which a methionyl initiator tRNA (Met-tRNA^{Met}) binds at the peptidyl (P) site of the mRNA. In elongation, aminoacyl tRNAs enter the acceptor (A) site and, if the correct tRNA is bound, the ribosome catalyzes the formation of a peptide bond. After the tRNAs and mRNA are translocated such that the next codon is moved into the A site, the process is repeated. If a stop codon is encountered, the translation process is terminated, releasing the peptide from the ribosome. The recycling step involves dissociation of the ribosome and release of mRNA and deacylated tRNA, thereby setting the stage for another round of translation initiation.

Translation initiation (rather than elongation or termination) is the key step in regulating protein synthesis events. Translational initiation requires at least nine eukaryotic initiation factors (eIFs) and comprises two steps: the formation of 48S initiation complexes with established codon–anticodon base pairing in the P-site of the 40S ribosomal subunits, and the joining of 60S subunits to 48S complexes to form the 80S complex. On capped mRNAs, 48S complexes are formed by the interaction of a 43S preinitiation complex (comprising a 40S subunit, the eIF2–GTP–Met-tRNA^{Meti} ternary complex, eIF3, eIF1, eIF1A, and probably eIF5) with the eIF4F complex (comprising eIF4E, a cap-binding protein, eIF4G, and eIF4A), which binds to the 5$'$ cap region of the mRNA. By unwinding the mRNA's 5$'$-terminal secondary structure primarily by eIF4A in the eIF4F complex, the 43S complex then scans the 5$'$ untranslated region (UTR) in the 5$'$–3$'$ direction to the initiation codon. After initiation codon recognition, eIF2 triggers GTP hydrolysis, which is facilitated by eIF5 and eIF5B, leading to the displacement of eIFs, and the joining of a 60S subunit to form the 80S complex (Kapp and Lorsch, 2004).

Several studies showed coronavirus (CoV)-mediated control/alteration at translational initiation step (see Section 4). In contrast, little is known as to whether CoVs also affect the elongation, termination, or recycling steps in host protein synthesis. We also do not know whether synthesis of CoV-encoded proteins is regulated at the translation termination and/or recycling steps in infected cells, while the critical role of a −1 ribosomal frameshift event during translation elongation to produce the ORF1a/ORF1b-encoded replicase polyprotein (pp) 1ab is well established and has been extensively characterized (see Section 2.4).

In addition to cap-dependent translation initiation, several viruses, such as picornaviruses (which lack a 5$'$-end cap structure) (Daijogo and Semler, 2011; Martinez-Salas et al., 2015), and some host mRNAs use cap-independent mechanisms for translation initiation. In contrast to cap-dependent translation, in which the 43S preinitiation complex binds to the 5$'$-terminal region of the mRNA through interaction between eIF3 in the 43S preinitiation complex and eIF4G in the eIF4F complex that is associated with the 5$'$-end of the mRNA, the 43S preinitiation complex (or the 40S ribosomal unit alone in some specific mRNA templates) directly binds to a specific region called internal ribosome entry site (IRES) of mRNAs that are translated by cap-independent translation mechanisms

(Kapp and Lorsch, 2004). The number and identities of translation initiation factors required for IRES-mediated translation initiation of specific mRNA species may vary significantly. For example, IRES-mediated translation of the picornavirus genome requires all translational initiation factors, except for eIF4E, 40S ribosome, and 60S ribosome subunits. However, other IRES elements, such as those present in the hepatitis C virus (HCV) genome or the cricket paralysis virus (CrPV) intergenic region, require a much smaller number of initiation factors compared to those required for cap-dependent translation. It is generally accepted that the vast majority of CoV proteins is synthesized by a cap-dependent translation mechanism and cap-independent translation initiation has only been reported for relatively few coronaviral mRNAs (see Section 2.7).

1.2 CoVs

CoVs are enveloped plus-strand RNA viruses that belong to the order *Nidovirales* in the subfamily *Coronavirinae* (family *Coronaviridae*) and are classified into four genera, *Alphacoronavirus*, *Betacoronavirus*, *Gammacoronavirus*, and *Deltacoronavirus* (de Groot et al., 2011; Gorbalenya et al., 2004; Snijder et al., 2003; Woo et al., 2010, 2012). CoVs cause primarily respiratory and/or enteric diseases and are found in many animal species, including wild animals, domestic animals, and humans (Weiss and Navas-Martin, 2005). While most human CoVs (HCoV) cause relatively mild upper respiratory tract infections (common cold), two zoonotic viruses called severe acute respiratory syndrome (SARS) CoV and Middle East respiratory syndrome (MERS) CoV are associated with severe lower respiratory tract infections and are major public health threats. SARS-CoV, MERS-CoV, and some HCoVs, including HCoV-OC43 and HCoV-HKU1, belong to the genus *Betacoronavirus*, while other HCoVs, HCoV-229E, and HCoV-NL63, belong to the genus *Alphacoronavirus*. Animal CoVs from the genera *Alpha-* and *Betacoronavirus* are mainly associated with infections in mammals, while viruses in the genera *Gamma-* and *Deltacoronavirus* primarily (but not exclusively) infect birds. There is now compelling evidence to suggest that bats are the natural reservoir involved in the evolution and spread of many mammalian CoVs, including SARS-CoV and MERS-CoV (Lau et al., 2005; Li et al., 2005; Memish et al., 2013).

The CoV particles have a spherical shape with a diameter of roughly 100 nm (Davies and Macnaughton, 1979; Wege et al., 1979). They carry three major structural proteins (S, M, and E) in the envelope and contain a helical nucleocapsid that is formed by the viral genomic RNA and the viral N protein. The viral S protein binds has receptor-binding and fusogenic functions (Heald-Sargent and Gallagher, 2012; Masters, 2006) and thus is essential for initiation of CoV infection.

1.3 Overview of CoV Genome Organization and Gene Expression Strategy

CoV genomes range between 27 and 32 kb, representing the largest RNA genome known to date. In common with typical mammalian mRNAs, the CoV genome has a 5′-terminal cap structure and a poly(A) sequence at the 3′-end (Masters, 2006) but, in contrast to most mammalian mRNAs, the CoV genome carries multiple open reading frames (ORFs) between the 5′- and 3′-terminal UTRs, both of which contain *cis*-acting signals involved in RNA replication (Brian and Baric, 2005; Masters, 2006). Genome regions upstream of these ORFs contain so-called transcription regulatory sequences (TRS) that are required for CoV transcription (Brian and Baric, 2005; Masters, 2006). All CoVs have two large ORFs, called ORF1a and ORF1b, that occupy the 5′-terminal two-thirds of the genome and are generally referred to as the viral replicase gene. Other ORFs located downstream of the replicase gene encode viral structural proteins and a varying number of accessory proteins, the latter being dispensable for virus replication in cell culture but involved in CoV pathogenicity (Liu et al., 2014; Narayanan et al., 2008b).

Following viral entry into the cell, the viral genome RNA undergoes translation to produce the viral proteins that are required for subsequent RNA replication and transcription. eIF4F and the 43S preinitiation complex access the 5′-end of the capped viral genome and the 43S preinitiation complex scans the 5′ UTR. The 80S complex is assembled at the translation initiation codon and protein synthesis starts and proceeds until the first termination codon is encountered and the ribosome dissociates from its mRNA template, resulting in polyprotein (pp) 1a. Production of pp1ab (encoded by both ORF1a and ORF1b) requires a −1 ribosomal frameshift before the translation stop codon is reached. This frameshift has been shown to occur in the overlap region between ORFs 1a and 1b. Viral proteins encoded downstream of ORFs 1a and 1b are not synthesized from the genome RNA but from a set of 5′-capped subgenomic mRNAs that carry the respective ORFs in their 5′-terminal regions. The two replicase polyproteins translated from ORFs 1a and 1b undergo proteolytic cleavage via viral-encoded proteinases encoded in ORF1a to generate 15–16 mature nonstructural proteins, termed nsp1 to nsp16. All of these nsps, except for nsp1 (Hurst-Hess et al., 2015) and nsp2 (Graham et al., 2005), are considered essential for transcription and replication of CoV RNA (Newman et al., 2014). CoV RNA synthesis occurs at double membrane vesicles that are derived from endoplasmic reticulum (ER) membranes (reviewed in Newman et al., 2014) in the cytoplasm.

Besides its role as an mRNA for replicase polyprotein expression, the genome RNA is packaged into progeny virus particles (in contrast to subgenomic mRNAs that are not packaged efficiently). The number of subgenomic mRNAs differs among CoVs. These CoV mRNAs share their 3′-terminal regions, constituting a 3′-coterminal nested set of RNAs (reviewed in Masters, 2006). The 5′-end of CoV genomic RNA carries a ~70-nt-long leader sequence. The same leader sequence is also found at the 5′-end of all CoV mRNAs. Subgenomic minus-strand RNAs, each of which corresponds to each subgenomic mRNA species, also accumulate in infected cells. These subgenomic minus-strand RNAs carry complement of the leader sequence (antileader) at the 3′-end. It has been proposed that subgenomic minus-strands are synthesized from intracellular genome-length RNA, in a process called discontinuous extension and involving base pairing interactions between the 3′-end of the nascent minus-strand and the leader TRS. Subsequently, these antileader-containing subgenomic minus-strand RNAs are used as templates to produce 5′ leader-containing mRNAs in which the 5′ leader sequence is fused to the mRNA body at the TRS (Masters, 2006; Sawicki and Sawicki, 1990; Sawicki et al., 2007). Like the genomic RNA, the CoV subgenomic mRNAs, except for the smallest mRNA, are polycistronic and, with very few exceptions, only the 5′-terminal ORF of each of these subgenomic mRNAs is translated into protein.

2. MECHANISMS AND CONTROL OF TRANSLATION OF CORONAVIRUS mRNAs

2.1 Evidence for Cap-Dependent Translation of CoV mRNAs

Because genomic and subgenomic CoV mRNAs have a 5′ cap structure, most CoV mRNAs are thought to undergo cap-dependent translation using eIF4F. Cencic et al. reported several compounds that inhibit eIF4F activity by preventing eIF4E–eIF4G interaction (Cencic et al., 2011b). The same group demonstrated that a molecule (4E2RCat) that prevents the interaction between eIF4E and eIF4G inhibits HCoV-229E replication (Cencic et al., 2011a), providing additional evidence to suggest that CoV mRNA translation depends on a 5′ cap structure being present on viral mRNAs. The authors also found that a certain concentration of 4E2RCat completely inhibited HCoV-229E replication, whereas it inhibited host protein synthesis by ~40%, indicating that HCoV-229E mRNAs show a higher dependency on eIF4F for ribosome recruitment compared to host mRNAs. To date, possible roles of eIF4F in viral mRNA translation have not reported for other CoVs.

2.2 Viral Enzymes Involved in Capping of CoV mRNAs

Formation of the cap structure of eukaryotic and eukaryotic viral mRNAs generally requires three successive enzymatic reactions (Furuichi et al., 1976). First, an RNA 5′-triphosphatase (TPase) removes the γ-phosphate group from the 5′-triphosphate end (pppN) of the nascent mRNA chain to generate the diphosphate 5′-ppN. Subsequently, an RNA guanylyltransferase transfers a GMP to the 5′-diphosphate end to produce the cap core structure (GpppN). Finally, an N7 methyltransferase (N7-MTase) methylates the attached GMP (cap) at the N7 position to produce a cap-0 structure (m7GpppN). Higher eukaryotes and viruses usually further methylate the cap-0 structure at the ribose 2′-O position of the first and second nucleotide of the mRNA by a ribose 2′-O-MTase to form cap-1 and cap-2 structures, respectively (Furuichi and Shatkin, 2000). Ribose 2′-O-methylation of viral RNA cap provides a mechanism for viruses to escape host immune recognition (Daffis et al., 2010; Zust et al., 2011).

The CoV genome encodes several RNA processing enzymes involved in RNA capping. The cap formation step by CoV has mainly been studied in SARS-CoV. In the first step of cap formation to generate diphosphate 5′-ppN, nsp13 may be involved because it has RNA 5′-TPase activity mediated by the NTPase active site of the nsp13-associated helicase domain (Ivanov and Ziebuhr, 2004; Ivanov et al., 2004). The next step is the formation of the cap core structure (GpppN) by an RNA guanylyltransferase. At present, it is not clear whether or not CoVs encode this enzyme; perhaps CoVs use cellular enzymes to perform this step. The third step is the methylation of the cap guanosine at the N7 position. This reaction is mediated by the C-terminal domain of CoV nsp14 (Chen et al., 2009) using S-adenosyl methionine (SAM) as a methyl group donor. Apparently, the enzyme is not specific for viral substrate RNAs (Bouvet et al., 2010). Conversion of cap-0 to cap-1 structures involves nsp16 that acts as a $2'-O$-MTase (Bouvet et al., 2010; Decroly et al., 2008) and forms a complex with nsp10 (Chen et al., 2011) that appears to be required for efficient binding to SAM and the RNA substrate. Interestingly, SARS-CoV nsp10 plays an essential role in the specific binding of nsp16 to m7GpppA-capped RNA (first nucleotide is adenine). Considering that both the genomic and subgenomic mRNAs of SARS-CoV start with an adenine, this feature appears beneficial for SARS-CoV replication. The crystal structure of the heterodimer of nsp16/nsp10 with bound methyl donor SAM showed that nsp10 may stabilize the SAM-binding pocket and extend the RNA-binding groove of nsp16 (Chen et al., 2011).

2.3 Changes in the Poly(A) Tail Length During CoV Replication

CoV genomic and subgenomic mRNAs carry a poly(A) tail at their 3' ends. Hofmann and Brian postulated that viral RNA-dependent RNA polymerase or a cellular cytoplasmic poly(A) polymerase synthesizes the poly(A) tail of CoV mRNAs (Hofmann and Brian, 1991). Wu et al. reported that the length of the poly(A) tail of bovine CoV (BCoV) mRNAs in infected human rectal tumor-18 cells varies at various times postinfection (p.i.), ranging from ∼45 nt immediately after virus entry to ∼65 nt at 6–9 h p.i. and ∼30 nt at 120–144 h p.i. (Wu et al., 2013). Differences in poly(A) length of viral mRNAs at different times p.i. was also observed in several other BCoV-infected cell lines and cells infected with different strains of infectious bronchitis virus (IBV) (Shien et al., 2014), indicating that changes in the poly(A) length during virus replication may be a common feature of CoVs. Factor and mechanisms involved in this process remain to be studied. Because the length of the poly(A) tail contributes to the efficiency of translation and replication of CoV defective interfering RNAs (Spagnolo and Hogue, 2000; Wu et al., 2013), the regulated changes in the length of the CoV poly(A) tail may affect efficiencies of viral translation and replication over the course of infection.

2.4 Ribosomal Frameshift in CoV Gene 1 Protein Expression

As mentioned earlier, two large polyproteins, one of which being translated from ORF1a and the other from ORFs 1a and 1b, are synthesized from the viral genome RNA. The synthesis of polyprotein 1ab involves a −1 ribosomal frameshift during the translational elongation step and occurs in the overlap region between ORFs 1a and 1b (Bekaert and Rousset, 2005; Brierley et al., 1987, 1989). Polyprotein 1ab is thus encoded by a (functionally) fused ORF produced from the two ORFs 1a and 1b. As most of the mature nsp proteins processed from these two large polyproteins are essential for CoV RNA synthesis, translation of polyprotein 1ab via −1 ribosomal frameshifting is an essential step in CoV replication.

The signals required for −1 ribosomal frameshifting in the ORF1a/1b overlap region were first identified in IBV (Brierley et al., 1987, 1989). Subsequently, putative ribosomal frameshift signals were also identified and characterized in other CoVs (Bekaert and Rousset, 2005). The −1 ribosomal frameshifting signals are composed of a slippery sequence "UUUAAAC" followed by a "stimulatory" RNA secondary structure. The −1 frameshifting occurs at this slippery sequence, where tRNAs are supposed to dissociate from the mRNA and then shift (by 1 nucleotide) to a codon in another reading frame, ORF1b (Plant and Dinman, 2008). The stimulatory structure is a complex RNA stem-loop structure RNA that varies among different CoVs. SARS-CoV's stimulatory structure contains three stem loops (Baranov et al., 2005; Plant et al., 2005; Su et al., 2005), and their disruption affects frameshift efficiency (Plant and Dinman, 2006; Plant et al., 2005). Recently, Ishimaru et al. showed that a homodimeric RNA complex formed by the SARS-CoV's stimulatory structure occurs within cells and that loop-to-loop kissing interactions involving stem3-loop2 modulate the −1 ribosome frameshift efficiency (Ishimaru et al., 2013). These reports indicate that an optimal secondary RNA structure and RNA–RNA interaction within the ribosomal frameshifting signal are important for efficient −1 ribosomal frameshifting.

Several studies showed that reduction of frameshifting efficiency affects virus infectivity and replication (Ishimaru et al., 2013; Plant et al., 2010, 2013). Plant et al. showed that the ratio of ORF1a- and ORF1b-encoded proteins plays a critical role in CoV replication efficiency (Plant et al., 2010). The authors proposed that CoVs have evolved to produce optimal levels of −1 ribosomal frame shift efficiency for efficient virus propagation. CoV frameshift signals characterized previously have −1 ribosomal frameshift efficiencies in the range of 20–45% (Baranov et al., 2005; Brierley et al., 1987; Herold and Siddell, 1993; Plant and Dinman, 2008). A recent study using ribosome profiling of mouse hepatitis virus (MHV)-infected cells suggests that the frameshift rate may be even slightly higher (Irigoyen et al., 2016).

2.5 Ribosomal Shunting Mechanism of Translation in CoVs

Ribosomal shunting is a translation initiation depending on cap-dependent discontinuous scanning, whereby ribosomes are loaded onto mRNA at the 5′ cap structure and scanning is started for a short distance before bypassing the large internal leader region and initiating at a downstream start site (Firth and Brierley, 2012). For some strains of transmissible gastroenteritis virus (TGEV), mRNA having ORF 3b as the first ORF is not produced and ORF 3b is present as a nonoverlapping second ORF on mRNA 3. O'Connor et al. proposed a possible ribosomal shunting in the translation of ORF 3b in TGEV (O'Connor and Brian, 2000). The basis for this proposal was as follows: (i) if 3b protein was translated by a leaky scanning mechanism, a modification of ORF 3a to generate a favorable Kozak sequence would be expected to diminish protein synthesis from ORF3b compared to ORF3a; however, optimization of the Kozak context for ORF 3a did not affect translation efficiency of 3b; (ii) the translation of 3b was shown to be cap dependent; and (iii) deletion analysis failed to provide evidence for an IRES within the ORF 3a sequence. In several viral mRNAs, the presence of a specific donor structure with a large stem-loop with a 5′-adjacent, short ORF appears to be required for ribosome shunting (Hemmings-Mieszczak and Hohn, 1999; Pooggin et al., 1999), whereas TGEV ORF3b shunting does not depend on such a donor structure. Given that Sendai virus mRNA (Latorre et al., 1998), avian orthoreovirus mRNA (Racine and Duncan, 2010), and avihepadnavirus mRNA (Cao and Tavis, 2011) undergo ribosomal shunting, with no apparent requirement for a donor structure, it seems reasonable to suggest that translation of the TGEV 3b protein may involve a ribosomal shunting-driven translation mechanism in the absence of a specific donor structure.

2.6 Leaky Scanning Translation Mechanism of CoV Internal ORFs

An internal CoV ORF was initially found within the 5′ half of the BCoV N gene ORF (Lapps et al., 1987). This internal ORF gene encodes the I protein and is likely translated by a leaky scanning mechanism in which ribosomes occasionally bypass the first AUG (the start codon for the N protein) and initiate translation from the AUG codon of the internal ORF encoding the I protein. The first AUG has a suboptimal Kozak context, while the downstream AUG (the start codon for the I protein) is in a more favorable Kozak context (Senanayake and Brian, 1997; Senanayake et al., 1992). The N genes of MHV-A59, HCoV-HKU1, HCoV-OC43, SARS-CoV, and MERS-CoV also have an internal ORF (Armstrong et al., 1983; Kamahora et al., 1989; Rota et al., 2003; van Boheemen et al., 2012; Woo et al., 2005); in HCoV-OC43, the AUG start codon of the internal ORF is changed to AUC.

Because many CoVs carry the internal ORF in the N gene, it is possible that the encoded proteins may have (an) important biological function(s) for virus replication/survival, whereas our knowledge of their biological functions is limited. Fischer et al. reported that the MHV I protein is not essential for virus replication in cultured cells and its natural host (Fischer et al., 1997). However, a MHV mutant lacking the I gene formed smaller plaques than did wild-type MHV, indicating that I protein expression could give some minor growth advantage to the virus. Shi et al. reported that the SARS-CoV internal ORF protein, called the 9b protein, which is encoded within the N gene ORF, evades host innate immunity by targeting the mitochondrial-associated adaptor molecule, MAVS (Shi et al., 2014), while a previous study using a SARS-CoV mutant lacking viral accessory genes, including 6, 7a, 7b, 8a, 8b, and 9b, showed that deletion of these genes had little effect on the pathogenicity of the virus in a mouse model (Dediego et al., 2008). Further investigation is required to better understand the role of the SARS-CoV 9b protein in viral pathogenicity.

2.7 IRES-Mediated Translation in CoVs

Some CoV proteins are synthesized by IRES-mediated translation. Liu and Inglis reported that mRNA 3 of IBV was functionally tricistronic, having the capacity to encode three proteins (3a, 3b, and 3c) from three ORFs (Liu and Inglis, 1992). Their study showed that (i) a synthetic mRNA whose peculiar 5′-end structure prevents translation of the 5′-proximal ORFs (3a and 3b) directs efficient synthesis of 3c, (ii) translation of 3c is not affected by the absence/presence of the 5′ cap analog as well as changes in the sequence contexts for initiation of ORF 3a and 3b translation, (iii) an mRNA in which the 3a/b/c coding region was placed downstream of the influenza A virus nucleocapsid protein gene directed the efficient synthesis of only 3c and nucleocapsid protein, and (iv) expression of the 3c ORF from this mRNA was abolished when the 3a and 3b coding region was deleted, the latter indicating that 3c translation initiation depends on upstream 3a/b sequences that serve as an IRES element. Further analyses showed that the sequence prior to the initiator AUG of 3c can form an RNA secondary structure comprised of five RNA stem-loops that can be modeled into a compact superstructure formed by interactions of two predicted pseudoknot structures. The proposed structure shares structural features with IRES elements conserved in picornavirus genome RNAs, and a base pairing model between mRNA 3 and 18S ribosomal RNA (rRNA) was suggested (Le et al., 1992, 1993). IRES-mediated translation was also reported in MHV. The unique region of MHV mRNA 5 has two ORFs, ORF 5a, and ORF 5b. Thiel and Siddell reported that a synthetic mRNA containing both ORFs is functionally bicistronic and expression of ORF 5b, but not ORF 5a, is maintained in a tricistronic mRNA containing an additional 5′-proximal ORF. These data suggested that initiation of protein synthesis from ORF 5b may be mediated by an internal entry of ribosomes (Thiel and Siddell, 1994). Further studies by Jendrach et al. identified the IRES element of mRNA 5, which contains ≤280 nucleotides including the ORF 5b initiation codon (Jendrach et al., 1999). The authors also showed that the IRES element of mRNA 5 interacts specifically with protein factors present in an L-cell lysate (Jendrach et al., 1999).

2.8 Presence of Upstream ORF in CoV Genomic RNAs

Short AUG-initiated upstream ORFs (uORF) located in the 5′ UTR of eukaryotic mRNAs are generally translated by 5′ cap-dependent ribosomal scanning, resulting in translation repression of the major downstream ORF expressed from this mRNA (Calvo et al., 2009; Somers et al., 2013). In CoVs, a uORF is found within the 5′ UTR of the genomic RNA, just downstream of the genomic leader sequence. Hence CoV genomic RNA, but not subgenomic mRNAs, carry a uORF in their 5′-terminal region(s) (Wu et al., 2014), with few exceptions (Hofmann et al., 1993). Wu et al. reported that the uORF encoding a potential polypeptide of 3 to 13 amino acids is found within the 5′ UTR of more than 75% of CoV genomes based on 38 reference strains (Wu et al., 2014), suggesting a functional significance for this sequence. The authors further showed by using an in vitro translation system that the MHV uORF is involved in a decreased rate of translation from the ORF1a start codon; wild-type MHV and mutants that lack the uORF showed similar growth properties in cell culture and, within 10 passages in cell culture, the uORF-disrupting mutations reverted back to the wild-type sequences or generated a new uORF sequence (Wu et al., 2014). These results suggest that the uORF represses ORF1a translation, yet plays a beneficial, but nonessential role in CoV replication in cell culture.

3. HOST AND VIRAL FACTORS THAT REGULATE CORONAVIRUS mRNA TRANSLATION

3.1 Factors That Bind to Viral UTRs

The 5′ and 3′ UTRs of CoV genomic RNA carry *cis*-acting elements that are important for viral RNA synthesis (Chen and Olsthoorn, 2010). A number of host factors have been identified to interact in vitro with *cis*-acting elements of the 5′ and 3′ UTRs; these include heterogeneous nuclear ribonucleoprotein (hnRNP) A1 and Q (Choi et al., 2004; Li et al., 1997), polypyrimidine tract-binding protein (PTB) (Li et al., 1999), mitochondrial aconitase (Nanda and Leibowitz, 2001), poly(A)-binding protein (PABP) (Spagnolo and Hogue, 2000), the glutamyl-prolyl-tRNA synthetase (Galan et al., 2009), the arginyl-tRNA synthetase (Galan et al., 2009), and the p100 transcriptional coactivator (Galan et al., 2009). Among these, hnRNP A1, hnRNP Q, and PTB have been shown to have a role in MHV RNA synthesis (Choi et al., 2002; Li et al., 1999; Shen and Masters, 2001; Shi et al., 2000). Likewise, PABP, hnRNP Q and EPRS proteins have been shown to play a positive role in TGEV infection (Galan et al., 2009). Currently, it is unclear whether any of these host proteins affect translation of viral proteins.

Senanayanke and Brian reported that replacement of the natural 77-nt 5′ UTR on synthetic transcripts of BCoV mRNA 7 (Senanayake and Brian, 1999), which encodes the N and I proteins, with the 210-nt 5′ UTR from BCoV mRNA 1 caused approximately twofold-less N protein in an in vitro translation system as well as in uninfected cells using a T7 RNA polymerase-driven transient transfection system. In infected cells, this difference became 12-fold as the result of both a stimulated translation from the 77-nt 5′ UTR and a repression of translation from the 210-nt 5′ UTR, demonstrating a differential 5′ UTR-directed regulation of translation in CoV-infected cells and suggesting that this regulation involves viral or virus-induced cellular factors that interact with *cis*-acting elements in the 5′ UTR.

3.2 N Protein-Mediated Enhancement of Viral Translation

Tahara et al. reported that capped chimeric mRNAs, in which the MHV 5′-leader sequence was positioned upstream of the human alpha-globin coding region, were translated three- to fourfold more efficiently in cell-free extracts derived from MHV-infected cells compared to extracts obtained from uninfected cells. In contrast, nonviral mRNAs were found to be translated equally efficiently in cell-free extracts derived from infected and uninfected cells (Tahara et al., 1994). The same group subsequently demonstrated that the MHV N protein (i) stimulates translation of a reporter mRNA carrying the MHV 5′-UTR (Tahara et al., 1998) and (ii) binds with high affinity to the leader sequence and TRS (Nelson et al., 2000), suggesting that CoV N proteins may act as *trans*-acting factors that enhance the translation efficiency of CoV mRNAs.

4. CORONAVIRUS-MEDIATED CONTROL OF HOST TRANSLATION

CoV replication in cultured cells is often linked to an inhibition of host protein synthesis (Hilton et al., 1986; Narayanan et al., 2008a), with some exceptions (Cencic et al., 2011a). MHV infection drastically inhibits host protein synthesis between 3 and 6 h p.i., while viral protein synthesis increases from 3 h p.i. and peaks at 6 h p.i. (Siddell et al., 1980, 1981a,b). An increased number of free 80S ribosomes and a shift to lighter polysomes during MHV infection suggest a potential suppression at the translation initiation phase (Anderson and Kedersha, 2009; Siddell et al., 1981b).

Banerjee et al. reported that MHV infection induces degradation of 28S rRNAs as early as 4 h p.i. and almost all 28S rRNAs are degraded by 24 h p.i., while 18S rRNA remains stable (Banerjee et al., 2000). Although the biological significance of MHV-induced 28S rRNA degradation is not entirely clear, it may contribute to the virus–induced translational suppression because the 28S rRNA is an integral component of the 60S ribosomal subunit. In addition to 28S rRNA, MHV infection was reported to stimulate degradation of specific host mRNAs (Kyuwa et al., 1994), thereby further limiting host protein expression.

4.1 CoV Proteins That Suppress Translation

One of the CoV proteins that have a role in inhibiting host translation was found to be nsp1, the most N-terminal pp1a/pp1ab-derived processing product. CoV nsp1 equivalents appear to employ divergent strategies to suppress host gene expression. For instance, SARS-CoV nsp1 inhibits mRNA translation and promotes host mRNA degradation by binding to the 40S ribosomal subunits (Kamitani et al., 2009), while nsp1 of MERS-CoV inhibits host mRNA translation and induces host mRNA cleavage without binding to the 40S ribosomal subunits (Lokugamage et al., 2015). Nsp1 of TGEV uses a yet another, currently unknown mechanism to suppress host protein synthesis without binding to the 40S ribosomal subunits or inducing mRNA degradation (Huang et al., 2011a).

SARS-CoV nsp1 inhibits cap-dependent and IRES-driven translation by suppressing multiple steps of translation initiation (Lokugamage et al., 2012). SARS-CoV nsp1 inhibits both 48S and 80S initiation complex formation with a stronger inhibition of the latter on cap-dependent and encephalomyocarditis virus (EMCV)-IRES-driven mRNA templates. SARS-CoV nsp1 inhibits different steps of initiation on mRNA templates that use the IRES-40S binary complex to initiate translation, such as CrPV and HCV-IRES-driven mRNAs. Nsp1 inhibits binary complex formation on CrPV-IRES and 48S complex formation on HCV-IRES (Lokugamage et al., 2012). In addition to the translation suppression function, SARS-CoV nsp1 promotes host mRNA degradation by inducing endonucleolytic RNA cleavage in the 5′-UTR of host mRNAs (Huang et al., 2011b). The internally cleaved mRNAs are then degraded by the cellular Xrn1-mediated 5′–3′ exonucleolytic mRNA degradation pathway (Gaglia et al., 2012; Huang et al., 2011b). SARS-CoV nsp1-induced degradation of host mRNAs also contributes to inhibition of host gene expression in infected cells (Narayanan et al., 2008a). While SARS-CoV nsp1 induces endonucleolytic cleavage of host mRNAs, viral mRNAs appear to be resistant to these cleavage mechanisms (Huang et al., 2011b; Kamitani et al., 2009; Lokugamage et al., 2012), most likely because the 5′ leader sequence present on these RNAs protects them from nsp1-induced endonucleolytic cleavage. SARS-CoV nsp1 was also shown to inhibit translation of viral mRNAs in cell-free extracts (Huang et al., 2011b) as well as infected cells (Narayanan et al., 2014). In contrast, Tanaka et al. postulated that specific interaction of SARS-CoV nsp1 with the stem-loop I structure in the 5′-UTR of the SARS-CoV genome facilitates efficient viral gene expression in infected cells (Tanaka et al., 2012).

Unlike SARS-CoV nsp1, which is a cytoplasmic protein (Kamitani et al., 2006), MERS-CoV nsp1 is localized in both the cytoplasm and the nucleus of cells transfected with suitable plasmid construct or virus-infected cells (Lokugamage et al., 2015). Interestingly, MERS-CoV nsp1 selectively inhibits translation and induces degradation of mRNAs that are transcribed in the nucleus and transported to the cytoplasm, whereas it does not suppress the expression of exogenous mRNAs directly introduced into the cytoplasm or viral mRNAs produced in the cytoplasm of infected cells (Lokugamage et al., 2015). As MERS-CoV mRNAs are of cytoplasmic origin, their expression remains unaffected by MERS-CoV-encoded nsp1 in infected cells (Lokugamage et al., 2015).

Xiao et al. reported that the S proteins of SARS-CoV and IBV interact with eIF3f, one of the subunits of eIF3, leading to the inhibition of host translation at a late point in the infection (Xiao et al., 2008). The authors suggested that the interaction between CoV S protein and eIF3f may play a functional role in controlling the expression of specific host genes, especially those that are induced by the viral infection.

Using a yeast two-hybrid human protein library screen, Zhou et al. (2008) found that the SARS-CoV N protein binds to eukaryotic translation elongation factor 1 alpha (eEF1α), an integral part of the translational machinery; GTP-bound EF1alpha plays a key role in recruiting aminoacyl-tRNA to the A site of the 80S translation initiation complex during translation. The authors showed that expression of SARS-CoV N protein induces the aggregation of eEF1α, resulting in the inhibition of host protein synthesis (Zhou et al., 2008).

Another CoV protein that affects host translation is the SARS-CoV 7a protein, a multifunctional protein that inhibits host translation, induces apoptosis, activates the p38 mitogen-activated protein kinase (Kopecky-Bromberg et al., 2006), and arrests the cell cycle at the G0/G1 phase (Yuan et al., 2006). Accumulating evidence suggests that the 7a protein inhibits cellular gene expression at the level of translation (Kopecky-Bromberg et al., 2006), whereas the precise mechanism of translation inhibition by the 7a protein remains to be determined.

4.2 CoV-Mediated Induction of ER Stress and Unfolded Protein Response

In eukaryotic cells, the ER plays a central role in the synthesis of secretory or transmembrane proteins, including folding of these proteins into their native conformation and mediating a range of posttranslational modifications. The ER maintains a homeostasis suitable to regulate the processing and prevent the aggregation of these proteins. If the accumulation of nascent, unfolded polypeptides exceeds the folding and processing capacity of the ER, the homeostasis is perturbed, resulting in ER stress and triggering the activation of the unfolded protein response (UPR). The latter contributes to restoring the normal function of ER or initiates apoptosis if the ER stress remains unchanged. UPR signaling activates three main branches: PKR-like ER kinase (PERK), activating transcription factor 6 (ATF6), and inositol-requiring protein-1 (IRE1). Once activated these pathways send signals to inhibit translation, degrade misfolded proteins through ER-associated degradation, express ER molecular chaperones, and expand the ER membrane to decrease the load of proteins and increase the protein-folding capacity in the ER. ER stress occurs under various conditions, such as nutrient deprivation, developmental processes, genetic changes, and invasion by pathogens, such as viruses and intracellular bacteria. Generally, in virus-infected cells, a copious amount of viral proteins is produced, often perturbing ER homeostasis and eventually causing ER stress (Schroder and Kaufman, 2005; Zhang and Wang, 2012).

Mounting evidence suggests that CoV replication causes ER stress that triggers UPR activation, thus possibly regulating host antiviral responses. Several studies have observed the induction of ER protein chaperones, such as the immunoglobulin heavy chain-binding protein (BiP), also known as glucose-regulated protein 94 (GRP94), which are considered indicators of ER stress, in SARS-CoV-infected cells (Chan et al., 2006; Yeung et al., 2008). In one of these studies, induction of luciferase reporter gene expression from BiP- or GRP94-driven promoters was observed in SARS-CoV-infected cells (Chan et al., 2006). Also other CoVs, such as MHV and IBV, have been shown to cause ER stress in infected cells (Fung and Liu, 2014; Versteeg et al, 2007). Cells overexpressing the S protein of SARS-CoV, MHV or HCoV-HKU1 also induce ER stress, suggesting that the S protein could be the main modulator of ER stress in CoV infections (Versteeg et al., 2007). Due to its large molecular weight and extensive glycosylation, the S protein requires ER protein chaperones for proper folding and maturation. ER stress has also been observed in cells that overexpress specific CoV accessory proteins, such as the 3a, 6, and 8ab proteins of SARS-CoV (Minakshi et al., 2009; Sung et al., 2009; Ye et al., 2008). In addition to the contribution of viral proteins in inducing ER stress, an increased overall ER burden, the formation of double membrane vesicle, and the depletion of ER lipids may contribute to ER stress, thereby affecting cellular mRNA expression in CoV-infected cells.

4.3 Status of Stress Granules and Processing Bodies in CoV Replication

In response to cellular stress caused by either environmental stimuli or viral infections, cells may reduce or even shut off their protein synthesis. Induction of translational shutoff also occurs by increased phosphorylation of eIF2α (de Haro et al., 1996), resulting in dissociation of polysomes which, in turn, leads to stalled 43S and 48S preinitiation complexes and nontranslating mRNAs that may be incorporated into stress granules (SGs), one of the two major cytoplasmic RNA granules. SGs are reversible foci and temporary store houses for these complexes. When the stress conditions are over, stored initiation complexes are rapidly released to resume translation. Under normal conditions, processing bodies (P-bodies), another type of cytoplasmic RNA granule, are present in cells. Typically, P-bodies increase in number and size under host shutoff conditions. Both SGs and P-bodies contain translationally incompetent mRNAs, while P-bodies contain exonucleases, deadenylases, and enzymes for decapping (Anderson and Kedersha, 2009; Beckham and Parker, 2008; Buchan and Parker, 2009). Although P-bodies are enriched with RNA decay machinery proteins, it is controversial as to whether RNA decay occurs within P-bodies. Although virus infections activate cellular stress responses, some viruses, such as poliovirus and CrPV, do not induce SGs in the course of infection. Other viruses, including the alphavirus Semliki Forest virus, rotavirus, dengue virus, and West Nile virus, may actively inhibit SG formation (Emara and Brinton, 2007; Khong and Jan, 2011; McInerney et al., 2005; Montero et al., 2008).

Increased formation of SGs and P-bodies has been shown to occur during replication of some CoVs. Both TGEV and MHV form SGs as the infection progresses (Raaben et al., 2007; Sola et al., 2011). During MHV infection, formation of SGs and P-bodies were observed concomitantly with increased eIF2α phosphorylation as early as 6 h p.i. when host translational shutoff and mRNA decay are apparent. These data led Raaben et al. to propose that MHV replication induces host translational shutoff by triggering an integrated stress response (Raaben et al., 2007). The authors further demonstrated that, when host translational shutoff was experimentally impaired, replication of MHV was not negatively affected but rather enhanced (Raaben et al., 2007). The appearance of SGs was also observed in TGEV infection, and these granules were not disassembled at later stages of the infection (Sola et al., 2011). However, when SG components were depleted, an increased replication of TGEV was observed, suggesting that SGs may restrict CoV replication.

4.4 Activation of PKR, PERK, and eIF2α Phosphorylation

Viral infections induce ER stress, thereby triggering phosphorylation of the serine residue at position 51 of eIF2α. Phosphorylation of eIF2α leads to translation inhibition. Two of the four protein kinases that are known to phosphorylate eIF2α at Ser51 are protein kinase RNA-activated (PKR) and PERK. Double-stranded RNA and ER stress activate PKR and PERK, respectively. Several studies examined the status of eIF2α phosphorylation and PKR/PERK activation in CoV-infected cells. Zorzitto et al. reported minimal transcriptional activation of PKR in MHV-1-infected cells at later time points (Zorzitto et al., 2006). Ye et al. found that PKR and eIF2α are not phosphorylated in MHV-A59-infected cells (Ye et al., 2007). In contrast, Bechill et al. observed significant phosphorylation of eIF2α in MHV-A59-infected cells and efficient translation of MHV mRNAs in the presence of phosphorylated eIF2α (Bechill et al., 2008). This discrepancy in the status of eIF2α phosphorylation and PKR activation in these studies could be due to the use of different cell types and virus strains. Krahling et al. reported that SARS-CoV infection not only activated PKR and PERK but also phosphorylated eIF2α at 8 and 24 h p.i. (Krahling et al., 2009), whereas activation of PKR and PERK or eIF2α phosphorylation did not impair SARS-CoV replication. Based on these data, the authors hypothesized that SARS-CoV has evolved a mechanism to overcome the inhibitory effects of phosphorylated eIF2α on viral mRNA translation. Wang et al. reported that that phosphorylation of eIF2α was severely suppressed in human and animal cells infected with IBV (Wang et al., 2009). The level of phosphorylated PKR was found to be greatly reduced in IBV-infected cells, and nsp2 was a weak PKR antagonist. Also GADD34, a component of the protein phosphatase 1 (PP1) complex that dephosphorylates eIF2α, was significantly induced in IBV-infected cells. Inhibition of the PP1 activity and over-expression of wild-type and mutant GADD34, eIF2α, and PKR provided evidence to suggest that these virus–modulated pathways play a synergistic role in facilitating IBV replication. It was also postulated that IBV may employ a combination of two mechanisms, i.e., blocking PKR activation and inducing GADD34 expression, to maintain de novo protein synthesis in infected cells to enhance viral replication (Wang et al., 2009).

5. CONCLUDING REMARKS

This chapter summarizes experimental studies of mechanisms that drive and control the translation of viral and host cell mRNAs in CoV-infected cells. There is now a large body of information to suggest that (i) CoV evolved a range of mechanisms to control viral gene expression at the posttranscriptional level and (ii) CoV replication results in cell stress, leading to the activation of stress-induced signaling pathways that suppress host translation. Also, recent studies have uncovered that a number of CoV-encoded proteins, such as nsp1, suppress host translation without severely affecting viral gene expression. Although studies of the posttranscriptional regulation of CoV gene expression have been making significant progress, our understanding of the translation mechanisms used by specific CoV mRNAs requires further investigation.

A major question to be addressed in more detail pertains to the mechanism that is used to ensure efficient CoV mRNA translation in cells in which host protein synthesis is suppressed. Replication of some CoVs, such as SARS-CoV, induces host mRNA decay (Lokugamage et al., 2015; Narayanan et al., 2008a). While the reduction of host mRNA levels is one of the factors involved in suppressing host translation in CoV-infected cells, it is unlikely to be the sole reason for the inhibition of host protein synthesis (Narayanan et al., 2008a). CoV mRNAs and host mRNAs share important structural features (such as the 5' cap structure and 3' poly(A) tail), suggesting that the translation of CoV and host mRNAs may be equally sensitive to the inhibition of factors that control specific translation steps, including eIF2α and its phosphorylated form. A possible mechanism for efficient CoV translation under conditions in which host translation is suppressed may be related to the extreme abundance of CoV mRNAs that, at least in part, may counterbalance effects of translational suppression. Alternatively, CoV mRNAs and host mRNAs may use a slightly different repertoire of translation factors. It has been reported that alphaviruses can use eIF2A, instead of eIF2α, in viral translation (Ventoso et al., 2006). If CoV mRNAs have similar requirements, then eIF2α phosphorylation may not or less profoundly affect the translation of CoV mRNAs.

Because CoV mRNAs are capped at the 5'-end and polyadenylated at the 3'-end, it has been postulated that most of the CoV mRNAs undergo cap-dependent translation. However, experimental data that support this hypothesis have only been provided for HCoV-229E-infected cells (Cencic et al., 2011b). Unlike many other CoVs, HCoV-229E replication does not inhibit translation of host proteins (Cencic et al., 2011b). Do mRNAs of other CoVs also undergo cap-dependent translation? Recent studies showed that m⁶A-modification in the 5' UTR of mammalian mRNA may stimulate cap-independent translation (Meyer et al., 2015; Zhou et al., 2015). If the 5' UTR of some CoV mRNAs are subject to m⁶A-modification in infected cells, these CoV mRNAs might use a cap-independent mechanism of translation.

Previous studies suggested that the CoV leader sequence that is attached to the 5'-end of all CoV mRNAs affects translation and stability of CoV mRNAs. For example, the MHV leader sequence was shown to promote translation in cell-free extracts from MHV-infected cells (Tahara et al.,

1994); the N protein was reported to bind to the MHV leader sequence and to facilitate translation (Nelson et al., 2000; Tahara et al., 1998); and the SARS-CoV leader sequence was shown to mediate viral mRNA escape from SARS-CoV nsp1-induced endonucleolytic RNA cleavage (Huang et al., 2011b). However, our understanding of the biological functions of the leader sequence of other CoVs at the posttranscriptional level remains limited. Also, it is possible that leader sequences of different CoVs have evolved to meet differential requirements imposed by specific (and different) target cells.

Mechanisms that control cellular gene expression at the posttranscriptional level remains a major research field, generating surprising and exiting new discoveries at an amazing speed and, undoubtedly, will influence our future studies of the posttranscriptional regulation of CoV gene expression.

ACKNOWLEDGMENTS

The work was partially supported by Public Health Service grants, AI99107 and AI114657 from the National Institutes of Health. K.N. was supported by the James W. McLaughlin fellowship fund.

REFERENCES

Anderson, P., Kedersha, N., 2009. RNA granules: post-transcriptional and epigenetic modulators of gene expression. Nat. Rev. Mol. Cell Biol. 10, 430–436.

Armstrong, J., Smeekens, S., Rottier, P., 1983. Sequence of the nucleocapsid gene from murine coronavirus MHV-A59. Nucleic Acids Res. 11, 883–891.

Banerjee, S., An, S., Zhou, A., Silverman, R.H., Makino, S., 2000. RNase L-independent specific 28S rRNA cleavage in murine coronavirus-infected cells. J. Virol. 74, 8793–8802.

Baranov, P.V., Henderson, C.M., Anderson, C.B., Gesteland, R.F., Atkins, J.F., Howard, M.T., 2005. Programmed ribosomal frameshifting in decoding the SARS-CoV genome. Virology 332, 498–510.

Bechill, J., Chen, Z., Brewer, J.W., Baker, S.C., 2008. Coronavirus infection modulates the unfolded protein response and mediates sustained translational repression. J. Virol. 82, 4492–4501.

Beckham, C.J., Parker, R., 2008. P bodies, stress granules, and viral life cycles. Cell Host Microbe 3, 206–212.

Bekaert, M., Rousset, J.P., 2005. An extended signal involved in eukaryotic −1 frameshifting operates through modification of the E site tRNA. Mol. Cell 17, 61–68.

Bouvet, M., Debarnot, C., Imbert, I., Selisko, B., Snijder, E.J., Canard, B., Decroly, E., 2010. In vitro reconstitution of SARS-coronavirus mRNA cap methylation. PLoS Pathog. 6, e1000863.

Brian, D.A., Baric, R.S., 2005. Coronavirus genome structure and replication. Curr. Top. Microbiol. Immunol. 287, 1–30.

Brierley, I., Boursnell, M.E., Binns, M.M., Bilimoria, B., Blok, V.C., Brown, T.D., Inglis, S.C., 1987. An efficient ribosomal frame-shifting signal in the polymerase-encoding region of the coronavirus IBV. EMBO J. 6, 3779–3785.

Brierley, I., Digard, P., Inglis, S.C., 1989. Characterization of an efficient coronavirus ribosomal frameshifting signal: requirement for an RNA pseudoknot. Cell 57, 537–547.

Buchan, J.R., Parker, R., 2009. Eukaryotic stress granules: the ins and outs of translation. Mol. Cell 36, 932–941.

Calvo, S.E., Pagliarini, D.J., Mootha, V.K., 2009. Upstream open reading frames cause widespread reduction of protein expression and are polymorphic among humans. Proc. Natl. Acad. Sci. U.S.A. 106, 7507–7512.

Cao, F., Tavis, J.E., 2011. RNA elements directing translation of the duck hepatitis B Virus polymerase via ribosomal shunting. J. Virol. 85, 6343–6352.

Cencic, R., Desforges, M., Hall, D.R., Kozakov, D., Du, Y., Min, J., et al., 2011a. Blocking eIF4E:eIF4G interaction as a strategy to impair coronavirus replication. J. Virol. 85, 6381–6389.

Cencic, R., Hall, D.R., Robert, F., Du, Y., Min, J., Li, L., et al., 2011b. Reversing chemoresistance by small molecule inhibition of the translation initiation complex eIF4F. Proc. Natl. Acad. Sci. U.S.A. 108, 1046–1051.

Chan, C.P., Siu, K.L., Chin, K.T., Yuen, K.Y., Zheng, B., Jin, D.Y., 2006. Modulation of the unfolded protein response by the severe acute respiratory syndrome coronavirus spike protein. J. Virol. 80, 9279–9287.

Chen, S.C., Olsthoorn, R.C., 2010. Group-specific structural features of the 5'-proximal sequences of coronavirus genomic RNAs. Virology 401, 29–41.

Chen, Y., Cai, H., Pan, J., Xiang, N., Tien, P., Ahola, T., et al., 2009. Functional screen reveals SARS coronavirus nonstructural protein nsp14 as a novel cap N7 methyltransferase. Proc. Natl. Acad. Sci. U.S.A. 106, 3484–3489.

Chen, Y., Su, C., Ke, M., Jin, X., Xu, L., Zhang, Z., et al., 2011. Biochemical and structural insights into the mechanisms of SARS coronavirus RNA ribose 2'-O-methylation by nsp16/nsp10 protein complex. PLoS Pathog. 7, e1002294.

Choi, K.S., Huang, P., Lai, M.M., 2002. Polypyrimidine-tract-binding protein affects transcription but not translation of mouse hepatitis virus RNA. Virology 303, 58–68.

Choi, K.S., Mizutani, A., Lai, M.M., 2004. SYNCRIP, a member of the heterogeneous nuclear ribonucleoprotein family, is involved in mouse hepatitis virus RNA synthesis. J. Virol. 78, 13153–13162.

Daffis, S., Szretter, K.J., Schriewer, J., Li, J., Youn, S., Errett, J., ... Diamond, M.S., 2010. 2'-O methylation of the viral mRNA cap evades host restriction by IFIT family members. Nature 468, 452–456.

Daijogo, S., Semler, B.L., 2011. Mechanistic intersections between picornavirus translation and RNA replication. Adv. Virus Res. 80, 1–24.

Davies, H.A., Macnaughton, M.R., 1979. Comparison of the morphology of three coronaviruses. Arch. Virol. 59, 25–33.

de Groot, R.J., Baric, R., Enjuanes, L., Gorbalenya, A.E., Holmes, K.V., Perlman, S., et al., 2011. Family Coronaviridae Virus Taxonomy. In: Classification and Nomenclature of Viruses: Ninth Report of the International Committee on Taxonomy of Viruses. Academic Press, London, United Kingdom, pp. 806–828.

de Haro, C., Mendez, R., Santoyo, J., 1996. The eIF-2alpha kinases and the control of protein synthesis. FASEB J. 10, 1378–1387.

Decroly, E., Imbert, I., Coutard, B., Bouvet, M., Selisko, B., Alvarez, K., et al., 2008. Coronavirus nonstructural protein 16 is a cap-0 binding enzyme possessing (nucleoside-2'O)-methyltransferase activity. J. Virol. 82, 8071–8084.

Dediego, M.L., Pewe, L., Alvarez, E., Rejas, M.T., Perlman, S., Enjuanes, L., 2008. Pathogenicity of severe acute respiratory coronavirus deletion mutants in hACE-2 transgenic mice. Virology 376, 379–389.

Emara, M.M., Brinton, M.A., 2007. Interaction of TIA-1/TIAR with West Nile and dengue virus products in infected cells interferes with stress granule formation and processing body assembly. FASEB J. 104, 9041–9046.

Firth, A.E., Brierley, I., 2012. Non-canonical translation in RNA viruses. J. Gen. Virol. 93, 1385–1409.

Fischer, F., Peng, D., Hingley, S.T., Weiss, S.R., Masters, P.S., 1997. The internal open reading frame within the nucleocapsid gene of mouse hepatitis virus encodes a structural protein that is not essential for viral replication. J. Virol. 71, 996–1003.

Fung, T.S., Liu, D.X., 2014. Coronavirus infection, ER stress, apoptosis and innate immunity. Front. Microbiol. 5, 296.

Furuichi, Y., Shatkin, A.J., 2000. Viral and cellular mRNA capping: past and prospects. Adv. Virus Res. 55, 135–184.

Furuichi, Y., Muthukrishnan, S., Tomasz, J., Shatkin, A.J., 1976. Mechanism of formation of reovirus mRNA 5′-terminal blocked and methylated sequence, m7GpppGmpC. J. Biol. Chem. 251, 5043–5053.

Gaglia, M.M., Covarrubias, S., Wong, W., Glaunsinger, B.A., 2012. A common strategy for host RNA degradation by divergent viruses. J. Virol. 86, 9527–9530.

Galan, C., Sola, I., Nogales, A., Thomas, B., Akoulitchev, A., Enjuanes, L., et al., 2009. Host cell proteins interacting with the 3′ end of TGEV coronavirus genome influence virus replication. Virology 391, 304–314.

Gorbalenya, A.E., Snijder, E.J., Spaan, W.J., 2004. Severe acute respiratory syndrome coronavirus phylogeny: toward consensus. J. Virol. 78, 7863–7866.

Graham, R.L., Sims, A.C., Brockway, S.M., Baric, R.S., Denison, M.R., 2005. The nsp2 replicase proteins of murine hepatitis virus and severe acute respiratory syndrome coronavirus are dispensable for viral replication. J. Virol. 79, 13399–13411.

Heald-Sargent, T., Gallagher, T., 2012. Ready, set, fuse! The coronavirus spike protein and acquisition of fusion competence. Viruses 4, 557–580.

Hemmings-Mieszczak, M., Hohn, T., 1999. A stable hairpin preceded by a short open reading frame promotes nonlinear ribosome migration on a synthetic mRNA leader. RNA 5, 1149–1157.

Herold, J., Siddell, S., 1993. An 'elaborated' pseudoknot is required for high frequency frameshifting during translation of HCV 229E polymerase mRNA. Nucleic Acids Res. 21, 5838–5842.

Hilton, A., Mizzen, L., MacIntyre, G., Cheley, S., Anderson, R., 1986. Translational control in murine hepatitis virus infection. J. Gen. Virol. 67, 923–932.

Hofmann, M.A., Brian, D.A., 1991. The 5′ end of coronavirus minus-strand RNAs contains a short poly(U) tract. J. Virol. 65, 6331–6333.

Hofmann, M.A., Chang, R.Y., Ku, S., Brian, D.A., 1993. Leader-mRNA junction sequences are unique for each subgenomic mRNA species in the bovine coronavirus and remain so throughout persistent infection. Virology 196, 163–171.

Huang, C., Lokugamage, K.G., Rozovics, J.M., Narayanan, K., Semler, B.L., Makino, S., 2011a. Alphacoronavirus transmissible gastroenteritis virus nsp1 protein suppresses protein translation in mammalian cells and in cell-free HeLa cell extracts but not in rabbit reticulocyte lysate. J. Virol. 85, 638–643.

Huang, C., Lokugamage, K.G., Rozovics, J.M., Narayanan, K., Semler, B.L., Makino, S., 2011b. SARS coronavirus nsp1 protein induces template-dependent endonucleolytic cleavage of mRNAs: viral mRNAs are resistant to nsp1-induced RNA cleavage. PLoS Pathog. 7, e1002433.

Hurst-Hess, K.R., Kuo, L., Masters, P.S., 2015. Dissection of amino-terminal functional domains of murine coronavirus nonstructural protein 3. J. Virol. 89, 6033–6047.

Irigoyen, N., Firth, A.E., Jones, J.D., Chung, B.Y., Siddell, S., Brierley, I., 2016. High-resolution analysis of coronavirus gene expression by RNA sequencing and ribosome profiling. PLoS Pathog. 12, e1005473.

Ishimaru, D., Plant, E.P., Sims, A.C., Yount Jr., B.L., Roth, B.M., Eldho, N.V., et al., 2013. RNA dimerization plays a role in ribosomal frameshifting of the SARS coronavirus. Nucleic Acids Res. 41, 2594–2608.

Ivanov, K.A., Ziebuhr, J., 2004. Human coronavirus 229E nonstructural protein 13: characterization of duplex-unwinding, nucleoside triphosphatase, and RNA 5′-triphosphatase activities. J. Virol. 78, 7833–7838.

Ivanov, K.A., Thiel, V., Dobbe, J.C., van der Meer, Y., Snijder, E.J., Ziebuhr, J., 2004. Multiple enzymatic activities associated with severe acute respiratory syndrome coronavirus helicase. J. Virol. 78, 5619–5632.

Jendrach, M., Thiel, V., Siddell, S., 1999. Characterization of an internal ribosome entry site within mRNA 5 of murine hepatitis virus. Arch. Virol. 144, 921–933.

Kamahora, T., Soe, L.H., Lai, M.M., 1989. Sequence analysis of nucleocapsid gene and leader RNA of human coronavirus OC43. Virus Res. 12, 1–9.

Kamitani, W., Narayanan, K., Huang, C., Lokugamage, K., Ikegami, T., Ito, N., et al., 2006. Severe acute respiratory syndrome coronavirus nsp1 protein suppresses host gene expression by promoting host mRNA degradation. Proc. Natl. Acad. Sci. U.S.A. 103, 12885–12890.

Kamitani, W., Huang, C., Narayanan, K., Lokugamage, K.G., Makino, S., 2009. A two-pronged strategy to suppress host protein synthesis by SARS coronavirus Nsp1 protein. Nat. Struct. Mol. Biol. 16, 1134–1140.

Kapp, L.D., Lorsch, J.R., 2004. The molecular mechanics of eukaryotic translation. Annu. Rev. Biochem. 73, 657–704.

Khong, A., Jan, E., 2011. Modulation of stress granules and P bodies during dicistrovirus infection. J. Virol. 85, 1439–1451.

Kopecky-Bromberg, S.A., Martinez-Sobrido, L., Palese, P., 2006. 7a protein of severe acute respiratory syndrome coronavirus inhibits cellular protein synthesis and activates p38 mitogen-activated protein kinase. J. Virol. 80, 785–793.

Krahling, V., Stein, D.A., Spiegel, M., Weber, F., Muhlberger, E., 2009. Severe acute respiratory syndrome coronavirus triggers apoptosis via protein kinase R but is resistant to its antiviral activity. J. Virol. 83, 2298–2309.

Kyuwa, S., Cohen, M., Nelson, G., Tahara, S.M., Stohlman, S.A., 1994. Modulation of cellular macromolecular synthesis by coronavirus: implication for pathogenesis. J. Virol. 68, 6815–6819.

Lapps, W., Hogue, B.G., Brian, D.A., 1987. Sequence analysis of the bovine coronavirus nucleocapsid and matrix protein genes. Virology 157, 47–57.

Latorre, P., Kolakofsky, D., Curran, J., 1998. Sendai virus Y proteins are initiated by a ribosomal shunt. Mol. Cell. Biol. 18, 5021–5031.

Lau, S.K., Woo, P.C., Li, K.S., Huang, Y., Tsoi, H.W., Wong, B.H., et al., 2005. Severe acute respiratory syndrome coronavirus-like virus in Chinese horseshoe bats. Proc. Natl. Acad. Sci. U.S.A. 102, 14040–14045.

Le, S.Y., Chen, J.H., Sonenberg, N., Maizel, J.V., 1992. Conserved tertiary structure elements in the 5′ untranslated region of human enteroviruses and rhinoviruses. Virology 191, 858–866.

Le, S.Y., Chen, J.H., Sonenberg, N., Maizel, J.V., 1993. Conserved tertiary structural elements in the 5′ nontranslated region of cardiovirus, aphthovirus and hepatitis A virus RNAs. Nucleic Acids Res. 21, 2445–2451.

Li, H.P., Zhang, X., Duncan, R., Comai, L., Lai, M.M., 1997. Heterogeneous nuclear ribonucleoprotein A1 binds to the transcription-regulatory region of mouse hepatitis virus RNA. Proc. Natl. Acad. Sci. U.S.A. 94, 9544–9549.

Li, H.P., Huang, P., Park, S., Lai, M.M., 1999. Polypyrimidine tract-binding protein binds to the leader RNA of mouse hepatitis virus and serves as a regulator of viral transcription. J. Virol. 73, 772–777.

Li, W., Shi, Z., Yu, M., Ren, W., Smith, C., Epstein, J.H., et al., 2005. Bats are natural reservoirs of SARS-like coronaviruses. Science 310, 676–679.

Liu, D.X., Inglis, S.C., 1992. Internal entry of ribosomes on a tricistronic mRNA encoded by infectious bronchitis virus. J. Virol. 66, 6143–6154.

Liu, D.X., Fung, T.S., Chong, K.K., Shukla, A., Hilgenfeld, R., 2014. Accessory proteins of SARS-CoV and other coronaviruses. Antiviral Res. 109, 97–109.

Lokugamage, K.G., Narayanan, K., Huang, C., Makino, S., 2012. Severe acute respiratory syndrome coronavirus protein nsp1 is a novel eukaryotic translation inhibitor that represses multiple steps of translation initiation. J. Virol. 86, 13598–13608.

Lokugamage, K.G., Narayanan, K., Nakagawa, K., Terasaki, K., Ramirez, S.I., Tseng, C.T., et al., 2015. Middle east respiratory syndrome coronavirus nsp1 inhibits host gene expression by selectively targeting mRNAs transcribed in the nucleus while sparing mRNAs of cytoplasmic origin. J. Virol. 89, 10970–10981.

Martinez-Salas, E., Francisco-Velilla, R., Fernandez-Chamorro, J., Lozano, G., Diaz-Toledano, R., 2015. Picornavirus IRES elements: RNA structure and host protein interactions. Virus Res. 206, 62–73.

Masters, P.S., 2006. The molecular biology of coronaviruses. Adv. Virus Res. 66, 193–292.

McInerney, G.M., Kedersha, N.L., Kaufman, R.J., Anderson, P., Liljestrom, P., 2005. Importance of eIF2alpha phosphorylation and stress granule assembly in alphavirus translation regulation. Mol. Biol. Cell 16, 3753–3763.

Memish, Z.A., Mishra, N., Olival, K.J., Fagbo, S.F., Kapoor, V., Epstein, J.H., et al., 2013. Middle East respiratory syndrome coronavirus in bats, Saudi Arabia. Emerg. Infect. Dis. 19, 1819–1823.

Meyer, K.D., Patil, D.P., Zhou, J., Zinoviev, A., Skabkin, M.A., Elemento, O., et al., 2015. 5′ UTR m(6)A promotes cap-independent translation. Cell 163, 999–1010.

Minakshi, R., Padhan, K., Rani, M., Khan, N., Ahmad, F., Jameel, S., 2009. The SARS Coronavirus 3a protein causes endoplasmic reticulum stress and induces ligand-independent downregulation of the type 1 interferon receptor. PLoS One 4, e8342.

Montero, H., Rojas, M., Arias, C.F., Lopez, S., 2008. Rotavirus infection induces the phosphorylation of eIF2alpha but prevents the formation of stress granules. J. Virol. 82, 1496–1504.

Nanda, S.K., Leibowitz, J.L., 2001. Mitochondrial aconitase binds to the 3′ untranslated region of the mouse hepatitis virus genome. J. Virol. 75, 3352–3362.

Narayanan, K., Huang, C., Lokugamage, K.G., Kamitani, W., Ikegami, T., Tseng, C.T., et al., 2008a. Severe acute respiratory syndrome coronavirus nsp1 suppresses host gene expression, including that of type I interferon, in infected cells. J. Virol. 82, 4471–4479.

Narayanan, K., Huang, C., Makino, S., 2008b. SARS coronavirus accessory proteins. Virus Res. 133, 113–121.

Narayanan, K., Ramirez, S.I., Lokugamage, K.G., Makino, S., 2014. Coronavirus nonstructural protein 1: common and distinct functions in the regulation of host and viral gene expression. Virus Res. 202, 89–100.

Nelson, G.W., Stohlman, S.A., Tahara, S.M., 2000. High affinity interaction between nucleocapsid protein and leader/intergenic sequence of mouse hepatitis virus RNA. J. Gen. Virol. 81, 181–188.

Newman, B.W., Chamberlain, P., Bowden, F., Joseph, J., 2014. Atlas of coronavirus replicase structure. Virus Res. 194, 49–66.

O'Connor, J.B., Brian, D.A., 2000. Downstream ribosomal entry for translation of coronavirus TGEV gene 3b. Virology 269, 172–182.

Plant, E.P., Dinman, J.D., 2006. Comparative study of the effects of heptameric slippery site composition on −1 frameshifting among different eukaryotic systems. RNA 12, 666–673.

Plant, E.P., Dinman, J.D., 2008. The role of programmed −1 ribosomal frameshifting in coronavirus propagation. Front. Biosci. 13, 4873–4881.

Plant, E.P., Perez-Alvarado, G.C., Jacobs, J.L., Mukhopadhyay, B., Hennig, M., Dinman, J.D., 2005. A three-stemmed mRNA pseudoknot in the SARS coronavirus frameshift signal. PLoS Biol. 3, e172.

Plant, E.P., Rakauskaite, R., Taylor, D.R., Dinman, J.D., 2010. Achieving a golden mean: mechanisms by which coronaviruses ensure synthesis of the correct stoichiometric ratios of viral proteins. J. Virol. 84, 4330–4340.

Plant, E.P., Sims, A.C., Baric, R.S., Dinman, J.D., Taylor, D.R., 2013. Altering SARS coronavirus frameshift efficiency affects genomic and subgenomic RNA production. Viruses 5, 279–294.

Pooggin, M.M., Futterer, J., Skryabin, K.G., Hohn, T., 1999. A short open reading frame terminating in front of a stable hairpin is the conserved feature in pregenomic RNA leaders of plant pararetroviruses. J. Gen. Virol. 80, 2217–2228.

Raaben, M., Groot Koerkamp, M.J., Rottier, P.J., de Haan, C.A., 2007. Mouse hepatitis coronavirus replication induces host translational shutoff and mRNA decay, with concomitant formation of stress granules and processing bodies. Cell. Microbiol. 9, 2218–2229.

Racine, T., Duncan, R., 2010. Facilitated leaky scanning and atypical ribosome shunting direct downstream translation initiation on the tricistronic S1 mRNA of avian reovirus. Nucleic Acids Res. 38, 7260–7272.

Rota, P.A., Oberste, M.S., Monroe, S.S., Nix, W.A., Campagnoli, R., Icenogle, J.P., et al., 2003. Characterization of a novel coronavirus associated with severe acute respiratory syndrome. Science 300, 1394–1399.

Sawicki, S.G., Sawicki, D.L., 1990. Coronavirus transcription: subgenomic mouse hepatitis virus replicative intermediates function in RNA synthesis. J. Virol. 64, 1050–1056.

Sawicki, S.G., Sawicki, D.L., Siddell, S., 2007. A contemporary view of coronavirus transcription. J. Virol. 81, 20–29.

Schroder, M., Kaufman, R.J., 2005. ER stress and the unfolded protein response. Mutat. Res. 569, 29–63.

Senanayake, S.D., Brian, D.A., 1997. Bovine coronavirus I protein synthesis follows ribosomal scanning on the bicistronic N mRNA. Virus Res. 48, 101–105.

Senanayake, S.D., Brian, D.A., 1999. Translation from the 5′ untranslated region (UTR) of mRNA 1 is repressed, but that from the 5′ UTR of mRNA 7 is stimulated in coronavirus-infected cells. J. Virol. 73, 8003–8009.

Senanayake, S.D., Hofmann, M.A., Maki, J.L., Brian, D.A., 1992. The nucleocapsid protein gene of bovine coronavirus is bicistronic. J. Virol. 66, 5277–5283.

Shen, X., Masters, P.S., 2001. Evaluation of the role of heterogeneous nuclear ribonucleoprotein A1 as a host factor in murine coronavirus discontinuous transcription and genome replication. Proc. Natl. Acad. Sci. U.S.A. 98, 2717–2722.

Shi, S.T., Huang, P., Li, H.P., Lai, M.M., 2000. Heterogeneous nuclear ribonucleoprotein A1 regulates RNA synthesis of a cytoplasmic virus. EMBO J. 19, 4701–4711.

Shi, C.S., Qi, H.Y., Boularan, C., Huang, N.N., Abu-Asab, M., Shelhamer, J.H., et al., 2014. SARS-coronavirus open reading frame-9b suppresses innate immunity by targeting mitochondria and the MAVS/TRAF3/TRAF6 signalosome. J. Immunol. 193, 3080–3089.

Shien, J.H., Su, Y.D., Wu, H.Y., 2014. Regulation of coronaviral poly(A) tail length during infection is not coronavirus species- or host cell-specific. Virus Genes 49, 383–392.

Siddell, S., Wege, H., Barthel, A., ter Meulen, V., 1980. Coronavirus JHM: cell-free synthesis of structural protein p60. J. Virol. 33, 10–17.

Siddell, S., Wege, H., Barthel, A., ter Meulen, V., 1981a. Coronavirus JHM: intracellular protein synthesis. J. Gen. Virol. 53, 145–155.

Siddell, S., Wege, H., Barthel, A., ter Meulen, V., 1981b. Intracellular protein synthesis and the in vitro translation of coronavirus JHM mRNA. Adv. Exp. Med. Biol. 142, 193–207.

Snijder, E.J., Bredenbeek, P.J., Dobbe, J.C., Thiel, V., Ziebuhr, J., Poon, L.L., et al., 2003. Unique and conserved features of genome and proteome of SARS-coronavirus, an early split-off from the coronavirus group 2 lineage. J. Mol. Biol. 331, 991–1004.

Sola, I., Galan, C., Mateos-Gomez, P.A., Palacio, L., Zuniga, S., Cruz, J.L., et al., 2011. The polypyrimidine tract-binding protein affects coronavirus RNA accumulation levels and relocalizes viral RNAs to novel cytoplasmic domains different from replication-transcription sites. J. Virol. 85, 5136–5149.

Somers, J., Poyry, T., Willis, A.E., 2013. A perspective on mammalian upstream open reading frame function. Int. J. Biochem. Cell Biol. 45, 1690–1700.

Spagnolo, J.F., Hogue, B.G., 2000. Host protein interactions with the 3' end of bovine coronavirus RNA and the requirement of the poly(A) tail for coronavirus defective genome replication. J. Virol. 74, 5053–5065.

Su, M.C., Chang, C.T., Chu, C.H., Tsai, C.H., Chang, K.Y., 2005. An atypical RNA pseudoknot stimulator and an upstream attenuation signal for −1 ribosomal frameshifting of SARS coronavirus. Nucleic Acids Res. 33, 4265–4275.

Sung, S.C., Chao, C.Y., Jeng, K.S., Yang, J.Y., Lai, M.M., 2009. The 8ab protein of SARS-CoV is a luminal ER membrane-associated protein and induces the activation of ATF6. Virology 387, 402–413.

Tahara, S.M., Dietlin, T.A., Bergmann, C.C., Nelson, G.W., Kyuwa, S., Anthony, R.P., et al., 1994. Coronavirus translational regulation: leader affects mRNA efficiency. Virology 202, 621–630.

Tahara, S.M., Dietlin, T.A., Nelson, G.W., Stohlman, S.A., Manno, D.J., 1998. Mouse hepatitis virus nucleocapsid protein as a translational effector of viral mRNAs. Adv. Exp. Med. Biol. 440, 313–318.

Tanaka, T., Kamitani, W., Dediego, M.L., Enjuanes, L., Matsuura, Y., 2012. Severe acute respiratory syndrome coronavirus nsp1 facilitates efficient propagation in cells through a specific translational shutoff of host mRNA. J. Virol. 86, 11128–11137.

Thiel, V., Siddell, S., 1994. Internal ribosome entry in the coding region of murine hepatitis virus mRNA 5. J. Gen. Virol. 75, 3041–3046.

van Boheemen, S., de Graaf, M., Lauber, C., Bestebroer, T.M., Raj, V.S., Zaki, A.M., et al., 2012. Genomic characterization of a newly discovered coronavirus associated with acute respiratory distress syndrome in humans. mBio 3. e00473-12.

Ventoso, I., Sanz, M.A., Molina, S., Berlanga, J.J., Carrasco, L., Esteban, M., 2006. Translational resistance of late alphavirus mRNA to eIF2alpha phosphorylation: a strategy to overcome the antiviral effect of protein kinase PKR. Genes Dev. 20, 87–100.

Versteeg, G.A., van de Nes, P.S., Bredenbeek, P.J., Spaan, W.J., 2007. The coronavirus spike protein induces endoplasmic reticulum stress and upregulation of intracellular chemokine mRNA concentrations. J. Virol. 81, 10981–10990.

Wang, X., Liao, Y., Yap, P.L., Png, K.J., Tam, J.P., Liu, D.X., 2009. Inhibition of protein kinase R activation and upregulation of GADD34 expression play a synergistic role in facilitating coronavirus replication by maintaining de novo protein synthesis in virus-infected cells. J. Virol. 83, 12462–12472.

Wege, H., Wege, H., Nagashima, K., ter Meulen, V., 1979. Structural polypeptides of the murine coronavirus JHM. J. Gen. Virol. 42, 37–47.

Weiss, S.R., Navas-Martin, S., 2005. Coronavirus pathogenesis and the emerging pathogen severe acute respiratory syndrome coronavirus. Microbiol. Mol. Biol. Rev. 69, 635–664.

Woo, P.C., Lau, S.K., Chu, C.M., Chan, K.H., Tsoi, H.W., Huang, Y., et al., 2005. Characterization and complete genome sequence of a novel coronavirus, coronavirus HKU1, from patients with pneumonia. J. Virol. 79, 884–895.

Woo, P.C., Huang, Y., Lau, S.K., Yuen, K.Y., 2010. Coronavirus genomics and bioinformatics analysis. Viruses 2, 1804–1820.

Woo, P.C., Lau, S.K., Lam, C.S., Lau, C.C., Tsang, A.K., Lau, J.H., et al., 2012. Discovery of seven novel Mammalian and avian coronaviruses in the genus deltacoronavirus supports bat coronaviruses as the gene source of alphacoronavirus and betacoronavirus and avian coronaviruses as the gene source of gammacoronavirus and deltacoronavirus. J. Virol. 86, 3995–4008.

Wu, H.Y., Ke, T.Y., Liao, W.Y., Chang, N.Y., 2013. Regulation of coronaviral poly(A) tail length during infection. PLoS One 8, e70548.

Wu, H.Y., Guan, B.J., Su, Y.P., Fan, Y.H., Brian, D.A., 2014. Reselection of a genomic upstream open reading frame in mouse hepatitis coronavirus 5′-untranslated-region mutants. J. Virol. 88, 846–858.

Xiao, H., Xu, L.H., Yamada, Y., Liu, D.X., 2008. Coronavirus spike protein inhibits host cell translation by interaction with eIF3f. PLoS One 3, e1494.

Ye, Y., Hauns, K., Langland, J.O., Jacobs, B.L., Hogue, B.G., 2007. Mouse hepatitis coronavirus A59 nucleocapsid protein is a type I interferon antagonist. J. Virol. 81, 2554–2563.

Ye, Z., Wong, C.K., Li, P., Xie, Y., 2008. A SARS-CoV protein, ORF-6, induces caspase-3 mediated, ER stress and JNK-dependent apoptosis. Biochim. Biophys. Acta 1780, 1383–1387.

Yeung, Y.S., Yip, C.W., Hon, C.C., Chow, K.Y., Ma, I.C., Zeng, F., et al., 2008. Transcriptional profiling of Vero E6 cells over-expressing SARS-CoV S2 subunit: insights on viral regulation of apoptosis and proliferation. Virology 371, 32–43.

Yuan, X., Wu, J., Shan, Y., Yao, Z., Dong, B., Chen, B., et al., 2006. SARS coronavirus 7a protein blocks cell cycle progression at G0/G1 phase via the cyclin D3/pRb pathway. Virology 346, 74–85.

Zhang, L., Wang, A., 2012. Virus-induced ER stress and the unfolded protein response. Front. Plant Sci. 3, 293.

Zhou, B., Liu, J., Wang, Q., Liu, X., Li, X., Li, P., et al., 2008. The nucleocapsid protein of severe acute respiratory syndrome coronavirus inhibits cell cytokinesis and proliferation by interacting with translation elongation factor 1alpha. J. Virol. 82, 6962–6971.

Zhou, J., Wan, J., Gao, X., Zhang, X., Jaffrey, S.R., Qian, S.B., 2015. Dynamic m(6)A mRNA methylation directs translational control of heat shock response. Nature 526, 591–594.

Zorzitto, J., Galligan, C.L., Ueng, J.J., Fish, E.N., 2006. Characterization of the antiviral effects of interferon-alpha against a SARS-like coronoavirus infection in vitro. Cell Res. 16, 220–229.

Zust, R., Cervantes-Barragan, L., Habjan, M., Maier, R., Neuman, B.W., Ziebuhr, J., et al., 2011. Ribose 2′-O-methylation provides a molecular signature for the distinction of self and non-self mRNA dependent on the RNA sensor Mda5. Nat. Immunol. 12, 137–143.

CHAPTER SIX

Feline Coronaviruses: Pathogenesis of Feline Infectious Peritonitis

G. Tekes[1], H.-J. Thiel

Institute of Virology, Faculty of Veterinary Medicine, Justus Liebig University Giessen, Giessen, Germany
[1]Corresponding author: e-mail address: gergely.tekes@vetmed.uni-giessen.de

Contents

Abstract

Feline infectious peritonitis (FIP) belongs to the few animal virus diseases in which, in the course of a generally harmless persistent infection, a virus acquires a small number of mutations that fundamentally change its pathogenicity, invariably resulting in a fatal outcome. The causative agent of this deadly disease, feline infectious peritonitis virus (FIPV), arises from feline enteric coronavirus (FECV). The review summarizes our current knowledge of the genome and proteome of feline coronaviruses (FCoVs), focusing on the viral surface (spike) protein S and the five accessory proteins. We also review the current classification of FCoVs into distinct serotypes and biotypes, cellular receptors of FCoVs and their presumed role in viral virulence, and discuss other aspects of FIPV-induced pathogenesis. Our current knowledge of genetic differences between FECVs and FIPVs has been mainly based on comparative sequence analyses that revealed "discriminatory" mutations that are present in FIPVs but not in FECVs. Most of these mutations result in amino acid substitutions in the S protein and these may have a critical role in the switch from FECV to FIPV. In most cases, the precise roles of these

Advances in Virus Research, Volume 96
ISSN 0065-3527
http://dx.doi.org/10.1016/bs.aivir.2016.08.002

193

mutations in the molecular pathogenesis of FIP have not been tested experimentally in the natural host, mainly due to the lack of suitable experimental tools including genetically engineered virus mutants. We discuss the recent progress in the development of FCoV reverse genetics systems suitable to generate recombinant field viruses containing appropriate mutations for in vivo studies.

1. FELINE CORONAVIRUSES

1.1 Taxonomy and Genome Organization

Together with the *Arteriviridae, Mesoniviridae,* and *Roniviridae,* the family *Coronaviridae* (subfamilies *Coronavirinae* and *Torovirinae*) make up the order *Nidovirales.* Coronaviruses belong to the subfamily *Coronavirinae* which has been divided into four genera: *Alpha-, Beta-, Gamma-,* and *Deltacoronavirus.* Within the genus *Alphacoronavirus,* feline coronaviruses (FCoVs) are part of the species *Alphacoronavirus 1,* the latter also containing a few other closely related viruses, such as canine coronaviruses (CCoVs) and the porcine transmissible gastroenteritis virus (TGEV). Other more distantly related species in the genus *Alphacoronavirus* include *Porcine epidemic diarrhea virus* (PEDV), *Human coronavirus 229E* (HCoV-229E), and *Human coronavirus NL63* (HCoV-NL63) (de Groot et al., 2012).

The positive-strand RNA genome of FCoVs has a size of approximately 29 kb and shows the typical genome organization of coronaviruses (Fig. 1). The 5′ untranslated region (UTR) comprises about 310 nucleotides (nts) and contains the leader sequence as well as the transcription regulatory sequence (TRS) with the core-TRS motif. This 5′-CUAAAC-3′ core-TRS motif is conserved in all FCoVs (de Groot et al., 1988; Dye and Siddell, 2005; Tekes et al., 2008). The 3′ UTR consists of around 275 nts and is followed by a poly(A) tail. The replicase gene covers around two-thirds of the genome and comprises open reading frames (ORFs) 1a and 1b. The translation of the FCoV replicase gene leads to the production of polyproteins (pp) 1a and pp1ab, which are processed by virus–encoded proteinases (Dye and Siddell, 2005; Ziebuhr et al., 2000). By analogy with other alphacoronaviruses, FCoV pp1a/pp1ab is thought to be cleaved by virus-encoded papain- and 3C-like proteases at 3 and 11 sites, respectively (Ziebuhr, 2005). Accordingly, proteolytic processing of the FCoV pp1a/1ab gives rise to 16 nonstructural proteins (nsps) that form the replication/transcription complex and, in some cases, are involved in interactions with host cell factors and functions. The 3′-terminal one-third of the FCoV

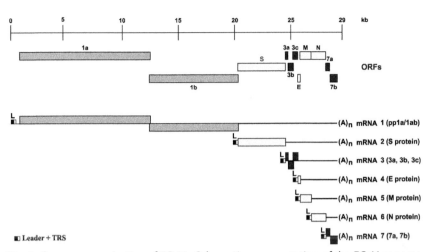

Fig. 1 Genome organization of FCoVs. Schematic representation of the FCoV genome, open reading frames (ORFs), and the characteristic set of subgenomic (sg) mRNAs. The predicted translated regions of each sg mRNA are indicated with *boxes*. The leader (L) sequence together with the transcription regulatory sequence (TRS) located at the 5' end of all mRNAs is depicted as a *black* and *white boxes*, respectively.

genome contains (i) four genes encoding the four structural proteins called spike (S), envelope (E), membrane (M), nucleocapsid (N) protein, respectively, and (ii) several accessory genes.

1.2 Accessory Genes and Proteins

FCoVs possess five accessory genes called 3a, 3b, 3c, 7a, and 7b. Like in other alphacoronaviruses, these genes are located at two different genome positions (Dye and Siddell, 2005; Haijema et al., 2007; Tekes et al., 2008). Between the S and E genes, FCoVs and CCoVs possess three ORFs (3a, 3b, and 3c), while TGEV contains only two ORFs (3a and 3b). Recently, an additional ORF named ORF3 was described in this region for CCoV (Lorusso et al., 2008). Other members of the *Alphacoronavirus* genus possess only one ORF 3. Sequence analyses suggest that FCoV ORF 3a is homologous to CCoV ORF 3a and TGEV ORF 3a, while the FCoV ORF 3c is a homolog of CCoV ORF 3c, TGEV ORF 3b, and ORF 3 of all other alphacoronaviruses (Narayanan et al., 2008). CCoV ORF 3b represents the only known homolog of FCoV ORF 3b. Furthermore, all members of the species *Alphacoronavirus 1* harbor various numbers of additional accessory genes downstream of the N gene. TGEV has only one ORF (called ORF 7), which is homologous to ORF 7a of FCoVs and CCoVs. The latter two contain yet another ORF, called 7b, which precedes the 3′ UTR. Deletions of the entire FCoV ORF 3 and 7 genome regions showed that the accessory genes are dispensable for viral growth in vitro; they were suggested to be important for virus replication and virulence in vivo (Haijema et al., 2004). However, the functions of the accessory proteins remain still to be investigated.

FCoV ORF 3a is predicted to encode a 72-amino acid(aa)–long protein without any known or predicted function. This protein is thought to be expressed from the subgenomic (sg) RNA 3, which has been detected in infected cells (Dye and Siddell, 2005; Tekes et al., 2008) and the synthesis of which involves the canonical core-TRS motif upstream of the translational start signal of 3a. However, until now, the expression of this protein in infected cells has not been demonstrated. ORF 3b overlaps with ORF 3a and is supposed to encode an approximately 9-kDa protein with currently unknown functions. Similar to 3a, expression of 3b protein has not been demonstrated in infected cells. It is generally thought that the 3b protein is translated by a noncanonical mechanism from the second ORF present in the 5′-unique region of sgRNA 3. ORF 3c is predicted to code for a protein of 238 amino acids which likely represents a membrane protein with three transmembrane regions. The predicted topology of the 3c protein transmembrane domains is similar to that of the viral M protein (Oostra et al., 2006). Thus far, the expression of 3c protein in infected cells could not be shown. Also, it is unclear whether the 3c protein is expressed from the same sgRNA3 (as predicted for the 3a and 3b proteins) or from a separate sgRNA that, however, has not been identified to date. The existence of an additional sgRNA from which 3c could be expressed receives some support by the observation that the genomes of most FCoV isolates contain a core-TRS motif or a very similar sequence immediately upstream of the ORF3c translational start codon. The transient expression of 3c in a cat cell line revealed a perinuclear localization (Hsieh et al., 2013). Based on the sequence analyses of FCoV field isolates, it has been proposed that 3c is essential for viral replication in the gut but dispensable for systemic infection (Chang et al., 2010). Furthermore, the FCoV 3c protein homologs conserved in PEDV and HCoV-229E were suggested to be incorporated into virus particles, to function as ion channels and to enhance virus production (Wang et al., 2012; Zhang et al., 2014). It remains to be determined whether FCoV 3c protein has similar functions.

Although the synthesis of the FCoV 7a protein in infected cells has not been confirmed experimentally, the protein is expected to be expressed from sgRNA 7 as was shown previously for its TGEV homolog. FCoV ORF7a is predicted to encode a 71-amino acid (~10 kDa) protein with an N-terminal signal sequence and a C-terminal transmembrane domain (Haijema et al., 2007). Using a plasmid construct expressing 7a with a C-terminal GFP tag, the 7a protein was shown to colocalize primarily with the endoplasmic reticulum (ER) and Golgi apparatus. Using the same plasmid construct and a recombinant virus lacking the entire ORF7, a specific function of the 7a protein in counteracting IFN-α-induced antiviral responses was suggested (Dedeurwaerder et al., 2014). In contrast to all other FCoV accessory proteins, the expression of 7b in infected cells has been confirmed experimentally and the detection of FCoV 7b-specific antibodies in sera obtained from infected cats indicates that the protein is produced in vivo (Herrewegh et al., 1995b; Kennedy et al., 2008; Vennema et al., 1992, 1993). Together with 7a, the 7b protein is expected to be expressed from sgRNA7; however, the translation mechanism used to initiate translation from this second ORF remains to be determined. The 7b protein has a molecular mass of ~26 kDa, it is secreted from the cell, and contains (i) an N-terminal signal sequence, (ii) a potential N-glycosylation site at aa position 68, and (iii) a C-terminal KDEL-like ER retention signal (Vennema et al., 1992). The presence of an internal stop codon or a deletion in the 7b gene has been suggested to indicate cell culture adaptation and a possible (partial) loss of virulence in vivo (Herrewegh et al., 1995b). The precise function of the 7b protein in the FCoV life cycle remains to be elucidated in further studies.

1.3 FCoV Serotypes and Cellular Receptor Usage

Based on serological properties, FCoVs are classified into two serotypes. The vast majority of the natural infections (80–95%) in Europe and America are caused by serotype I FCoVs, while serotype II FCoVs are less common in the field (Benetka et al., 2004; Kummrow et al., 2005). Furthermore, serotype II FCoVs have predominantly been observed in Asia and they were reported to be responsible for up to 25% of the natural infections in those countries (Amer et al., 2012; An et al., 2011; Sharif et al., 2010). There is consistent evidence from independent studies that serotype II viruses emerge via double homologous recombination between serotype I FCoV and CCoV (Decaro and Buonavoglia, 2008; Haijema et al., 2007; Herrewegh et al., 1998; Lin et al., 2013; Lorusso et al., 2008; Terada et al., 2014). As a consequence of the recombination, approximately one-third (~10 kb) of the serotype I FCoV genome including the S gene and the neighboring regions are replaced with the equivalent parts of the CCoV genome (Decaro and Buonavoglia, 2008; Haijema et al., 2007; Herrewegh et al., 1998; Lin et al., 2013; Lorusso et al., 2008; Terada et al., 2014). Detailed sequence analyses of numerous serotype II FCoVs revealed that the 5'-recombination event occurs in the polymerase gene while the 3'-recombination site is located in the E or M genes. However, the exact locations of these recombination sites vary in the different isolates, indicating that serotype II FCoVs continuously arise through independent recombination events (Haijema et al., 2007; Herrewegh et al., 1998; Lin et al., 2013; Terada et al., 2014). It is considered likely that the described recombination occurs in cats that are coinfected with serotype I FCoV and CCoV. However, the exact source of serotype II FCoVs is unclear. It is believed that serotype II FCoVs are more virulent (Lin et al., 2013; Wang et al., 2013).

The most important biological consequence of the recombination is the integration of the CCoV S gene into serotype I FCoV. The coronaviral S protein is the major determinant for viral attachment and host cell type specificity. While the S1 domain of the S protein is responsible for receptor binding, the S2 domain is required for fusion of the viral and cellular membranes (Bosch et al., 2003; Kubo et al., 1994; Yoo et al., 1991). The poor sequence identity (~30%) of the S1 domains of FCoVs serotype I and II strongly suggests that the two serotypes use different receptors for cell entry. Early studies showed that serotype II FCoVs employ as a cellular receptor the feline aminopeptidase N (fAPN) (Tresnan and Holmes, 1998; Tresnan et al., 1996), a 150-kDa glycoprotein with metalloprotease activity that is expressed in many host tissues, including epithelial cells from the intestinal brush border (Kenny and Maroux, 1982; Look et al., 1989; Semenza, 1986). These early studies suggested that fAPN may also facilitate the entry of serotype I FCoVs, albeit less efficiently. Subsequent experiments showed that an fAPN-specific monoclonal antibody is able to block infection by serotype II FCoVs (as well as CCoV and TGEV), but not by serotype I FCoVs (Hohdatsu et al., 1998), suggesting that the two serotypes use different receptors for cell entry. This hypothesis was supported by experiments using pseudotyped retroviruses containing the spike protein of FCoV serotypes I and II, respectively, to transduce different continuous cat cell lines. The data obtained in this study provided evidence that serotype I spike fails to recognize fAPN as a receptor for attachment and entry, suggesting that fAPN is not a functional receptor for serotype I FCoVs (Dye et al., 2007). In line with this, recombinant serotype I FCoVs generated by reverse genetics and expressing serotype I and serotype II S proteins, respectively, were used to demonstrate that the S protein alone is responsible for the different receptor usage of serotype I and serotype II FCoVs (Tekes et al., 2010). It is now generally accepted that serotype I FCoVs employ another cellular receptor. Other studies suggest that feline C-type lectin dendritic cell-specific intercellular adhesion molecule-3-grabbing nonintegrin (fDC-SIGN) has a role in cellular attachment and may serve as a coreceptor for both FCoV serotypes in vitro (Regan and Whittaker, 2008; Regan et al., 2010; Van Hamme et al., 2011). The identification of the cellular receptor for serotype I FCoVs remains an important topic in FCoV research.

The usage of different cellular receptors by the FCoV serotypes is reflected in the characteristics of these viruses in vitro. Whereas serotype II FCoVs replicate well in feline tissue culture cells in vitro, serotype I FCoVs grow poorly, if at all, in cell culture, except for a few cell culture-adapted isolates. Accordingly, in the last decade, most studies on FCoVs were based on serotype II viruses, while the more prevalent serotype I FCoVs were largely neglected (de Haan et al., 2005; Dye and Siddell, 2005; Haijema et al., 2003, 2004; Rottier et al., 2005; Tekes et al., 2012).

2. INFECTION WITH FELINE CORONAVIRUSES

2.1 Feline Enteric Coronavirus

FCoVs can cause infections in domestic and wild *Felidae* worldwide (Hofmann-Lehmann et al., 1996; Leutenegger et al., 1999; Munson et al., 2004; Paul-Murphy et al., 1994). Approximately 20–60% of domestic cats are seropositive, with seropositivity rates approaching 90% in animal shelters or multi-cat households (Hohdatsu et al., 1992; Pedersen, 2009, 2014). As pointed out earlier, most of the natural infections are caused by serotype I FCoVs (Addie et al., 2003; Hohdatsu et al., 1992; Kennedy et al., 2002; Kummrow et al., 2005). According to pathogenicity, FCoVs are separated into two biotypes that are generally referred to as feline enteric coronavirus (FECV) and feline infectious peritonitis virus (FIPV). These two biotypes exist in both serotypes I and II.

The vast majority of FECV infections are benign and they either remain undetected or cause a mild diarrhea. However, FECVs can occasionally induce severe enteritis (Kipar et al., 1998b). Convincing evidence for persistent infections caused by FECVs was first provided in the late 1990s (Herrewegh et al., 1997). In these experiments, naturally infected cats were isolated and monitored for virus shedding in the feces. In several cases, FECVs remained detectable in the feces of the cats for more than 15 weeks, although with decreasing viral loads. To investigate the course of infection in more detail, FECV infection experiments have also been performed under controlled conditions (Desmarets et al., 2016; Kipar et al., 2010; Pedersen et al., 2008; Vogel et al., 2010). These studies showed that FECVs induce symptomless persistent infections similar to natural infections. The virus could be detected a few days after infection in the feces, and virus shedding was confirmed to last for several months (Pedersen et al., 1981b; Vogel et al., 2010). Furthermore, similar to natural infections viral RNA was also found in the blood (Gunn-Moore et al., 1998; Herrewegh et al., 1995a, 1997; Kipar et al., 2006a,b; Meli et al., 2004; Simons et al., 2005; Vogel et al., 2010). Seroconversion of the animals started approximately 10 days postinfection and the antibody titers remained at a relatively low level. Postmortem analyses showed that, in acute infections, FECVs have a tropism to the apical epithelium of the intestinal villi from the lower part of the small intestines to the caecum (Pedersen et al., 1981b). Although coronaviral RNA can be detected in persistently infected cats in the entire gastrointestinal tract, blood, and different tissues, experimental infections revealed that the lower part of the gastrointestinal tract is the major site for viral replication and FECV persistence (Herrewegh et al., 1997; Kipar et al., 2010; Vogel et al., 2010). These observations confirm that FECVs are primarily associated with the gastrointestinal tract but they are also capable of infecting monocytes, albeit less efficiently, and thereby spread throughout the body (Dewerchin et al., 2005; Kipar et al., 2006a, 2010; Meli et al., 2004; Porter et al., 2014).

FECVs are highly contagious and are transmitted horizontally via the fecal–oral route (Pedersen, 2009, 2014; Pedersen et al., 1981b). Usually, kittens become infected with FECVs at a young age in the litter, most probably through viruses in the feces of the mother (Addie and Jarrett, 1990, 1992; Pedersen et al., 1981b). Since persistently infected cats shed the virus in their feces for extended periods of time, they play a central role in spreading and maintaining FECVs in cat populations and therefore represent a threat to other animals.

2.2 Feline Infectious Peritonitis Virus

In sharp contrast to FECVs, FIPV causes a lethal disease called feline infectious peritonitis (FIP). The disease is characterized by fibrinous and granulomatous serositis, protein-rich serous effusion in body cavities, and/or granulomatous lesions (pyogranulomas) (Hayashi et al., 1977; Kipar and

Meli, 2014; Kipar et al., 1998a, 2005; Pedersen, 1987, 2009; Weiss and Scott, 1981a,b). The cellular composition, the level of viral antigen expression and the distribution of the FIP-characteristic lesions in different organs can vary in individual cases (Kipar and Meli, 2014). For the development of these lesions, FIPV-infected monocytes and macrophages have been identified as major target cells of FIPVs and are assumed to play a pivotal role (Haijema et al., 2007). FIPVs are able to efficiently infect and replicate in monocytes/macrophages (Dewerchin et al., 2005; Rottier et al., 2005; Stoddart and Scott, 1989) and to trigger an activation of these cells (Regan et al., 2009). Circulating activated monocytes heavily express cytokines such as tumor necrosis factor -α, IL-1β, and adhesion molecules (e.g., CD11b and CD18) (Kipar et al., 2006b; Kiss et al., 2004; Regan et al., 2009; Takano et al., 2009, 2007a,b); the latter facilitate the interaction of monocytes with activated endothelial cells in the small- and medium-sized veins. Moreover, it has been suggested that the increased expression of enzymes such as matrix metalloproteinase-9 by the activated monocytes contributes to endothelial barrier dysfunction and subsequent extravasation of monocytes (Kipar and Meli, 2014; Kipar et al., 2005). Furthermore, the production of vascular endothelial growth factor produced in FIPV-infected monocytes and macrophages was proposed to induce increased vascular permeability and hence effusion in body cavities (Takano et al., 2011). Although leukocytes are not susceptible to FIPV infection, they appear to become activated during FIPV infection by as-yet-unknown mechanisms, thereby probably contributing to endothelial cell damage and the development of FIP lesions (Olyslaegers et al., 2013).

Based on the presence or absence of protein-rich effusions in the abdominal and pleural cavities, wet (effusive), dry (noneffusive), and a combination of these two clinical forms (mixed form) of FIP can be distinguished (Drechsler et al., 2011; Hartmann, 2005; Kipar and Meli, 2014; Pedersen, 2009). In natural infections, the wet form seems to be more prevalent than the dry and mixed form, respectively (Pedersen, 2009). The development of the various clinical forms is believed to be dependent on the host immune response. Although the underlying mechanisms are not completely understood, it is generally accepted that the balance between cellular and humoral immune responses in infected animals critically determines the clinical progression of the disease. While strong cellular immune responses may control the disease (Pedersen, 2009, 2014), weak cellular but vigorous B cell responses have been associated with the wet form and somewhat stronger T cell immune responses are thought to cause the dry form of FIP

(Pedersen, 2009, 2014). It has been observed in field cases of FIP that the wet form often develops during the terminal stage of dry FIP, probably reflecting a collapse of the immune system (Pedersen, 2009, 2014).

Factors that may trigger the progression of the disease have been described for both naturally and experimentally infected cats. It was shown that stress or superinfections with feline leukemia virus and feline immuno-deficiency virus, respectively, increase the risk for FIP development (Poland et al., 1996). The underlying mechanism for this phenomenon is not completely understood, but it is assumed that immunosuppression favors the generation of escape mutants and thereby, the probability of clinical manifestation of FIP. Furthermore, genetic predisposition to FIP was suggested (Golovko et al., 2013; Hsieh and Chueh, 2014; Pedersen, 2009; Pesteanu-Somogyi et al., 2006; Wang et al., 2014; Worthing et al., 2012).

Despite the existence of serotype I and II FIPVs, the characteristics of the disease caused by these serotypes appear to be very similar. The incubation time for naturally occurring FIP cases is difficult to assess, but a number of studies using experimentally infected specific pathogen-free (SPF) cats revealed incubation times of 2–14 days for the wet form and several weeks for the dry form (Kiss et al., 2004; Pedersen and Black, 1983; Pedersen et al., 1981a, 1984; Tekes et al., 2012). Following experimental infections with the prototype serotype II FIPV 79-1146 strain or a recombinant form of this virus, respectively, cats were shown to develop fever after a few days and lost weight rapidly. Shortly after infection, viral RNA became detectable in the feces and blood; serum antibody titers increased rapidly and remained at a high level during the entire course of infection. In some cases, infected animals seemed to recover after the first week of clinical signs, but subsequently developed pronounced clinical signs of the disease including fever, lack of appetite, weight loss and a progressively worsening condition. Although most of the cats died within 4–5 weeks after experimental exposure to serotype II FIPV strain 79-1146, a low number of animals survived for a few months and succumbed to the disease at a later time point (de Groot-Mijnes et al., 2005). The overall survival time of cats can vary significantly, depending on the amount and virulence of the virus used for the experimental infection (Kiss et al., 2004; Pedersen and Black, 1983; Pedersen et al., 1981a; Tekes et al., 2012). It should also be noted that experimental infections of cats with FIPVs always lead to clinical signs, but approximately 20% of the animals can survive and recover (de Groot-Mijnes et al., 2005; Dean et al., 2003; Kipar and Meli, 2014; Tekes et al., 2012).

Naturally occurring FIP usually affects cats at a young age of less than 2 years; the incidence of FIP dramatically decreases with increasing age (Foley et al., 1997; Pedersen, 2009). It is generally accepted that FIP occurs sporadically. Unlike FCoVs, FIPVs are usually not transmitted horizontally from cat to cat, even though FIPV shedding into feces has been detected under experimental infection conditions (Bank-Wolf et al., 2014; Pedersen, 2009; Pedersen et al., 2012; Tekes et al., 2012; Thiel et al., 2014). However, FIPV shedding in the feces does not lead to FIP in contact animals (Pedersen et al., 2012) and horizontal transmission of FIPV resulting in FIP is thought to have no epidemiological role.

2.3 Origin of FIPV

FIP develops in approximately 5% of cats that are persistently infected with FECV (Chang et al., 2011; Haijema et al., 2007; Pedersen, 2009). Over many years, the origin of FIPVs was unclear and discussed quite controversially. In early investigations, FECVs and FIPVs were considered different virus species. In subsequent studies, FECVs and FIPVs were proposed to be closely related viruses with distinct virulence properties. Sequence analyses of both biotypes revealed much higher sequence similarity of FIPV and FECV isolates collected in the same cattery compared to FCoV sequences from distinct catteries/geographical regions (Herrewegh et al., 1995b; Pedersen et al., 1981b; Poland et al., 1996; Vennema et al., 1998). These observations led to the hypothesis that FIPV evolves from FECV by specific mutations occurring in the viral genomes in individually infected cats. This "internal mutation" hypothesis received further support from a series of animal experiments (Poland et al., 1996; Vennema et al., 1998). Based on other data, an alternative "circulating virulent–avirulent FCoV" hypothesis that contradicted the widely accepted theory was also proposed (Brown et al., 2009). The study suggested the independent coexistence of virulent and avirulent FCoVs in a cat population. The authors claimed that cats develop FIP only upon infection with the virulent FCoV type. However, this hypothesis has failed to receive any backing, and since then additional experiments and further analyses have strengthened the "internal mutation" theory (Bank-Wolf et al., 2014; Barker et al., 2013; Chang et al., 2010, 2011, 2012; Lewis et al., 2015; Licitra et al., 2013; Pedersen et al., 2012; Porter et al., 2014). It is now widely accepted that FIPV emerges during persistent infection through mutations from the harmless FECV. However, it is not understood which mutation(s) occur(s) at which stage during the development of FIP. As mentioned earlier, FECVs show a pronounced tropism toward epithelial cells in the gut, but they are also able to infect monocytes, albeit inefficiently. It was suggested that in monocytes—rather than in intestinal epithelial cells—FECVs acquire mutations that can convert them into FIPVs (Pedersen et al., 2012). The resulting FIPVs display an altered cell tropism; they infect and replicate efficiently in monocytes and macrophages. This property is considered a key step in the development of FIP.

3. MOLECULAR PATHOGENESIS OF FIP

3.1 Differences Between FECV and FIPV

In the past decades, many studies were aimed at identifying mutations responsible for the biotype switch. Mutations in accessory genes and the S gene of FCoVs have been associated with FIP development. In this regard, accessory gene 3c was one major focus. Early studies showed that FECVs always contain an intact 3c gene, while more than two-thirds of FIPV-derived 3c sequences were found to contain mutations (e.g., deletion or point mutation) that prevented translation of an intact full-length protein. Therefore, mutations in 3c were initially thought to be a general virulence marker indicative of FIP (Pedersen et al., 2009; Vennema et al., 1998). More recent studies confirmed these earlier observations, with 3c being heavily mutated in the majority of FIPV isolates and possibly involved in FIP development. Comprehensive sequence analyses of FECVs isolated from the gut and FIPVs isolated from the gut, organ lesions, and effusions, respectively, suggested that an intact 3c gene is required for viral replication in the gut but nonessential for systemic replication of FIPVs (Bank–Wolf et al., 2014; Chang et al., 2010, 2011, 2012; Pedersen et al., 2012). It is currently considered likely that the mutations in 3c are no virulence markers for FIP, but rather a consequence of systemic spread and enhanced replication of FIPVs. Nevertheless, it cannot be excluded that (frameshift and other) mutations that affect 3c protein expression contribute to an increased viral fitness in monocytes/macrophages and, thereby, to the development of FIP.

Based on sequence analyses of the 7a gene of FECV/FIPVs obtained from Persian cats, one study proposed that deletions in the 7a gene are associated with the development of FIP (Kennedy et al., 2001). However, these data did not receive support from others. Mutations in the 7a gene are not currently considered to be crucial for the biotype switch.

Deletions in the 7b gene have also been proposed to play an important role in FIP development. However, consecutive analyses revealed that deletions in 7b primarily evolve during cell culture adaptation and are associated with loss of virulence (Herrewegh et al., 1995b, 1998; Takano et al., 2011). The existence of deletions in the 7b gene in naturally occurring FECVs argues against a major involvement of mutations in 7b in FIP development (Lin et al., 2009).

Lately, the focus of research on FIP pathogenesis shifted toward the investigation of the S gene. The coronaviral S protein is crucial for receptor binding and virus entry. Since the FECV–FIPV transition involves a switch of target cell tropism, mutations in the S gene alone or in combination with changes in other genes may contribute to the biotype switch. To address this possibility, recent studies investigated the involvement of S gene mutations in FIP pathogenesis. An analysis of 11 FECV and 11 FIPV full-length genome sequences identified two point mutations in the S gene that can distinguish the vast majority of FIPVs from FECVs (Chang et al., 2012). To confirm this observation, the same research group investigated additional FECV and FIPV S gene sequences; the outcome was basically identical. The analyses showed that either one or both mutations were present in approximately 96% of the FIPV sequences while they were absent in all examined FECVs, providing strong evidence to suggest that these mutations correlate with the occurrence of FIP. One of the mutations leads to a Met-to-Leu substitution at amino acid position 1058 in the S protein (M1058L) and the other causes a Ser-to-Ala substitution (S1060A) (Chang et al., 2012). Since the affected residues are located in the putative fusion peptide of the S protein, it is tempting to speculate that amino acid changes in this region affect the cellular tropism of the virus, resulting in enhanced monocyte/macrophage tropism, a hypothesis that remains to be confirmed in additional experiments. Based on the observed sequence differences between the S genes of FECVs and FIPVs, a diagnostic assay for FIP diagnosis has been developed.

Porter et al. sequenced a short fragment of the S gene derived from fecal and tissue samples of both FECVs and FIPVs. In the majority of the fecal samples, the authors found methionine at position 1058 and in the majority of tissue samples leucine at position 1058, regardless of whether the cats were infected with FECV or FIPV. They concluded that the M1058L substitution represents a marker for systemic FCoV infection rather than a marker for FIP (Porter et al., 2014).

Another study investigating FECV–FIPV discriminatory mutations determined 3 FECV and 3 FIPV full-length genome sequences (Lewis et al., 2015). Similar to the observations described earlier (Chang et al., 2012), M1058L was identified as a fully discriminatory mutation between FECVs and FIPVs because it was exclusively present in the analyzed FIPV but not in any of the FECV samples. Interestingly, this work identified one more substitution suitable to discriminate between FECVs and FIPVs, an Ile-to-Thr substitution at position 1108 (I1108T) in the heptad repeat 1 (HR1) region, which was exclusively found in FIPVs. Amino acid substitutions in the HR1 region of FIPVs (but not of FECVs) have also been described by others (Bank-Wolf et al., 2014). As discussed by Lewis et al., it seems plausible that changes in the HR 1 region result in an altered fusogenic activity of the S protein which may affect the cellular tropism of the virus. However, it remains unclear whether the described mutations are relevant to FIP development.

In a recent study, the furin cleavage site located between the S1 and S2 domains of the S protein was investigated in FECV and FIPV samples (Licitra et al., 2013). While all FECVs were found to contain an intact and functional furin cleavage motif, as many as 10 out of 11 FIPVs contained amino acid substitutions at the cleavage site itself or in close proximity to the furin cleavage site. Fluorogenic peptide assays showed that the mutations identified in FIPVs affect the efficiency of furin-mediated S protein cleavage. Because the fusion activity of the coronaviral S protein generally requires activation by cellular proteases, substitutions at the protease cleavage site may indirectly affect viral spread and, thus, disease progression and the development of FIP (Bosch and Rottier, 2008).

The point mutations described earlier were detected in different regions of the S gene. Obviously, there is a strong correlation between the genetic changes and the occurrence of FIPV. However, it is important to emphasize that the FECV–FIPV substitutions were identified only via comparative sequence analyses and, so far, none of the assumed functional changes concerning cell tropism and biotype switch for FIP pathogenesis have been proved experimentally.

4. REVERSE GENETICS OF FELINE CORONAVIRUSES

Reverse genetics approaches are extremely valuable tools to produce recombinant FCoVs containing genetic changes suitable to investigate the role of specific viral proteins in the molecular pathogenesis of FIP. So far, three different reverse genetics systems have been described for FCoVs.

The very first system is based on targeted RNA recombination and was established for the highly virulent serotype II FIPV strain 79-1146 (Haijema et al., 2003). This system proved to be a very useful tool for modifying the

FCoV genome but, for technical reasons, only the 3′-terminal third of the genome is amenable to mutagenesis via this approach. By deleting the entire ORF3 (FIPV-Δ3), the entire ORF7 (FIPV-Δ7), or both (FIPV-Δ3Δ7), recombinant viruses were generated that displayed similar properties in cell culture to the parental virus. However, all of these recombinant viruses were attenuated in vivo and did not induce FIP. Furthermore, cats inoculated with FIPV-Δ3 or FIPV-Δ7 mutants were protected against a challenge with the parental virus, demonstrating that the accessory genes are dispensable for viral growth in vitro but contribute to virulence in vivo (Haijema et al., 2004). The same group investigated the genetic determinants for macrophage tropism of FIPV strain 79-1146. Parts of the S gene or the entire S gene of the FIPV strain 79-1146 were replaced with the corresponding S gene sequences derived from a cell culture-adapted serotype II FECV. Infection of macrophages with the recombinant viruses and the parental virus, respectively, revealed that the S protein alone was responsible for efficient macrophage infection and replication. Moreover, the C-terminal domain of the S protein was suggested as a key determinant for target cell tropism (Rottier et al., 2005).

Others have also used the recombinant viruses FIPV-Δ3, FIPV-Δ7, and FIPV-Δ3Δ7 to study certain aspects of FCoV biology. Dedeurwaerder et al. investigated the role of ORF3 and ORF7 for replication of FIPV in peripheral blood monocytes (Dedeurwaerder et al., 2013). They were able to show that only the FIPV-Δ3 and the parental virus but not FIPV-Δ3Δ7 and FIPV-Δ7 were able to maintain replication in monocytes. Accordingly, it was suggested that ORF7 is crucial for FIPV replication in monocytes and macrophages. In another study, this group addressed the question of whether 7a and 7b proteins interfere with the cellular innate immune system. Using the recombinant FIPV-Δ7 virus and a 7a-expressing plasmid construct, data were obtained to suggest a function of the 7a protein in counteracting IFN-α-induced antiviral responses (Dedeurwaerder et al., 2014).

Balint et al. established a bacterial artificial chromosome (BAC)-based reverse genetic system for serotype II FIPV strain DF2 (Balint et al., 2012). In contrast to the targeted RNA recombination system, the BAC system is suitable to mutagenize the entire FCoV genome. However, instability of the cloned FCoV cDNA in BAC may hinder the efficient generation of recombinant FCoVs. Strain DF2 is the cell culture-adapted variant of strain 79-1146, which is used in the only available vaccine against FIP (Kipar and Meli, 2014; Pedersen, 2009). The strain DF2 contains a 338-nt-long deletion in ORF3, resulting in truncated ORF3a and ORF3c proteins and deletion of the entire ORF3b. Using the BAC-based reverse genetic system, the authors generated a recombinant DF2 identical to the parental virus and a virus with fully restored ORF3 derived from CCoV. All recombinant viruses showed similar characteristics in established cell lines. However, after infection of peripheral blood monocytes, the virus with the fully restored ORF3 showed significantly lower replication compared to the virus in which ORF3 was deleted, suggesting that the ORF3 deletion may promote efficient virus replication in monocytes. In follow-up studies, the virulence of the recombinant viruses and the role of a fully restored ORF3 were assessed in experimental infections (Balint et al., 2014a,b). Only the recombinant virus containing an intact ORF3 was associated mainly with the gut and did not cause systemic infection. Accordingly, a pivotal role of ORF3 in establishing efficient infection of the intestine in vivo was suggested.

The third type of reverse genetic system relies on the integration of the entire coronaviral genome as a cDNA into the vaccinia virus genome, with the resulting recombinant vaccinia viruses serving as vectors to clone and manipulate the coronavirus cDNA insert (Casais et al., 2001; Coley et al., 2005; Thiel et al., 2001). The advantages of this system are that desired changes can be introduced at any position of the viral genome and, in contrast to *Escherichia coli*-based cloning systems, genetic instabilities of the full-length FCoV cDNA insert in the vaccinia virus genome have never been observed. However, complex and time-consuming procedures are required to manipulate the FCoV cDNA by vaccinia virus-mediated recombination and to produce (wild-type or mutant) genome-length FCoV RNA to be transfected into susceptible cells. Such a vaccinia virus-based system was reported for the serotype I FCoV strain Black (Tekes et al., 2008). This virus was isolated from a cat with FIP (Black, 1980). Interestingly, the recombinant serotype I FCoV strain Black and the virus isolate that was used to assemble the FCoV Black sequence did not induce FIP in SPF cats. Most likely, the propagation of the virus in tissue culture led to adaptive mutations that resulted in a nonpathogenic virus. The serotype I FCoV strain Black contains a stop codon in the accessory gene 7b that was thought to be an adaptive mutation responsible for the nonvirulent phenotype of this virus. Accordingly, a recombinant virus with a fully restored ORF 7b was generated and used for animal experiments; however, FIP was still not induced. In an attempt to generate a recombinant FCoV that reproducibly induces FIP, increasing portions of the recombinant FCoV Black genome were replaced with the homologous genome regions derived from the highly virulent serotype II FIPV strain 79-1146 (Tekes et al., 2012; Thiel et al., 2014). One of the chimeric viruses contained the S-3abc region, and another additionally possessed most of the ORF 1b of the serotype II FIPV in the serotype I virus backbone. Both viruses led to a systemic infection and induced high serum antibody titers, but FIP was not induced. Accordingly, the introduced parts of the serotype II genome were not sufficient to convert the nonpathogenic virus into a FIP-inducing virus, possibly because the virus contained additional attenuating mutations elsewhere in the serotype I backbone. Another explanation is an "incompatibility" of specific genome regions derived from serotype I FCoV strain Black and serotype II FIPV strain 79-1146, respectively, in inducing FIP. Finally, in order to generate a full-length recombinant serotype II FIPV strain 79-1146, the remaining parts of the serotype I backbone were replaced with the homologous regions of the serotype II FIPV. As expected, the recombinant serotype II FIPV and its parental virus induced FIP in experimentally infected SPF cats. In both the recombinant and the parental serotype II FIPV strain 79-1146, the ORF 3c contained a mutation causing a premature translational termination of this protein. Remarkably, sequence analyses of viral RNA originating from tissues of diseased cats revealed the restoration of ORF 3c. At first glance, this finding contradicts previous reports since an intact 3c is thought to be required for replication in the gut but dispensable for systemic infection (Chang et al., 2010; Pedersen et al., 2012). Interestingly, FIPV RNA with intact ORF 3c was detected in the fecal samples shortly before the cats succumbed to the disease, suggesting that FIPV replication also took place in the gut prior to death. It remains to be determined whether the restoration of ORF 3c is required to induce FIP after experimental infection with FIPV strain 79-1146.

5. PERSPECTIVES

Over the past few years, many aspects of FCoV biology have been studied, providing interesting new insight into FIP pathogenesis; however, a number of important questions remain to be addressed. This also applies to the emergence of FIPVs. It is now generally accepted that FIPV evolves from FECV through mutations being acquired in persistently infected animals. However, mutations that can convert a nonpathogenic FECV into a deadly virus have not been determined unambiguously, for example, by using reverse genetics approaches to produce and characterize mutants containing specific genetic changes predicted (by previous sequence analyses of

FCEV/FIPV pairs) to convey a FIP-inducing phenotype. In the past, various reverse genetic systems have been developed, but all of them are based either on cell culture-adapted serotype I or serotype II FCoVs that are not suitable for detecting mutations responsible for the biotype switch. Accordingly, there is an urgent need for a robust reverse genetic system that allows the production, characterization, and manipulation of serotype I field isolates.

There are two promising strategies for identifying mutations that can turn a nonpathogenic FECV into a FIP-inducing virus. One possibility is to introduce the described FECV–FIPV discriminatory mutations into a FECV field isolate using reverse genetics. So far, the most relevant data concerning FECV–FIPV discriminatory mutations originate from extensive comparative sequence analyses of serotype I FECV and FIPV field isolates. The identified mutations correlate with the switch of biotype and primarily concern the S gene. Accordingly, these discriminatory mutations should be introduced into the S gene of a serotype I FECV field isolate using reverse genetics and the virulence of the resulting viruses should be assessed in animal experiments.

We hypothesize that mutations in one gene or a combination of changes in different parts of the FCoV genome can lead to the emergence of FIPV. This assumption is supported by published data, which locate the mutations in four different regions of the S protein (predicted fusion peptide, HR1 region, furin cleavage site, and C-terminal region). All of these S gene mutations may lead to an altered cell tropism and finally to the development of FIP. However, sequence analyses also revealed substitutions located at many positions throughout the genomes of the corresponding FECV and FIPV pairs. Accordingly, it should be possible to identify mutations responsible for the biotype switch by generating and studying chimeric FCoVs. The starting materials for these viruses are "infectious clones" of genetically defined FECV–FIPV pairs. By replacing increasing parts of the FECV genome with the corresponding FIPV genome segments, the region of the FIPV genome that is able to convert the nonvirulent FECV into a FIP-inducing virus will be localized. However, this approach requires the growth of field viruses in standard cell culture systems, which has not been achieved so far. There is only one recently described feline enterocyte cell line that apparently allows propagation of serotype I FCoVs (Desmarets et al., 2013). It remains to be seen whether such a cell line can be used to grow FCoV field isolates to high titers or even for the identification of the cellular receptor(s) of serotype I FCoVs.

Another intriguing question relates to the role of accessory proteins in the FCoV life cycle. Due to the lack of appropriate tools, our knowledge about these proteins remains limited. So far, only the expression of 7b gene could be demonstrated in infected cells (Herrewegh et al., 1995b; Vennema et al., 1992, 1993). Furthermore, one publication suggested a role for the 7a protein as an interferon antagonist (Dedeurwaerder et al., 2014). There is increasing evidence that the accessory proteins are important for virulence in vivo, but the underlying molecular mechanisms are not understood. Also, FCoV accessory proteins may be required for viral persistence in specific cell types. Future studies are required to elucidate the functions of the accessory proteins and are expected to provide interesting insight into the molecular mechanisms that determine the pathogenesis, progression, and outcome of FCoV-induced diseases.

REFERENCES

Addie, D.D., Jarrett, O., 1990. Control of feline coronavirus infection in kittens. Vet. Rec. 126, 164.

Addie, D.D., Jarrett, O., 1992. A study of naturally occurring feline coronavirus infections in kittens. Vet. Rec. 130, 133–137.

Addie, D.D., Schaap, I.A., Nicolson, L., Jarrett, O., 2003. Persistence and transmission of natural type I feline coronavirus infection. J. Gen. Virol. 84, 2735–2744.

Amer, A., Siti Suri, A., Abdul Rahman, O., Mohd, H.B., Faruku, B., Saeed, S., et al., 2012. Isolation and molecular characterization of type I and type II feline coronavirus in Malaysia. Virol. J. 9, 278.

An, D.J., Jeoung, H.Y., Jeong, W., Park, J.Y., Lee, M.H., Park, B.K., 2011. Prevalence of Korean cats with natural feline coronavirus infections. Virol. J. 8, 455.

Balint, A., Farsang, A., Zadori, Z., Hornyak, A., Dencso, L., Almazan, F., et al., 2012. Molecular characterization of feline infectious peritonitis virus strain DF-2 and studies of the role of ORF3abc in viral cell tropism. J. Virol. 86, 6258–6267.

Balint, A., Farsang, A., Szeredi, L., Zadori, Z., Belak, S., 2014a. Recombinant feline coronaviruses as vaccine candidates confer protection in SPF but not in conventional cats. Vet. Microbiol. 169, 154–162.

Balint, A., Farsang, A., Zadori, Z., Belak, S., 2014b. Comparative in vivo analysis of recombinant type II feline coronaviruses with truncated and completed ORF3 region. PLoS One 9, e88758.

Bank-Wolf, B.R., Stallkamp, I., Wiese, S., Moritz, A., Tekes, G., Thiel, H.J., 2014. Mutations of 3c and spike protein genes correlate with the occurrence of feline infectious peritonitis. Vet. Microbiol. 173, 177–188.

Barker, E.N., Tasker, S., Gruffydd-Jones, T.J., Tuplin, C.K., Burton, K., Porter, E., et al., 2013. Phylogenetic analysis of feline coronavirus strains in an epizootic outbreak of feline infectious peritonitis. J. Vet. Intern. Med. 27, 445–450.

Benetka, V., Kubber-Heiss, A., Kolodziejek, J., Nowotny, N., Hofmann-Parisot, M., Mostl, K., 2004. Prevalence of feline coronavirus types I and II in cats with histopathologically verified feline infectious peritonitis. Vet. Microbiol. 99, 31–42.

Black, J.W., 1980. Recovery and in vitro cultivation of a coronavirus from laboratory-induced cases of feline infectious peritonitis (FIP). Vet. Med. Small Anim. Clin. 75, 811–814.

Bosch, B.J., Rottier, P.J., 2008. Nidovirus entry into cells. In: Perlman, S., Gallagher, T., Snijder, E. (Eds.), Nidoviruses. ASM Press, Washington, DC, pp. 157–178.

Bosch, B.J., van der Zee, R., de Haan, C.A., Rottier, P.J., 2003. The coronavirus spike protein is a class I virus fusion protein: structural and functional characterization of the fusion core complex. J. Virol. 77, 8801–8811.

Brown, M.A., Troyer, J.L., Pecon-Slattery, J., Roelke, M.E., O'Brien, S.J., 2009. Genetics and pathogenesis of feline infectious peritonitis virus. Emerg. Infect. Dis. 15, 1445–1452.

Casais, R., Thiel, V., Siddell, S.G., Cavanagh, D., Britton, P., 2001. Reverse genetics system for the avian coronavirus infectious bronchitis virus. J. Virol. 75, 12359–12369.

Chang, H.W., de Groot, R.J., Egberink, H.F., Rottier, P.J., 2010. Feline infectious peritonitis: insights into feline coronavirus pathobiogenesis and epidemiology based on genetic analysis of the viral 3c gene. J. Gen. Virol. 91, 415–420.

Chang, H.W., Egberink, H.F., Rottier, P.J., 2011. Sequence analysis of feline coronaviruses and the circulating virulent/avirulent theory. Emerg. Infect. Dis. 17, 744–746.

Chang, H.W., Egberink, H.F., Halpin, R., Spiro, D.J., Rottier, P.J., 2012. Spike protein fusion peptide and feline coronavirus virulence. Emerg. Infect. Dis. 18, 1089–1095.

Coley, S.E., Lavi, E., Sawicki, S.G., Fu, L., Schelle, B., Karl, N., et al., 2005. Recombinant mouse hepatitis virus strain A59 from cloned, full-length cDNA replicates to high titers in vitro and is fully pathogenic in vivo. J. Virol. 79, 3097–3106.

de Groot, R.J., Andeweg, A.C., Horzinek, M.C., Spaan, W.J., 1988. Sequence analysis of the 3'-end of the feline coronavirus FIPV 79-1146 genome: comparison with the genome of porcine coronavirus TGEV reveals large insertions. Virology 167, 370–376.

de Groot, R.J., Baker, S.C., Baric, R., Enjuanes, L., Gorbalenya, A.E., Holmes, K.V., et al., 2012. Family Coronaviridae. In: King, A.M.Q., Adams, M.J., Carstens, E.B., Lefkowitz, E.J. (Eds.), Virus Taxonomy. Elsevier, Amsterdam, pp. 806–828.

de Groot-Mijnes, J.D., van Dun, J.M., van der Most, R.G., de Groot, R.J., 2005. Natural history of a recurrent feline coronavirus infection and the role of cellular immunity in survival and disease. J. Virol. 79, 1036–1044.

de Haan, C.A., Haijema, B.J., Boss, D., Heuts, F.W., Rottier, P.J., 2005. Coronaviruses as vectors: stability of foreign gene expression. J. Virol. 79, 12742–12751.

Dean, G.A., Olivry, T., Stanton, C., Pedersen, N.C., 2003. In vivo cytokine response to experimental feline infectious peritonitis virus infection. Vet. Microbiol. 97, 1–12.

Decaro, N., Buonavoglia, C., 2008. An update on canine coronaviruses: viral evolution and pathobiology. Vet. Microbiol. 132, 221–234.

Dedeurwaerder, A., Desmarets, L.M., Olyslaegers, D.A., Vermeulen, B.L., Dewerchin, H.L., Nauwynck, H.J., 2013. The role of accessory proteins in the replication of feline infectious peritonitis virus in peripheral blood monocytes. Vet. Microbiol. 162, 447–455.

Dedeurwaerder, A., Olyslaegers, D.A., Desmarets, L.M., Roukaerts, I.D., Theuns, S., Nauwynck, H.J., 2014. ORF7-encoded accessory protein 7a of feline infectious peritonitis virus as a counteragent against IFN-alpha-induced antiviral response. J. Gen. Virol. 95, 393–402.

Desmarets, L.M., Theuns, S., Olyslaegers, D.A., Dedeurwaerder, A., Vermeulen, B.L., Roukaerts, I.D., et al., 2013. Establishment of feline intestinal epithelial cell cultures for the propagation and study of feline enteric coronaviruses. Vet. Res. 44, 71.

Desmarets, L.M., Vermeulen, B.L., Theuns, S., Conceicao-Neto, N., Zeller, M., Roukaerts, I.D., et al., 2016. Experimental feline enteric coronavirus infection reveals an aberrant infection pattern and shedding of mutants with impaired infectivity in enterocyte cultures. Sci. Rep. 6, 20022.

Dewerchin, H.L., Cornelissen, E., Nauwynck, H.J., 2005. Replication of feline coronaviruses in peripheral blood monocytes. Arch. Virol. 150, 2483–2500.

Drechsler, Y., Alcaraz, A., Bossong, F.J., Collisson, E.W., Diniz, P.P., 2011. Feline coronavirus in multicat environments. Vet. Clin. North Am. Small Anim. Pract. 41, 1133–1169.

Dye, C., Siddell, S.G., 2005. Genomic RNA sequence of Feline coronavirus strain FIPV WSU-79/1146. J. Gen. Virol. 86, 2249–2253.

Dye, C., Temperton, N., Siddell, S.G., 2007. Type I feline coronavirus spike glycoprotein fails to recognize aminopeptidase N as a functional receptor on feline cell lines. J. Gen. Virol. 88, 1753–1760.

Foley, J.E., Poland, A., Carlson, J., Pedersen, N.C., 1997. Risk factors for feline infectious peritonitis among cats in multiple-cat environments with endemic feline enteric coronavirus. J. Am. Vet. Med. Assoc. 210, 1313–1318.

Golovko, L., Lyons, L.A., Liu, H., Sorensen, A., Wehnert, S., Pedersen, N.C., 2013. Genetic susceptibility to feline infectious peritonitis in Birman cats. Virus Res. 175, 58–63.

Gunn-Moore, D.A., Gruffydd-Jones, T.J., Harbour, D.A., 1998. Detection of feline coronaviruses by culture and reverse transcriptase-polymerase chain reaction of blood samples from healthy cats and cats with clinical feline infectious peritonitis. Vet. Microbiol. 62, 193–205.

Haijema, B.J., Volders, H., Rottier, P.J., 2003. Switching species tropism: an effective way to manipulate the feline coronavirus genome. J. Virol. 77, 4528–4538.

Haijema, B.J., Volders, H., Rottier, P.J., 2004. Live, attenuated coronavirus vaccines through the directed deletion of group-specific genes provide protection against feline infectious peritonitis. J. Virol. 78, 3863–3871.

Haijema, B.J., Rottier, P.J.M., de Groot, R.J., 2007. Feline Coronaviruses: a tale of two-faced types. In: Thiel, V. (Ed.), Coronaviruses: Molecular and Cellular Biology. Caister Academic Press, Norfolk, pp. 182–203.

Hartmann, K., 2005. Feline infectious peritonitis. Vet. Clin. North Am. Small Anim. Pract. 35, 39–79.

Hayashi, T., Goto, N., Takahashi, R., Fujiwara, K., 1977. Systemic vascular lesions in feline infectious peritonitis. Nihon Juigaku Zasshi 39, 365–377.

Herrewegh, A.A., de Groot, R.J., Cepica, A., Egberink, H.F., Horzinek, M.C., Rottier, P.J., 1995a. Detection of feline coronavirus RNA in feces, tissues, and body fluids of naturally infected cats by reverse transcriptase PCR. J. Clin. Microbiol. 33, 684–689.

Herrewegh, A.A., Vennema, H., Horzinek, M.C., Rottier, P.J., de Groot, R.J., 1995b. The molecular genetics of feline coronaviruses: comparative sequence analysis of the ORF7a/7b transcription unit of different biotypes. Virology 212, 622–631.

Herrewegh, A.A., Mahler, M., Hedrich, H.J., Haagmans, B.L., Egberink, H.F., Horzinek, M.C., et al., 1997. Persistence and evolution of feline coronavirus in a closed cat-breeding colony. Virology 234, 349–363.

Herrewegh, A.A., Smeenk, I., Horzinek, M.C., Rottier, P.J., de Groot, R.J., 1998. Feline coronavirus type II strains 79-1683 and 79-1146 originate from a double recombination between feline coronavirus type I and canine coronavirus. J. Virol. 72, 4508–4514.

Hofmann-Lehmann, R., Fehr, D., Grob, M., Elgizoli, M., Packer, C., Martenson, J.S., et al., 1996. Prevalence of antibodies to feline parvovirus, calicivirus, herpesvirus, coronavirus, and immunodeficiency virus and of feline leukemia virus antigen and the interrelationship of these viral infections in free-ranging lions in east Africa. Clin. Diagn. Lab. Immunol. 3, 554–562.

Hohdatsu, T., Okada, S., Ishizuka, Y., Yamada, H., Koyama, H., 1992. The prevalence of types I and II feline coronavirus infections in cats. J. Vet. Med. Sci. 54, 557–562.

Hohdatsu, T., Izumiya, Y., Yokoyama, Y., Kida, K., Koyama, H., 1998. Differences in virus receptor for type I and type II feline infectious peritonitis virus. Arch. Virol. 143, 839–850.

Hsieh, L.E., Chueh, L.L., 2014. Identification and genotyping of feline infectious peritonitis-associated single nucleotide polymorphisms in the feline interferon-gamma gene. Vet. Res. 45, 57.

Hsieh, L.E., Huang, W.P., Tang, D.J., Wang, Y.T., Chen, C.T., Chueh, L.L., 2013. 3C protein of feline coronavirus inhibits viral replication independently of the autophagy pathway. Res. Vet. Sci. 95, 1241–1247.

Kennedy, M., Boedeker, N., Gibbs, P., Kania, S., 2001. Deletions in the 7a ORF of feline coronavirus associated with an epidemic of feline infectious peritonitis. Vet. Microbiol. 81, 227–234.

Kennedy, M., Citino, S., McNabb, A.H., Moffatt, A.S., Gertz, K., Kania, S., 2002. Detection of feline coronavirus in captive Felidae in the USA. J. Vet. Diagn. Invest. 14, 520–522.

Kennedy, M., Abd-Eldaim, M., Zika, S.E., Mankin, J.M., Kania, S.A., 2008. Evaluation of antibodies against feline coronavirus 7b protein for diagnosis of feline infectious peritonitis in cats. Am. J. Vet. Res. 69, 1179–1182.

Kenny, A.J., Maroux, S., 1982. Topology of microvillar membrance hydrolases of kidney and intestine. Physiol. Rev. 62, 91–128.

Kipar, A., Meli, M.L., 2014. Feline infectious peritonitis: still an enigma? Vet. Pathol. 51, 505–526.

Kipar, A., Bellmann, S., Kremendahl, J., Kohler, K., Reinacher, M., 1998a. Cellular composition, coronavirus antigen expression and production of specific antibodies in lesions in feline infectious peritonitis. Vet. Immunol. Immunopathol. 65, 243–257.

Kipar, A., Kremendahl, J., Addie, D.D., Leukert, W., Grant, C.K., Reinacher, M., 1998b. Fatal enteritis associated with coronavirus infection in cats. J. Comp. Pathol. 119, 1–14.

Kipar, A., May, H., Menger, S., Weber, M., Leukert, W., Reinacher, M., 2005. Morphologic features and development of granulomatous vasculitis in feline infectious peritonitis. Vet. Pathol. 42, 321–330.

Kipar, A., Baptiste, K., Barth, A., Reinacher, M., 2006a. Natural FCoV infection: cats with FIP exhibit significantly higher viral loads than healthy infected cats. J. Feline Med. Surg. 8, 69–72.

Kipar, A., Meli, M.L., Failing, K., Euler, T., Gomes-Keller, M.A., Schwartz, D., et al., 2006b. Natural feline coronavirus infection: differences in cytokine patterns in association with the outcome of infection. Vet. Immunol. Immunopathol. 112, 141–155.

Kipar, A., Meli, M.L., Baptiste, K.E., Bowker, L.J., Lutz, H., 2010. Sites of feline coronavirus persistence in healthy cats. J. Gen. Virol. 91, 1698–1707.

Kiss, I., Poland, A.M., Pedersen, N.C., 2004. Disease outcome and cytokine responses in cats immunized with an avirulent feline infectious peritonitis virus (FIPV)-UCD1 and challenge-exposed with virulent FIPV-UCD8. J. Feline Med. Surg. 6, 89–97.

Kubo, H., Yamada, Y.K., Taguchi, F., 1994. Localization of neutralizing epitopes and the receptor-binding site within the amino-terminal 330 amino acids of the murine coronavirus spike protein. J. Virol. 68, 5403–5410.

Kummrow, M., Meli, M.L., Haessig, M., Goenczi, E., Poland, A., Pedersen, N.C., et al., 2005. Feline coronavirus serotypes 1 and 2: seroprevalence and association with disease in Switzerland. Clin. Diagn. Lab. Immunol. 12, 1209–1215.

Leutenegger, C.M., Hofmann-Lehmann, R., Riols, C., Liberek, M., Worel, G., Lups, P., et al., 1999. Viral infections in free-living populations of the European wildcat. J. Wildl. Dis. 35, 678–686.

Lewis, C.S., Porter, E., Matthews, D., Kipar, A., Tasker, S., Helps, C.R., et al., 2015. Genotyping coronaviruses associated with feline infectious peritonitis. J. Gen. Virol. 96, 1358–1368.

Licitra, B.N., Millet, J.K., Regan, A.D., Hamilton, B.S., Rinaldi, V.D., Duhamel, G.E., et al., 2013. Mutation in spike protein cleavage site and pathogenesis of feline coronavirus. Emerg. Infect. Dis. 19, 1066–1073.

Lin, C.N., Su, B.L., Huang, H.P., Lee, J.J., Hsieh, M.W., Chueh, L.L., 2009. Field strain feline coronaviruses with small deletions in ORF7b associated with both enteric infection and feline infectious peritonitis. J. Feline Med. Surg. 11, 413–419.

Lin, C.N., Chang, R.Y., Su, B.L., Chueh, L.L., 2013. Full genome analysis of a novel type II feline coronavirus NTU156. Virus Genes 46, 316–322.

Look, A.T., Ashmun, R.A., Shapiro, L.H., Peiper, S.C., 1989. Human myeloid plasma membrane glycoprotein CD13 (gp150) is identical to aminopeptidase N. J. Clin. Invest. 83, 1299–1307.

Lorusso, A., Decaro, N., Schellen, P., Rottier, P.J., Buonavoglia, C., Haijema, B.J., et al., 2008. Gain, preservation, and loss of a group 1a coronavirus accessory glycoprotein. J. Virol. 82, 10312–10317.

Meli, M., Kipar, A., Muller, C., Jenal, K., Gonczi, E., Borel, N., et al., 2004. High viral loads despite absence of clinical and pathological findings in cats experimentally infected with feline coronavirus (FCoV) type I and in naturally FCoV-infected cats. J. Feline Med. Surg. 6, 69–81.

Munson, L., Marker, L., Dubovi, E., Spencer, J.A., Evermann, J.F., O'Brien, S.J., 2004. Serosurvey of viral infections in free-ranging Namibian cheetahs (Acinonyx jubatus). J. Wildl. Dis. 40, 23–31.

Narayanan, K., Huang, C., Makino, S., 2008. Coronavirus accessory proteins. In: Perlman, S., Gallagher, T., Snijder, E. (Eds.), Nidoviruses. ASM Press, Washington, DC, pp. 235–244.

Olyslaegers, D.A., Dedeurwaerder, A., Desmarets, L.M., Vermeulen, B.L., Dewerchin, H.L., Nauwynck, H.J., 2013. Altered expression of adhesion molecules on peripheral blood leukocytes in feline infectious peritonitis. Vet. Microbiol. 166, 438–449.

Oostra, M., de Haan, C.A., de Groot, R.J., Rottier, P.J., 2006. Glycosylation of the severe acute respiratory syndrome coronavirus triple-spanning membrane proteins 3a and M. J. Virol. 80, 2326–2336.

Paul-Murphy, J., Work, T., Hunter, D., McFie, E., Fjelline, D., 1994. Serologic survey and serum biochemical reference ranges of the free-ranging mountain lion (Felis concolor) in California. J. Wildl. Dis. 30, 205–215.

Pedersen, N.C., 1987. Virologic and immunologic aspects of feline infectious peritonitis virus infection. Adv. Exp. Med. Biol. 218, 529–550.

Pedersen, N.C., 2009. A review of feline infectious peritonitis virus infection: 1963–2008. J. Feline Med. Surg. 11, 225–258.

Pedersen, N.C., 2014. An update on feline infectious peritonitis: virology and immunopathogenesis. Vet. J. 201, 123–132.

Pedersen, N.C., Black, J.W., 1983. Attempted immunization of cats against feline infectious peritonitis, using avirulent live virus or sublethal amounts of virulent virus. Am. J. Vet. Res. 44, 229–234.

Pedersen, N.C., Boyle, J.F., Floyd, K., 1981a. Infection studies in kittens, using feline infectious peritonitis virus propagated in cell culture. Am. J. Vet. Res. 42, 363–367.

Pedersen, N.C., Boyle, J.F., Floyd, K., Fudge, A., Barker, J., 1981b. An enteric coronavirus infection of cats and its relationship to feline infectious peritonitis. Am. J. Vet. Res. 42, 368–377.

Pedersen, N.C., Evermann, J.F., McKeirnan, A.J., Ott, R.L., 1984. Pathogenicity studies of feline coronavirus isolates 79-1146 and 79-1683. Am. J. Vet. Res. 45, 2580–2585.

Pedersen, N.C., Allen, C.E., Lyons, L.A., 2008. Pathogenesis of feline enteric coronavirus infection. J. Feline Med. Surg. 10, 529–541.

Pedersen, N.C., Liu, H., Dodd, K.A., Pesavento, P.A., 2009. Significance of coronavirus mutants in feces and diseased tissues of cats suffering from feline infectious peritonitis. Viruses 1, 166–184.

Pedersen, N.C., Liu, H., Scarlett, J., Leutenegger, C.M., Golovko, L., Kennedy, H., et al., 2012. Feline infectious peritonitis: role of the feline coronavirus 3c gene in intestinal tropism and pathogenicity based upon isolates from resident and adopted shelter cats. Virus Res. 165, 17–28.

Pesteanu-Somogyi, L.D., Radzai, C., Pressler, B.M., 2006. Prevalence of feline infectious peritonitis in specific cat breeds. J. Feline Med. Surg. 8, 1–5.

Poland, A.M., Vennema, H., Foley, J.E., Pedersen, N.C., 1996. Two related strains of feline infectious peritonitis virus isolated from immunocompromised cats infected with a feline enteric coronavirus. J. Clin. Microbiol. 34, 3180–3184.

Porter, E., Tasker, S., Day, M.J., Harley, R., Kipar, A., Siddell, S.G., et al., 2014. Amino acid changes in the spike protein of feline coronavirus correlate with systemic spread of virus from the intestine and not with feline infectious peritonitis. Vet. Res. 45, 49.

Regan, A.D., Whittaker, G.R., 2008. Utilization of DC-SIGN for entry of feline coronaviruses into host cells. J. Virol. 82, 11992–11996.

Regan, A.D., Cohen, R.D., Whittaker, G.R., 2009. Activation of p38 MAPK by feline infectious peritonitis virus regulates pro-inflammatory cytokine production in primary blood-derived feline mononuclear cells. Virology 384, 135–143.

Regan, A.D., Ousterout, D.G., Whittaker, G.R., 2010. Feline lectin activity is critical for the cellular entry of feline infectious peritonitis virus. J. Virol. 84, 7917–7921.

Rottier, P.J., Nakamura, K., Schellen, P., Volders, H., Haijema, B.J., 2005. Acquisition of macrophage tropism during the pathogenesis of feline infectious peritonitis is determined by mutations in the feline coronavirus spike protein. J. Virol. 79, 14122–14130.

Semenza, G., 1986. Anchoring and biosynthesis of stalked brush border membrane proteins: glycosidases and peptidases of enterocytes and renal tubuli. Annu. Rev. Cell Biol. 2, 255–313.

Sharif, S., Arshad, S.S., Hair-Bejo, M., Omar, A.R., Zeenathul, N.A., Fong, L.S., et al., 2010. Descriptive distribution and phylogenetic analysis of feline infectious peritonitis virus isolates of Malaysia. Acta Vet. Scand. 52, 1.

Simons, F.A., Vennema, H., Rofina, J.E., Pol, J.M., Horzinek, M.C., Rottier, P.J., et al., 2005. A mRNA PCR for the diagnosis of feline infectious peritonitis. J. Virol. Methods 124, 111–116.

Stoddart, C.A., Scott, F.W., 1989. Intrinsic resistance of feline peritoneal macrophages to coronavirus infection correlates with in vivo virulence. J. Virol. 63, 436–440.

Takano, T., Hohdatsu, T., Hashida, Y., Kaneko, Y., Tanabe, M., Koyama, H., 2007a. A "possible" involvement of TNF-alpha in apoptosis induction in peripheral blood lymphocytes of cats with feline infectious peritonitis. Vet. Microbiol. 119, 121–131.

Takano, T., Hohdatsu, T., Toda, A., Tanabe, M., Koyama, H., 2007b. TNF-alpha, produced by feline infectious peritonitis virus (FIPV)-infected macrophages, upregulates expression of type II FIPV receptor feline aminopeptidase N in feline macrophages. Virology 364, 64–72.

Takano, T., Azuma, N., Satoh, M., Toda, A., Hashida, Y., Satoh, R., et al., 2009. Neutrophil survival factors (TNF-alpha, GM-CSF, and G-CSF) produced by macrophages in cats infected with feline infectious peritonitis virus contribute to the pathogenesis of granulomatous lesions. Arch. Virol. 154, 775–781.

Takano, T., Ohyama, T., Kokumoto, A., Satoh, R., Hohdatsu, T., 2011. Vascular endothelial growth factor (VEGF), produced by feline infectious peritonitis (FIP) virus-infected monocytes and macrophages, induces vascular permeability and effusion in cats with FIP. Virus Res. 158, 161–168.

Tekes, G., Hofmann-Lehmann, R., Stallkamp, I., Thiel, V., Thiel, H.J., 2008. Genome organization and reverse genetic analysis of a type I feline coronavirus. J. Virol. 82, 1851–1859.

Tekes, G., Hofmann-Lehmann, R., Bank-Wolf, B., Maier, R., Thiel, H.J., Thiel, V., 2010. Chimeric feline coronaviruses that encode type II spike protein on type I genetic background display accelerated viral growth and altered receptor usage. J. Virol. 84, 1326–1333.

Tekes, G., Spies, D., Bank-Wolf, B., Thiel, V., Thiel, H.J., 2012. A reverse genetics approach to study feline infectious peritonitis. J. Virol. 86, 6994–6998.

Terada, Y., Matsui, N., Noguchi, K., Kuwata, R., Shimoda, H., Soma, T., et al., 2014. Emergence of pathogenic coronaviruses in cats by homologous recombination between feline and canine coronaviruses. PLoS One 9, e106534.

Thiel, V., Herold, J., Schelle, B., Siddell, S.G., 2001. Infectious RNA transcribed in vitro from a cDNA copy of the human coronavirus genome cloned in vaccinia virus. J. Gen. Virol. 82, 1273–1281.

Thiel, V., Thiel, H.J., Tekes, G., 2014. Tackling feline infectious peritonitis via reverse genetics. Bioengineered 5, 396–400.

Tresnan, D.B., Holmes, K.V., 1998. Feline aminopeptidase N is a receptor for all group I coronaviruses. Adv. Exp. Med. Biol. 440, 69–75.

Tresnan, D.B., Levis, R., Holmes, K.V., 1996. Feline aminopeptidase N serves as a receptor for feline, canine, porcine, and human coronaviruses in serogroup I. J. Virol. 70, 8669–8674.

Van Hamme, E., Desmarets, L., Dewerchin, H.L., Nauwynck, H.J., 2011. Intriguing interplay between feline infectious peritonitis virus and its receptors during entry in primary feline monocytes. Virus Res. 160, 32–39.

Vennema, H., Rossen, J.W., Wesseling, J., Horzinek, M.C., Rottier, P.J., 1992. Genomic organization and expression of the 3' end of the canine and feline enteric coronaviruses. Virology 191, 134–140.

Vennema, H., Heijnen, L., Rottier, P.J., Horzinek, M.C., Spaan, W.J., 1993. A novel glycoprotein of feline infectious peritonitis coronavirus contains a KDEL-like endoplasmic reticulum retention signal. Adv. Exp. Med. Biol. 342, 209–214.

Vennema, H., Poland, A., Foley, J., Pedersen, N.C., 1998. Feline infectious peritonitis viruses arise by mutation from endemic feline enteric coronaviruses. Virology 243, 150–157.

Vogel, L., Van der Lubben, M., te Lintelo, E.G., Bekker, C.P., Geerts, T., Schuijff, L.S., et al., 2010. Pathogenic characteristics of persistent feline enteric coronavirus infection in cats. Vet. Res. 41, 71.

Wang, K., Lu, W., Chen, J., Xie, S., Shi, H., Hsu, H., et al., 2012. PEDV ORF3 encodes an ion channel protein and regulates virus production. FEBS Lett. 586, 384–391.

Wang, Y.T., Su, B.L., Hsieh, L.E., Chueh, L.L., 2013. An outbreak of feline infectious peritonitis in a Taiwanese shelter: epidemiologic and molecular evidence for horizontal transmission of a novel type II feline coronavirus. Vet. Res. 44, 57.

Wang, Y.T., Hsieh, L.E., Dai, Y.R., Chueh, L.L., 2014. Polymorphisms in the feline TNFA and CD209 genes are associated with the outcome of feline coronavirus infection. Vet. Res. 45, 123.

Weiss, R.C., Scott, F.W., 1981a. Antibody-mediated enhancement of disease in feline infectious peritonitis: comparisons with dengue hemorrhagic fever. Comp. Immunol. Microbiol. Infect. Dis. 4, 175–189.

Weiss, R.C., Scott, F.W., 1981b. Pathogenesis of feline infetious peritonitis: pathologic changes and immunofluorescence. Am. J. Vet. Res. 42, 2036–2048.

Worthing, K.A., Wigney, D.I., Dhand, N.K., Fawcett, A., McDonagh, P., Malik, R., et al., 2012. Risk factors for feline infectious peritonitis in Australian cats. J. Feline Med. Surg. 14, 405–412.

Yoo, D.W., Parker, M.D., Babiuk, L.A., 1991. The S2 subunit of the spike glycoprotein of bovine coronavirus mediates membrane fusion in insect cells. Virology 180, 395–399.

Zhang, R., Wang, K., Lv, W., Yu, W., Xie, S., Xu, K., et al., 2014. The ORF4a protein of human coronavirus 229E functions as a viroporin that regulates viral production. Biochim. Biophys. Acta 1838, 1088–1095.

Ziebuhr, J., 2005. The coronavirus replicase. Curr. Top. Microbiol. Immunol. 287, 57–94.

Ziebuhr, J., Snijder, E.J., Gorbalenya, A.E., 2000. Virus-encoded proteinases and proteolytic processing in the Nidovirales. J. Gen. Virol. 81, 853–879.

CHAPTER SEVEN

Interaction of SARS and MERS Coronaviruses with the Antiviral Interferon Response

E. Kindler*,†, V. Thiel*,†, F. Weber‡,1
*University of Bern, Bern, Switzerland
†Institute of Virology and Immunology, Bern and Mittelhäusern, Switzerland
‡Institute of Virology, Faculty of Veterinary Medicine, Justus Liebig University Giessen, Giessen, Germany
1Corresponding author: e-mail address: friedemann.weber@vetmed.uni-giessen.de

Contents

Abstract

Severe Acute Respiratory Syndrome (SARS) and Middle East Respiratory Syndrome (MERS) are the most severe coronavirus (CoV)-associated diseases in humans. The causative agents, SARS-CoV and MERS-CoV, are of zoonotic origin but may be transmitted to humans, causing severe and often fatal respiratory disease in their new host. The two coronaviruses are thought to encode an unusually large number of factors that allow them to thrive and replicate in the presence of efficient host defense mechanisms, especially the antiviral interferon system. Here, we review the recent progress in our understanding of the strategies that highly pathogenic coronaviruses employ to escape, dampen, or block the antiviral interferon response in human cells.

Advances in Virus Research, Volume 96
ISSN 0065-3527
http://dx.doi.org/10.1016/bs.aivir.2016.08.006
219

1. INTRODUCTION

Coronaviruses have made a remarkable career. Originally recognized as viral pathogens of veterinary importance but little medical (i.e., human) relevance, the appearance of SARS-CoV causing a worldwide epidemic with a large number of fatalities has changed everything. In 2003, the virus emerged in Chinese animal markets to circle the world in just a few weeks, teaching us important new lessons on perceived "differences" between animal and human pathogens. Just in case someone did not get the message, MERS-CoV repeated the coronavirus wake-up call 10 years later, providing yet another example for how easily animal viruses may be transmitted and adapt to new hosts including humans. Often, the tricks and strategies that viruses evolved to propagate in specific animal hosts may only need some fine-tuning (if at all) to enter the wide world of human crowds, air travel, and camel races. Here, we will summarize the insights gathered so far on an important aspect of virulence and host adaptation, the interactions of SARS-CoV and MERS-CoV with antiviral interferon (IFN) responses of human cells.

2. THE CORONAVIRUS GENOME

The coronavirus genome is composed of a linear, single-stranded, monopartite RNA with a cap structure at its $5'$ end and a polyA tail at the $3'$ end (Fehr and Perlman, 2015). The $5'$-terminal two-thirds of the CoV genome contain the open reading frames (ORF) 1a and 1b that together constitute the viral replicase gene. Translation is initiated at the start codon of ORF1a and may continue to ORF1b via a ribosomal frameshift mechanism, ultimately giving rise to two overlapping replicase polyproteins pp1a and pp1ab (Fehr and Perlman, 2015; Perlman and Netland, 2009; Snijder et al., 2003; Thiel et al., 2003). Virus-encoded proteinases, namely two papain-like cysteine proteases (PL1^{pro} and PL2^{pro}), residing in non-structural protein (nsp) 3 and a 3C-like cysteine protease (3CL^{pro}) associated with nsp5, proteolytically process the polyproteins into nsps 1–16 (Anand et al., 2003; Schiller et al., 1998; Thiel et al., 2003; Ziebuhr et al., 2007). A multitude of functions and enzymatic activities associated with specific nsps have been identified over the past years (for reviews, see Masters and Perlman, 2013; Ziebuhr, 2005). Moreover, ORF1b harbors several RNA-processing enzymes, including a $3'$–$5'$ exonuclease and a guanosine

N7-methyltransferase (associated with the N- and C-terminal domains, respectively, of nsp14), an endoribonuclease (nsp15) and a 2'-O-methyltransferase (nsp16) (Chen et al., 2009; Decroly et al., 2008; Fehr and Perlman, 2015; Ivanov et al., 2004; Kindler and Thiel, 2014; Minskaia et al., 2006; Perlman and Netland, 2009; Snijder et al., 2003; Thiel et al., 2003; Zust et al., 2011). The 3' ORFs are translated from a set of subgenomic (sg) RNAs and yield on one hand four canonical structural proteins like the spike protein (S), the envelope (E), the membrane (M), and the nucleoprotein (N). Moreover, sgRNAs express accessory genes, which vary in function and number between different CoV strains and are interspersed between the structural genes (Fehr and Perlman, 2015; Perlman and Netland, 2009; Snijder et al., 2003; Thiel et al., 2003). Specifically, the genome of SARS-CoV expresses eight different accessory genes (3a, 3b, 6, 7a, 7b, 8a, 8b, and 9b), while MERS-CoV encodes five accessory genes (3, 4a, 4b, 5, and 8b). The schematic overview of the genome organization of SARS-CoV and MERS-CoV is depicted in Fig. 1.

The CoV life cycle starts with the attachment of the viral spike protein to particular cellular receptors, subsequently leading to fusion between the viral envelope and the plasma membrane or the endosome membrane of the host. CoV uses a range of receptors, with SARS-CoV employing angiotensin-converting enzyme 2 (ACE2) and MERS-CoV employing dipeptyl peptidase 4 (DPP4) (Li et al., 2003; Raj et al., 2013). Following membrane fusion, the viral RNA genome is delivered into the host cytoplasm, where

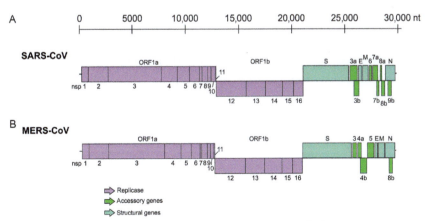

Fig. 1 Coronavirus genomes. Schematic representation of the genome regions encoding nonstructural (nsp), structural, and accessory proteins of SARS-CoV (A) and MERS-CoV (B).

translation of the two 5′-terminal ORFs 1a and 1b is accomplished by the cellular translation machinery. Most of the newly synthesized nsps assemble with the N protein into a replicase–transcriptase complex (RTC) responsible for viral genome replication and transcription. At the site of replicative organelles (Knoops et al., 2008), the RTC initiates minus-strand synthesis using the full-length genome as template, thereby either copying the entire template to generate full-length minus strands or to move discontinuously along the template to produce a nested set of sgRNAs with negative polarity. The minus strands of genomic and sgRNAs are subsequently used as templates to synthesize positive sense strands (mRNAs), specifically the genomic RNA (genome replication) and sg mRNAs (transcription) (Sawicki et al., 2007). The N protein then encapsidates the newly synthetized RNA genome and thereby forms a helical nucleocapsid. Virion assembly is triggered by the action of the M protein, which assists in incorporating the nucleocapsid, the envelope and the spike into virus particles. Budding takes place between the endoplasmic reticulum and the Golgi and new viruses are released by exocytosis (McBride et al., 2014; Neuman et al., 2011; Ruch and Machamer, 2012; Ujike and Taguchi, 2015).

3. THE TYPE I IFN SYSTEM

3.1 Types of IFNs and Their Signaling Pathways

The antiviral IFN (IFN-alpha/beta) system confers an important part of the innate immune defense in chordates (tenOever, 2016). IFNs are cytokines that are produced and secreted by cells encountering viruses or parts thereof (Fig. 2). Humans are able to express one IFN-beta, 13 subtypes of IFN-alphas, and one each of IFN-kappa and IFN-omega (Schneider et al., 2014). All nucleated cells are able to respond to them as they express the IFN receptor (composed of the two subunits IFNAR-1 and IFNAR-2) on their surface (Bekisz et al., 2004). The docking of IFN-alpha/beta onto its cognate receptor activates the so-called JAK–STAT pathway. Thereby, the Janus kinases JAK1 and TYK2 are waiting to be activated on the cytoplasmic side of IFNAR2 and 1, respectively. The activated kinases then phosphorylate the signal transducers and transcription factors STAT1 and STAT2, which form a complex with IRF9 (ISGF3) that enters the nucleus to transactivate promoters of an antiviral gene expression program. Genes that are specifically upregulated by IFNs are collectively called ISGs (IFN-stimulated genes).

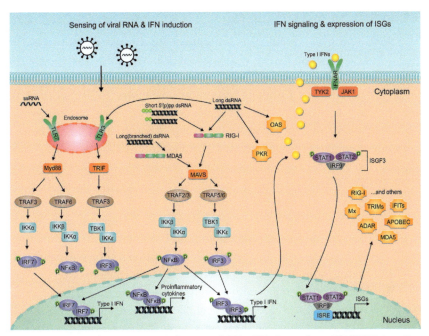

Fig. 2 The antiviral IFN system. Induction of IFNs, IFN-dependent JAK/STAT signaling, and ISG expression is depicted. For details, see text.

The alpha/beta-IFNs are classified as type I IFNs, since they had been discovered first (Isaacs and Lindenmann, 1957). The type II IFNs use a different receptor and consist of only one member, IFN-gamma. IFN-gamma also confers some antiviral activity but is regarded more of an immunoregulator (produced by specialized immune cells) than a general antiviral mediator (Schneider et al., 2014). It signals through a JAK/STAT pathway that partially overlaps with one of the type I IFNs. IFN-gamma will not be further discussed here as it is not to the core of the antiviral IFN response to coronaviruses.

Recently, the IFN family was extended by the newly discovered type III IFNs, consisting of IFN–lambda 1–4 (Schneider et al., 2014). Type III IFNs resemble type I IFNs in that they also trigger STAT1/2 phosphorylation via JAK1 and TYK2. They employ however a different receptor which is only expressed by epithelial cells (Sommereyns et al., 2008). Thus, type I and type III IFNs trigger largely overlapping sets of ISGs, but while the former constitute a major, general antiviral cytokine system, the latter are mainly restricted to mucosal sites (Galani et al., 2015).

3.2 Induction of Type I IFNs

The molecular events leading to the upregulation of type I IFNs are well established. As indicated earlier, molecular structures that are specific for virus infections (often called PAMPs, for pathogen-associated molecular patterns) are sensed by pathogen recognition receptors (PRRs) of the host, that in turn are triggering the upregulation of IFN genes. A prototypical PAMP relevant for coronaviruses is double-stranded RNA (dsRNA), a by-product of genome replication and transcription (Weber et al., 2006; Zielecki et al., 2013). dsRNA can be sensed by toll-like receptor 3 (TLR3) in the endosome, and in the cytoplasm by the RNA helicases RIG-I (retinoic acid-inducible gene I) and MDA5 (melanoma differentiation antigen 5), as well as by the kinase PKR (protein kinase, RNA-activated) (Rasmussen et al., 2009; Yim and Williams, 2014; Yoneyama et al., 2016). RIG-I is thereby specific for long dsRNA molecules and short dsRNAs bearing a tri- or di-phosphorylated 5′ end, whereas MDA5 senses long dsRNAs, preferentially with a higher-order structure (Binder et al., 2011; Goubau et al., 2014; Pichlmair et al., 2009; Schlee, 2013). PKR is activated by dsRNA as well as by short stem-loop RNAs bearing a 5′ triphosphate end (Dabo and Meurs, 2012; Nallagatla et al., 2011). Also specific single-stranded RNAs (ssRNAs) can act as PAMPs, either if they are in the wrong location or if they display particular features. TLR7 senses GU-rich ssRNA in the endosome (Heil et al., 2004). RIG-I can be triggered by polyU/UC rich or 3′ monophosphorylated ssRNAs stretches, and MDA5 was found to bind ssRNA stretches of negative-sense RNA viruses and hypomethylated 5′ capped mRNAs (Luthra et al., 2011; Malathi et al., 2007, 2010; Rasmussen et al., 2009; Runge et al., 2014; Saito et al., 2008; Zust et al., 2011).

Depending on the particular PRR, various-partially cross-talking-signaling pathways lead to the transactivation of promoters for antiviral genes (O'Neill et al., 2013). The endosomal PRR TLR3 engages the intracellular adapters TRIF (TIR domain-containing adapter protein inducing IFN-beta) and TRAF3 (TNF receptor-associated factor 3) to activate the kinases TBK1 (TANK-binding kinase 1) and IKKepsilon (inhibitor of NF-kappaB kinase epsilon). The kinases TBK1 and IKKepsilon then phosphorylate IRF3 (IFN regulatory factor 3), a transcription factor that activates genes for IFNs and other immunoregulatory cytokines. Signaling by TLR7, by contrast, requires the adaptor proteins MyD88 (myeloid differentiation primary-response protein 88) and TRAF3 to channel to the kinase

IKKalpha. This kinase then phosphorylates IRF7, a transcription factor that covers a gene spectrum similar to IRF3. TLR7/MyD88 also recruits the adaptor protein TRAF6 that eventually activates the transcription factor NF-kappaB via the kinases IKKalpha and IKKbeta. NF-kappaB drives transcription of genes for proinflammatory cytokines but also enhances IFN gene expression.

In the cytoplasm, RNA sensing by the two PRRs RIG-I and MDA5 (collectively termed RIG-I-like receptors, RLRs) converges on the adaptor protein MAVS (mitochondrial antiviral signaling protein) that uses various TRAFs (TRAF 2, 3, 5, 6) to trigger TBK1/IKKepsilon and IKKalpha/IKKbeta. These kinases then activate IRF3 and NF-kappaB, respectively (Belgnaoui et al., 2011; Liu et al., 2013). Besides the RLRs, PKR contributes to IFN induction in the cytoplasm. PKR is a master regulator of mRNA translation (see later), but several lines of evidence indicate a role in activation of NF-kappaB and IRF3 via TRAF2/6, IKKalpha/beta, antiviral stress granule formation, and IFN-alpha/beta mRNA stability (Gil et al., 2004; Onomoto et al., 2012; Pfaller et al., 2011; Pham et al., 2016; Schulz et al., 2010; Zamanian-Daryoush et al., 2000).

Thus, several types of PRRs are constantly surveying the extracellular and intracellular space to detect virus infections in a timely and sensitive manner. Importantly, TLRs are preferentially expressed by immune cells, especially myeloid dendritic cells (mDCs) and plasmacytoid cells (pDCs), whereas RLRs and PKR are thought to be active in all nucleated cells. Detection of viral RNA in mDCs is mainly mediated by TLR3 (and some TLR7), and in pDCs by TLR7 (and TLR8 in human pDCs) (Schreibelt et al., 2010).

PAMP sensing by PRRs eventually culminates in activation of IRF3, IRF7, and NF-kappaB, as described earlier, the transcription factors driving the expression of genes for IFN-beta, IFN-alpha, and various proinflammatory and immunomodulatory cytokines (Belgnaoui et al., 2011).

3.3 IFN-Stimulated Gene Expression

Signaling by both type I and type III IFNs triggers the formation of ISGF3 (see Section 3.1), the heterotrimeric transcription factor complex consisting of phosphorylated STAT1 and STAT2, and IRF9 (Schneider et al., 2014). ISGF3 binds to the ISREs (for IFN-stimulated response element), specific promoter sequences of the so-called ISGs. Of note, there are actually several types of "ISREs" that are responding to different types of triggers and transcription factors. First, there are the ISREs that purely respond to IFN signaling and ISGF3, as it would be expected from the name. A prominent example is given by the promoter of the human antiviral protein MxA (Holzinger et al., 2007). Second, there are the—somewhat mislabeled—ISREs that do not respond to IFN at all, but only to the IRF3-, IRF7-, and NF-kappaB-related signal transduction that occurs much earlier, directly after virus infection has triggered a PRR. The IFN-beta promoter belongs to this class of ISREs (Freaney et al., 2013; Schmid et al., 2010). Third, there are mixed-type ISREs that can be activated by both virus infection and IFNs. An example is the promoter of the gene for the antiviral protein IFIT1 (also known as ISG56) (Fensterl and Sen, 2015). The different ISRE classes can be distinguished as IRF-specific ISREs (responding only to PRR signaling), ISGF3-specific ISREs (responding only to type I or type III IFNs), and universal ISREs (responding to both infection and IFNs) (Schmid et al., 2010). Many ISGs are controlled by additional promoter elements ensuring basal levels of expression already in the absence of IFN. Moreover, low levels of IFN itself are constitutively secreted by many tissues (tonic IFN), ensuring physiological homeostasis and priming of cells for a rapid response against pathogens (Gough et al., 2012).

It is estimated that, depending on the IFN subtype, dose, and cell type, IFNs regulate hundreds, if not thousands of genes (Rusinova et al., 2013). Many of the ISGs (i.e., those genes that are upregulated by IFNs) are known to have antiviral, immunomodulatory, or antiproliferative function (Samuel, 2001; Stark and Darnell, 2012). The broad antiviral activity of IFNs occurs on several levels, namely virus entry, viral polymerase function, host cell translation, RNA availability, RNA stability, particle budding, apoptosis, or general boosting of innate and adaptive immune responses.

4. ANTIVIRAL ACTION OF IFNs AGAINST HUMAN CORONAVIRUSES

High-dose IFN treatment (type I and type III) has clear effects against SARS-CoV and MERS-CoV in cell culture (Chan et al., 2013; Cinatl et al., 2003; Falzarano et al., 2013; Kindler et al., 2013; Spiegel et al., 2004; Stroher et al., 2004; Zielecki et al., 2013), in animal experiments (Channappanavar et al., 2016; Frieman et al., 2010; Haagmans et al., 2004; Mahlakoiv et al., 2012; Mordstein et al., 2008), and possibly also in patients (Loutfy et al.,

2003; Omrani et al., 2014; Strayer et al., 2014). Remarkably, MERS-CoV was found to be substantially more IFN sensitive than SARS-CoV in cell culture (Zielecki et al., 2013).

The cellular basis for the relatively low (SARS-CoV) and high (MERS-CoV) IFN sensitivity is currently unknown. Several prominent (i.e., potent) ISG products were studied in the context of human pathogenic coronaviruses, but only some of them were found to have an effect. The IFN-induced transmembrane (IFITMs) proteins 1, 2, and 3 restrict the entry of many enveloped viruses including SARS-CoV (Huang et al., 2011b) as well as reoviruses (Bailey et al., 2014). They act by altering the site of membrane fusion, but the exact mechanism remains to be elucidated (Bailey et al., 2014). Strikingly, while IFITMs are inhibitory for the highly pathogenic SARS-CoV, they appear to boost infection with the related, low pathogenic coronavirus HCoV-OC43 (Zhao et al., 2014). In particular, IFITM2 or IFITM3 acts as entry factor for HCoV-OC43 by facilitating—rather than impeding—membrane fusion. Human MxA (for Myxovirus resistance protein A) is a well-known antiviral host factor with activity against a wide range of (mostly) RNA viruses (Haller et al., 2015). It blocks early replication steps of influenza viruses but was found have no effect on SARS-CoV (Spiegel et al., 2004). The kinase PKR is an ISG product acting as a signaling PRR on one hand (see earlier), but its main function in antiviral defense is the inhibition of protein synthesis. After binding viral dsRNA, PKR undergoes autophosphorylation to activate itself, and subsequently phosphorylates eIF-2alpha that is thereby converted from a translation initiation factor to a translation inhibitor (Yim and Williams, 2014). PKR has a broad antiviral spectrum. Nonetheless, PKR has no bearing on the replication of SARS-CoV, although it is involved in virally induced apoptosis (Krahling et al., 2009). Also the $2'-5'$ oligoadenylate synthetase (OAS) family members are triggered by viral dsRNA (Chakrabarti et al., 2011). In the dsRNA-bound state they synthesize short chains of $2'-5'$ oligoadenylates that activate the latent RNase L. RNase L then cleaves virus and host ssRNAs, predominantly at single-stranded UA and UU dinucleotides (Wreschner et al., 1981). Interestingly, the small $3'$-monophosphorylated cleavage products of RNase L are recognized by the PRRs RIG-I and MDA5, thus amplifying the IFN response in an infection-dependent manner (Malathi et al., 2007). Polymorphisms of the OAS-1 gene might affect susceptibility to SARS-CoV (Hamano et al., 2005), but to our knowledge, there is no direct data on antiviral effects of the OAS/RNase L system on human coronaviruses. For the mouse coronavirus MHV-A59, however, it was shown that mutants deficient in the ns2 gene are highly sensitive against RNase L (Zhao et al., 2012) (see also later).

As mentioned, there are several hundreds of ISGs, of which about 40 were characterized as being antiviral (Schneider et al., 2014). It is in a way remarkable that relatively little is known about ISGs that impede human pathogenic coronaviruses. Most likely, active and passive evasion mechanisms such as the ones described later are responsible for the relative insensitivity of at least SARS-CoV against IFN and potent antiviral ISGs. Although our review will focus on the human pathogenic coronaviruses SARS-CoV and MERS-CoV, we will draw additional conclusions from well investigated other coronaviruses whenever adequate.

5. EVASION STRATEGIES OF CORONAVIRUSES

Viral evasion strategies against the IFN response can act on several levels, namely the induction of IFN, IFN signaling, or antiviral action of individual ISG products (Gack, 2014; Kindler and Thiel, 2014; Vijay and Perlman, 2016; Weber and Weber, 2014; Wong et al., 2016; Zinzula and Tramontano, 2013). The viruses can thereby actively sequester or destroy key regulators, or otherwise interfere with the IFN system. Moreover, several aspects of the viral replication cycle can be regarded as a passive IFN evasion. The strategies described later are also summarized in three tables.

5.1 Inhibition of IFN Induction

Both SARS-CoV and MERS-CoV induce very little—if any—IFN in most cell types (Chan et al., 2013; Cheung et al., 2005; Kindler et al., 2013; Lau et al., 2013; Menachery et al., 2014a; Spiegel et al., 2005; Zhou et al., 2014; Ziegler et al., 2005; Zielecki et al., 2013). In fact, it was recently shown in a mouse model of SARS that the delay in IFN induction is responsible for the activation of proinflammatory monocyte-macrophages and cytokines in the lung, resulting in vascular leakage and impaired adaptive immune responses (Channappanavar et al., 2016). Thus, the high levels of dsRNA that are produced during replication (Weber et al., 2006; Zielecki et al., 2013) do not result in an adequate IFN induction. One of the reasons (besides the active measures described later) is certainly the storage of coronaviral dsRNA inside double-membrane vesicles (Knoops et al., 2008; van Hemert et al., 2008; Versteeg et al., 2007). Moreover, the N protein sequesters IFN-inducing RNA PAMPs (Kopecky-Bromberg et al., 2007; Lu et al., 2011). However, the fact that infection with coronaviruses activates the cytosolic dsRNA-sensing host factors PKR and OAS (Birdwell et al., 2016; Krahling et al., 2009; Zhao et al., 2012), as well as the existence of numerous mechanisms dedicated to suppress dsRNA-dependent IFN induction (see later) strongly suggest that dsRNA stashing alone is not sufficient and that some dsRNA or other PAMPs are exposed to PRRs, thus necessitating the presence of additional, active mechanisms.

While most cell types remain IFN-silent after infection, a notable exception are pDCs, which express high levels of IFN-alpha/beta in response to infection with both SARS-CoV and MERS-CoV (Cervantes-Barragan et al., 2007; Channappanavar et al., 2016; Scheuplein et al., 2015). For the mouse coronavirus MHV-A59 it was shown that IFN induction in pDCs occurs through TLR7 (Cervantes-Barragan et al., 2007), suggesting the same to be true for SARS-CoV and MERS-CoV. Indeed, GU-rich ssRNAs from the SARS-CoV genome were shown to activate an excessive innate immune response via TLR7 (Li et al., 2013). Moreover, the membrane (M) protein and the envelope (E) protein of SARS-CoV are able to activate a TLR-like pathway and NF-kappaB signaling, respectively (DeDiego et al., 2014; Wang and Liu, 2016).

The mouse coronavirus MHV-A59 also naturally induces IFN in brain macrophages/microglia, with MDA5 being the responsible PRR (Birdwell et al., 2016; Roth-Cross et al., 2008). Also in oligodendrocytes IFN induction by MHV occurs through both MDA5 and RIG-I (Li et al., 2010). Interestingly, a general (i.e., not restricted to particular cell types) MDA5-dependent IFN induction can be obtained by ablating the ribose 2'-O-methylation activity of the nsp16. As it was shown for MHV-A59, SARS-CoV, and the mildly human pathogenic coronavirus HCoV-229E, nsp16-mediated 2'-O-methylation of viral mRNA cap structures prevents recognition by MDA5 (Menachery et al., 2014b; Zust et al., 2011).

Besides these "hiding" or "disguising" strategies, active mechanisms targeting specific host factors are in place (Table 1). SARS-CoV was shown to inhibit IRF3 by preventing its hyperphosphorylation, dimerization, and interaction with the cofactor CBP (Spiegel et al., 2005). Curiously, IRF3 initially enters the nucleus of infected cells, but later returns to the cytoplasm. SARS-CoV also inhibits the nuclear import of the related transcription factor IRF7 (Kuri et al., 2009). In this context, the papain-like protease (PL^{pro}) domain of nsp3 (the largest coronaviral protein) of SARS-CoV and the mildly pathogenic HCoV-NL63 both interact with IRF3 and block its activation (Devaraj et al., 2007; Frieman et al., 2009). Moreover, PL^{pro} was shown to drive the deubiquitination (or inhibit ubiquitination) of RIG-I,

Table 1 Mechanisms and Factors of Human Coronaviruses to Counteract IFN Induction

Virus	Viral Protein or Function	Mechanism	References
SARS-CoV (MHV-A59)	Storage of dsRNA inside double-membrane vesicles	Prevents exposure of dsRNA to PRRs	Knoops et al. (2008), van Hemert et al. (2008), and Versteeg et al. (2007)
SARS-CoV	N	Sequesters IFN-inducing RNA PAMPs	Kopecky-Bromberg et al. (2007) and Lu et al. (2011)
SARS-CoV, HCoV-229E (MHV-A59)	nsp16	Ribose 2′-O-methylation of viral mRNA cap structures prevents recognition by MDA5	Menachery et al. (2014b) and Zust et al. (2011)
SARS-CoV, NL63	PL^{pro}	Interacts with IRF3, inhibits IRF3 activation, deubiquitinates RIG-I, TBK1, IRF3	Clementz et al. (2010), Devaraj et al. (2007), Frieman et al. (2009), and Sun et al. (2012)
SARS-CoV	M	Inhibits TRAF3/TBK1 complex formation	Siu et al. (2009)
SARS-CoV	nsp7, nsp15, ORF3b, ORF6	Mechanism unclear	Frieman et al. (2009) and Kopecky-Bromberg et al. (2007)
SARS-CoV	nsp1	Mediates host mRNA degradation	Huang et al. (2011a) and Narayanan et al. (2008)
SARS-CoV	nsp1	Blocks host mRNA translation	Narayanan et al. (2008) and Tanaka et al. (2012)
SARS-CoV	ORF9b protein	Proteasomal degradation of MAVS, TRAF3, and TRAF6	Shi et al. (2014)
MERS-CoV	ORF4a protein	Interacts with dsRNA and the RLR cofactor PACT	Niemeyer et al. (2013)
MERS-CoV	ORF4a protein	Interacts with the RLR cofactor PACT	Siu et al. (2014)

Table 1 Mechanisms and Factors of Human Coronaviruses to Counteract IFN Induction—cont'd

Virus	Viral Protein or Function	Mechanism	References
MERS-CoV	ORF4a, 4b, and ORF5 proteins, M	Prevent IRF3 translocation	Yang et al. (2013)
MERS-CoV	ORF4b protein	Binds TBK1 and IKKepsilon	Matthews et al. (2014) and Yang et al. (2015)
MERS-CoV	PL^{pro}	Deubiquitination	Bailey-Elkin et al. (2014) and Mielech et al. (2014)
MERS-CoV	nsp1	Degrades host mRNAs	Lokugamage et al. (2015)
MERS-CoV	Unknown	Repressive histone modifications	Menachery et al. (2014a)

TBK1, and IRF3 (Clementz et al., 2010; Devaraj et al., 2007; Frieman et al., 2009; Sun et al., 2012). IRF3 activation is also prevented by the M protein of SARS-CoV through inhibiting complex formation between TRAF3 and TBK1 (Siu et al., 2009). Since M was also found to activate a TLR-like signaling pathway (Wang and Liu, 2016), a final picture of M protein function in the context of IFN induction/inhibition remains to be provided. IFN induction is also disturbed by the SARS-CoV nsp1, nsp7, nsp15, ORF3b, ORF6, and ORF9b proteins, respectively (Frieman et al., 2009; Kopecky-Bromberg et al., 2007; Shi et al., 2014; Zust et al., 2007). The anti-IFN function of nsp1 is based on its ability to mediate host mRNA degradation, while sparing viral mRNAs at the same time, and to block host mRNA translation (Huang et al., 2011a; Narayanan et al., 2008; Tanaka et al., 2012). Nsp1 also has a function in evasion from IFN signaling (see later), providing a possible reason why nsp1 mutants are particularly IFN sensitive (Wathelet et al., 2007; Zust et al., 2007). While the mechanisms of other SARS-CoV IFN induction antagonists like nsp7, nsp15, ORF3b, and ORF6 proteins remain to be characterized, for the ORF9b protein it was shown that it drives degradation of MAVS, TRAF3, and TRAF6 by interacting with the host factors PCBP2 and the E3 ubiquitin ligase AIP4 (Shi et al., 2014).

Also for MERS-CoV, the reason for the low levels of IFN produced by infected cells (Chan et al., 2013; Kindler et al., 2013; Lau et al., 2013;

Menachery et al., 2014a; Zhou et al., 2014; Zielecki et al., 2013) was further investigated. The ORF4a protein inhibits IFN induction by interaction with dsRNA and the RLR cofactor PACT (Niemeyer et al., 2013; Siu et al., 2014). Like the ORF4a, the ORF4b, 5, and M proteins of MERS-CoV were shown to prevent IRF3 translocation (Yang et al., 2013). The ORF4b protein, in particular, inhibits IFN induction by binding to TBK1 and IKKepsilon (Matthews et al., 2014; Yang et al., 2015). In agreement with the data on SARS-CoV, the PL^{pro} of MERS-CoV has deubiquitinating activity and inhibits IFN induction (Bailey-Elkin et al., 2014; Mielech et al., 2014), and the nsp1 mediates host mRNA degradation (Lokugamage et al., 2015). In contrast to SARS-CoV, however, infection with MERS-CoV additionally activates repressive histone modifications that downregulate ISG expression (Menachery et al., 2014a).

5.2 Inhibition of IFN Signaling

Several proteins of SARS-CoV and MERS-CoV were found to interfere with the signal transduction chain that leads from IFN docking onto its receptor to the upregulation of ISGs by ISGF3, the STAT1/STAT2/IRF9 complex (Table 2). The ORF3a protein was shown to decrease levels of IFNAR, most probably by ubiquitination and proteolytic degradation

Table 2 Mechanisms and Factors of Human Coronaviruses to Counteract IFN-Stimulated Gene Expression

Virus	Viral Protein or Function	Mechanism	References
SARS–CoV	ORF3a protein	Proteolytic degradation of IFNAR	Minakshi et al. (2009)
	ORF6 protein	Inhibits STAT1 nuclear import by sequestering karyopherin alpha 2 to intracellular membranes	Frieman et al. (2007) and Kopecky-Bromberg et al. (2007)
SARS–CoV	nsp1	Decreases phosphorylation of STAT1	Wathelet et al. (2007)
MERS–CoV	ORF4a, and ORF4b proteins, M	Inhibit ISRE activation after stimulation with IFN, mechanism unknown	Yang et al. (2013)
MERS–CoV	Unknown	Repressive histone modifications	Menachery et al. (2014a)

(Minakshi et al., 2009). The ORF6 protein was the first factor described for SARS-CoV that affects IFN signaling in infected cells, disrupting nuclear import of STAT1 (Frieman et al., 2007; Kopecky-Bromberg et al., 2007). The ORF6 protein binds to the nuclear import factor karyopherin alpha 2 and tethers it (together with karyopherin beta 1) to intracellular membranes (Frieman et al., 2007). There, they become unavailable for their normal cellular function, the import of, e.g., STAT1. The phosphorylation of STAT1 is impeded by the multifunctional nsp1 protein of SARS-CoV, which otherwise drives degradation of host mRNAs and inhibits translation (see earlier) (Wathelet et al., 2007). For MERS-CoV, the ORF4a, 4b, and M proteins inhibit ISRE activation after stimulation with IFN (Yang et al., 2013). The mechanisms are currently unknown. The ORF4a protein, which also acts as an inhibitor of IFN induction (see earlier), had the strongest activity. Lastly, the repressive modifications that are imposed by MERS-CoV onto the cellular histones are also a strategy to dampen ISG expression (Menachery et al., 2014a).

5.3 Increasing IFN Resistance

Despite having some sensitivity toward IFN, especially MERS-CoV (see Section 4), viral strategies to increase IFN resistance are also in place (Table 3). The sequestration of viral dsRNA in DMVs (Knoops et al., 2008; van Hemert et al., 2008; Versteeg et al., 2007) not only reduces cytoplasmic exposure to PRRs and hence IFN induction but also limits activation of antiviral dsRNA-responsive ISG products like PKR. However, PKR is eventually activated by SARS-CoV infection, but has no effect on viral replication (Krahling et al., 2009). Interestingly, other coronaviruses

Table 3 Mechanisms and Factors of Human Coronaviruses to Increase IFN Resistance

Virus	Viral Protein or Function	Mechanism	References
SARS-CoV (MHV-1)	Storage of dsRNA inside double-membrane vesicles	Prevents exposure of dsRNA to PKR and OAS	Knoops et al. (2008), van Hemert et al. (2008), and Versteeg et al. (2007)
SARS-CoV	Unknown	Insensitivity to activated PKR	Krahling et al. (2009)
SARS-CoV	ADP-ribose-1″-monophosphatase domain of nsp3	Unknown	Kuri et al. (2011)

cope differently with PKR. The avian infectious bronchitis virus (IBV) expresses a weak inhibitor or PKR (nsp2) and additionally upregulates the phosphatase subunit GADD34 to reduce phosphorylation of the PKR downstream target eIF-2alpha (Wang et al., 2009). By contrast, the porcine reproductive and respiratory syndrome virus (PRRSV; a member of the *Arteriviridae* that are related to the *Coronaviridae* and other nidoviruses) does not inhibit but rather requires PKR for optimal replication and gene expression (Wang et al., 2016). Thus, the interactions and interdependencies of coronaviruses with PKR are complex and far from being understood. With respect to the antiviral OAS/RNase L system that is also activated by dsRNA, the mouse coronavirus MHV-A59 was shown to expresses an ns2 protein that antagonizes by degrading the product of the OAS enzyme, $2'-5'$ oligoadenylate that would activate RNase L (Zhao et al., 2012). SARS-CoV and MERS-CoV do not possess an ns2 homolog (Silverman and Weiss, 2014), but the MERS-CoV ns4b was recently demonstrated to cleave $2'-5'$ oligoadenylate (Thornbrough et al., 2016). Although ns4b-mutated MERS-CoV was not attenuated in cell culture, it provoked increased RNAse L activity in infected cells (Thornbrough et al., 2016). A critical factor for IFN resistance of SARS-CoV (and of the low pathogenic HCoV-229E) is the ADP-ribose-1″-monophosphatase (ADRP) domain that is contained within the nsp3 protein (Kuri et al., 2011). Virus mutants lacking a functional ADRP domain (also called macrodomain) display an increased IFN sensitivity. ADRP-like macrodomains are encoded by other coronaviruses and several other positive-strand RNA viruses (Gorbalenya et al., 1991). Also for MHV-A59, a role of the ADRP domain in pathogenesis was shown (Eriksson et al., 2008; Fehr et al., 2015), but this seems not be related to IFN sensitivity.

6. CONCLUSIONS AND OUTLOOK

The last 10+ years have seen tremendous progress toward the identification of IFN antagonists of human coronaviruses (De Diego et al., 2014; Gralinski and Baric, 2015; Kindler and Thiel, 2014; Perlman and Netland, 2009; Thiel and Weber, 2008; Totura and Baric, 2012; Vijay and Perlman, 2016; Wong et al., 2016). For SARS-CoV, the catalogue of IFN antagonists may be nearly complete by now and that of MERS-CoV may follow soon. Nonetheless, we are still far from comprehensively understanding the manifold interactions of human pathogenic coronaviruses with the IFN system. Many of the factors described here were identified by overexpression studies, and still lack the final biological assessment through generation and characterization of adequate virus mutants. It would also be interesting to see at which infections stage, in which subcellular compartment, and with which comparative intensity the IFN antagonists act, and whether and how they interact with each other. It is however safe to state that coronaviruses, which have the largest RNA genome known to date, do not rely on single virulence factors but employ several layers of anti-IFN strategies. Otherwise they would not be able to exist, thrive, and even expand to new hosts in the presence of powerful antiviral IFN responses.

ACKNOWLEDGMENTS

F.W. is supported by the SFB 1021 and Grant We 2616/7-1 (SPP 1596) of the Deutsche Forschungsgemeinschaft. E.K. and V.T. were supported by the Swiss National Science Foundation (SNF Grant 149784).

Disclosures: No conflicts of interest declared.

REFERENCES

Anand, K., Ziebuhr, J., Wadhwani, P., Mesters, J.R., Hilgenfeld, R., 2003. Coronavirus main proteinase (3CLpro) structure: basis for design of anti-SARS drugs. Science 300, 1763–1767.

Bailey, C.C., Zhong, G.C., Huang, I.C., Farzan, M., 2014. IFITM-family proteins: the cell's first line of antiviral defense. Annu. Rev. Virol. 1 (1), 261–283.

Bailey-Elkin, B.A., Knaap, R.C.M., Johnson, G.G., Dalebout, T.J., Ninaber, D.K., van Kasteren, P.B., et al., 2014. Crystal structure of the middle east respiratory syndrome coronavirus (MERS-CoV) papain-like protease bound to ubiquitin facilitates targeted disruption of deubiquitinating activity to demonstrate its role in innate immune suppression. J. Biol. Chem. 289, 34667–34682.

Bekisz, J., Schmeisser, H., Hernandez, J., Goldman, N.D., Zoon, K.C., 2004. Human interferons alpha, beta and omega. Growth Factors 22, 243–251.

Belgnaoui, S.M., Paz, S., Hiscott, J., 2011. Orchestrating the interferon antiviral response through the mitochondrial antiviral signaling (MAVS) adapter. Curr. Opin. Immunol. 23, 564–572.

Binder, M., Eberle, F., Seitz, S., Mucke, N., Huber, C.M., Kiani, N., et al., 2011. Molecular mechanism of signal perception and integration by the innate immune sensor retinoic acid-inducible gene-I (RIG-I). J. Biol. Chem. 286, 27278–27287.

Birdwell, L.D., Zalinger, Z.B., Li, Y., Wright, P.W., Elliott, R., Rose, K.M., et al., 2016. Activation of RNase L by murine coronavirus in myeloid cells is dependent on basal OAS gene expression and independent of virus-induced interferon. J. Virol. 90, 3160–3172.

Cervantes-Barragan, L., Zust, R., Weber, F., Spiegel, M., Lang, K.S., Akira, S., et al., 2007. Control of coronavirus infection through plasmacytoid dendritic-cell-derived type I interferon. Blood 109, 1131–1137.

Chakrabarti, A., Jha, B.K., Silverman, R.H., 2011. New insights into the role of RNase L in innate immunity. J. Interferon Cytokine Res. 31, 49–57.

Chan, R.W.Y., Chan, M.C.W., Agnihothram, S., Chan, L.L.Y., Kuok, D.I.T., Fong, J.H.M., et al., 2013. Tropism of and innate immune responses to the novel human betacoronavirus lineage C virus in human ex vivo respiratory organ cultures. J. Virol. 87, 6604–6614.

Channappanavar, R., Fehr, A.R., Vijay, R., Mack, M., Zhao, J., Meyerholz, D.K., et al., 2016. Dysregulated type I interferon and inflammatory monocyte-macrophage responses cause lethal pneumonia in SARS-CoV-infected mice. Cell Host Microbe 19, 181–193.

Chen, Y., Cai, H., Pan, J., Xiang, N., Tien, P., Ahola, T., et al., 2009. Functional screen reveals SARS coronavirus nonstructural protein nsp14 as a novel cap N7 methyl-transferase. Proc. Natl. Acad. Sci. U.S.A. 106, 3484–3489.

Cheung, C.Y., Poon, L.L., Ng, I.H., Luk, W., Sia, S.F., Wu, M.H., et al., 2005. Cytokine responses in severe acute respiratory syndrome coronavirus-infected macrophages in vitro: possible relevance to pathogenesis. J. Virol. 79, 7819–7826.

Cinatl, J., Morgenstern, B., Bauer, G., Chandra, P., Rabenau, H., Doerr, H.W., 2003. Treatment of SARS with human interferons. Lancet 362, 293–294.

Clementz, M.A., Chen, Z., Banach, B.S., Wang, Y., Sun, L., Ratia, K., et al., 2010. Deubiquitinating and interferon antagonism activities of coronavirus papain-like prote-ases. J. Virol. 84, 4619–4629.

Dabo, S., Meurs, E.F., 2012. dsRNA-dependent protein kinase PKR and its role in stress, signaling and HCV infection. Viruses 4, 2598–2635.

De Diego, M.L., Nieto-Torres, J.L., Jimenez-Guardeno, J.M., Regla-Nava, J.A., Castano-Rodriguez, C., Fernandez-Delgado, R., et al., 2014. Coronavirus virulence genes with main focus on SARS-CoV envelope gene. Virus Res. 194, 124–137.

Decroly, E., Imbert, I., Coutard, B., Bouvet, M.L., Selisko, B., Alvarez, K., et al., 2008. Coronavirus nonstructural protein 16 is a cap-0 binding enzyme possessing (nucleo-side-2'O)-methyltransferase activity. J. Virol. 82, 8071–8084.

DeDiego, M.L., Nieto-Torres, J.L., Regla-Nava, J.A., Jimenez-Guardeno, J.M., Fernandez-Delgado, R., Fett, C., et al., 2014. Inhibition of NF-kappaB-mediated inflammation in severe acute respiratory syndrome coronavirus-infected mice increases survival. J. Virol. 88, 913–924.

Devaraj, S.G., Wang, N., Chen, Z., Tseng, M., Barretto, N., Lin, R., et al., 2007. Regu-lation of IRF-3-dependent innate immunity by the papain-like protease domain of the severe acute respiratory syndrome coronavirus. J. Biol. Chem. 282, 32208–32221.

Eriksson, K.K., Cervantes-Barragan, L., Ludewig, B., Thiel, V., 2008. Mouse hepatitis virus liver pathology is dependent on ADP-ribose-1''-phosphatase, a viral function conserved in the alpha-like supergroup. J. Virol. 82, 12325–12334.

Falzarano, D., de Wit, E., Martellaro, C., Callison, J., Munster, V.J., Feldmann, H., 2013. Inhibition of novel beta coronavirus replication by a combination of interferon-alpha2b and ribavirin. Sci. Rep. 3, 1686.

Fehr, A.R., Perlman, S., 2015. Coronaviruses: an overview of their replication and patho-genesis. Methods Mol. Biol. 1282, 1–23.

Fehr, A.R., Athmer, J., Channappanavar, R., Phillips, J.M., Meyerholz, D.K., Perlman, S., 2015. The nsp3 macrodomain promotes virulence in mice with coronavirus-induced encephalitis. J. Virol. 89, 1523–1536.

Fensterl, V., Sen, G.C., 2015. Interferon-induced IFIT proteins: their role in viral pathogen-esis. J. Virol. 89, 2462–2468.

Freaney, J.E., Kim, R., Mandhana, R., Horvath, C.M., 2013. Extensive cooperation of immune master regulators IRF3 and NFkappaB in RNA Pol II recruitment and pause release in human innate antiviral transcription. Cell Rep. 4, 959–973.

Frieman, M., Yount, B., Heise, M., Kopecky-Bromberg, S.A., Palese, P., Baric, R.S., 2007. Severe acute respiratory syndrome coronavirus ORF6 antagonizes STAT1 function by sequestering nuclear import factors on the rough endoplasmic reticulum/Golgi mem-brane. J. Virol. 81, 9812–9824.

Frieman, M., Ratia, K., Johnston, R.E., Mesecar, A.D., Baric, R.S., 2009. Severe acute respiratory syndrome coronavirus papain-like protease ubiquitin-like domain and catalytic domain regulate antagonism of IRF3 and NF-kappaB signaling. J. Virol. 83, 6689–6705.

Frieman, M.B., Chen, J., Morrison, T.E., Whitmore, A., Funkhouser, W., Ward, J.M., et al., 2010. SARS-CoV pathogenesis is regulated by a STAT1 dependent but a type I, II and III interferon receptor independent mechanism. PLoS Pathog. 6, e1000849.

Gack, M.U., 2014. Mechanisms of RIG-I-like receptor activation and manipulation by viral pathogens. J. Virol. 88, 5213–5216.

Galani, I.E., Koltsida, O., Andreakos, E., 2015. Crossroads between innate and adaptive immunity V. In: Type III interferons (IFNs): Emerging Master Regulators of Immunity. Advances in Experimental Medicine and Biology, vol. 850. Springer International Publishing, pp. 1–15.

Gil, J., Garcia, M.A., Gomez-Puertas, P., Guerra, S., Rullas, J., Nakano, H., et al., 2004. TRAF family proteins link PKR with NF-kappa B activation. Mol. Cell. Biol. 24, 4502–4512.

Gorbalenya, A.E., Koonin, E.V., Lai, M.M.C., 1991. Putative papain-related thiol proteases of positive-strand RNA viruses—identification of rubivirus and aphthovirus proteases and delineation of a novel conserved domain associated with proteases of rubivirus, alpha- and coronaviruses. FEBS Lett. 288, 201–205.

Goubau, D., Schlee, M., Deddouche, S., Pruijssers, A.J., Zillinger, T., Goldeck, M., et al., 2014. Antiviral immunity via RIG-I-mediated recognition of RNA bearing 5'-diphosphates. Nature 514, 372.

Gough, D.J., Messina, N.L., Clarke, C.J.P., Johnstone, R.W., Levy, D.E., 2012. Constitutive type I interferon modulates homeostatic balance through tonic signaling. Immunity 36, 166–174.

Gralinski, L.E., Baric, R.S., 2015. Molecular pathology of emerging coronavirus infections. J. Pathol. 235, 185–195.

Haagmans, B.L., Kuiken, T., Martina, B.E., Fouchier, R.A.M., Rimmelzwaan, G.F., van Amerongen, G., et al., 2004. Pegylated interferon-alpha protects type 1 pneumocytes against SARS coronavirus infection in macaques. Nat. Med. 10, 290–293.

Haller, O., Staeheli, P., Schwemmle, M., Kochs, G., 2015. Mx GTPases: dynamin-like antiviral machines of innate immunity. Trends Microbiol. 23, 154–163.

Hamano, E., Hijikata, M., Itoyama, S., Quy, T., Phi, N.C., Long, H.T., et al., 2005. Polymorphisms of interferon-inducible genes OAS-1 and MxA associated with SARS in the Vietnamese population. Biochem. Biophys. Res. Commun. 329, 1234–1239.

Heil, F., Hemmi, H., Hochrein, H., Ampenberger, F., Kirschning, C., Akira, S., et al., 2004. Species-specific recognition of single-stranded RNA via toll-like receptor 7 and 8. Science 303, 1526–1529.

Holzinger, D., Jorns, C., Stertz, S., Boisson-Dupuis, S., Thimme, R., Weidmann, M., et al., 2007. Induction of MxA gene expression by influenza A virus requires type I or type III interferon signaling. J. Virol. 81, 7776–7785.

Huang, C., Lokugamage, K.G., Rozovics, J.M., Narayanan, K., Semler, B.L., Makino, S., 2011a. SARS coronavirus nsp1 protein induces template-dependent endonucleolytic cleavage of mRNAs: viral mRNAs are resistant to nsp1-induced RNA cleavage. PLoS Pathog. 7, e1002433.

Huang, I.C., Bailey, C.C., Weyer, J.L., Radoshitzky, S.R., Becker, M.M., Chiang, J.J., et al., 2011b. Distinct patterns of IFITM-mediated restriction of filoviruses, SARS coronavirus, and influenza A virus. PLoS Pathog. 7, e1001258.

Isaacs, A., Lindenmann, J., 1957. Virus interference. I. The interferon. Proc. R. Soc. Lond. B Biol. Sci. 147, 258–267.

Ivanov, K.A., Hertzig, T., Rozanov, M., Bayer, S., Thiel, V., Gorbalenya, A.E., et al., 2004. Major genetic marker of nidoviruses encodes a replicative endoribonuclease. Proc. Natl. Acad. Sci. U.S.A. 101, 12694–12699.

Kindler, E., Thiel, V., 2014. To sense or not to sense viral RNA—essentials of coronavirus innate immune evasion. Curr. Opin. Microbiol. 20, 69–75.

Kindler, E., Jonsdottir, H.R., Muth, D., Hamming, O.J., Hartmann, R., Rodriguez, R., et al., 2013. Efficient replication of the novel human betacoronavirus EMC on primary human epithelium highlights its zoonotic potential. mBio 4, e00611–e00612.

Knoops, K., Kikkert, M., Worm, S.H., Zevenhoven-Dobbe, J.C., van der Meer, Y., Koster, A.J., et al., 2008. SARS-coronavirus replication is supported by a reticulovesicular network of modified endoplasmic reticulum. PLoS Biol. 6, e226.

Kopecky-Bromberg, S.A., Martinez-Sobrido, L., Frieman, M., Baric, R.A., Palese, P., 2007. Severe acute respiratory syndrome coronavirus open reading frame (ORF) 3b, ORF 6, and nucleocapsid proteins function as interferon antagonists. J. Virol. 81, 548–557.

Krahling, V., Stein, D.A., Spiegel, M., Weber, F., Muhlberger, E., 2009. Severe acute respiratory syndrome coronavirus triggers apoptosis via protein kinase R but is resistant to its antiviral activity. J. Virol. 83, 2298–2309.

Kuri, T., Zhang, X., Habjan, M., Martinez-Sobrido, L., Garcia-Sastre, A., Yuan, Z., et al., 2009. Interferon priming enables cells to partially overturn the SARS coronavirus-induced block in innate immune activation. J. Gen. Virol. 90, 2686–2694.

Kuri, T., Eriksson, K.K., Putics, A., Zust, R., Snijder, E.J., Davidson, A.D., et al., 2011. The ADP-ribose-1″-monophosphatase domains of severe acute respiratory syndrome coronavirus and human coronavirus 229E mediate resistance to antiviral interferon responses. J. Gen. Virol. 92, 1899–1905.

Lau, S.K.P., Lau, C.C.Y., Chan, K.H., Li, C.P.Y., Chen, H.L., Jin, D.Y., et al., 2013. Delayed induction of proinflammatory cytokines and suppression of innate antiviral response by the novel Middle East respiratory syndrome coronavirus: implications for pathogenesis and treatment. J. Gen. Virol. 94, 2679–2690.

Li, W., Moore, M.J., Vasilieva, N., Sui, J., Wong, S.K., Berne, M.A., et al., 2003. Angiotensin-converting enzyme 2 is a functional receptor for the SARS coronavirus. Nature 426, 450–454.

Li, J., Liu, Y., Zhang, X., 2010. Murine coronavirus induces type I interferon in oligodendrocytes through recognition by RIG-I and MDA5. J. Virol. 84, 6472–6482.

Li, Y., Chen, M., Cao, H.W., Zhu, Y.F., Zheng, J., Zhou, H., 2013. Extraordinary GU-rich single-strand RNA identified from SARS coronavirus contributes an excessive innate immune response. Microbes Infect. 15, 88–95.

Liu, S., Chen, J., Cai, X., Wu, J., Chen, X., Wu, Y.T., et al., 2013. MAVS recruits multiple ubiquitin E3 ligases to activate antiviral signaling cascades. elife 2, e00785.

Lokugamage, K.G., Narayanan, K., Nakagawa, K., Terasaki, K., Ramirez, S.I., Tseng, C.T., et al., 2015. Middle East respiratory syndrome coronavirus nsp1 inhibits host gene expression by selectively targeting mRNAs transcribed in the nucleus while sparing mRNAs of cytoplasmic origin. J. Virol. 89, 10970–10981.

Loutfy, M.R., Blatt, L.M., Siminovitch, K.A., Ward, S., Wolff, B., Lho, H., et al., 2003. Interferon alfacon-1 plus corticosteroids in severe acute respiratory syndrome—a preliminary study. J. Am. Med. Assoc. 290, 3222–3228.

Lu, X., Pan, J., Tao, J., Guo, D., 2011. SARS-CoV nucleocapsid protein antagonizes IFN-beta response by targeting initial step of IFN-beta induction pathway, and its C-terminal region is critical for the antagonism. Virus Genes 42, 37–45.

Luthra, P., Sun, D.Y., Silverman, R.H., He, B.A., 2011. Activation of IFN-beta expression by a viral mRNA through RNase L and MDA5. Proc. Natl. Acad. Sci. U.S.A. 108, 2118–2123.

Mahlakoiv, T., Ritz, D., Mordstein, M., DeDiego, M.L., Enjuanes, L., Muller, M.A., et al., 2012. Combined action of type I and type III interferon restricts initial replication of severe acute respiratory syndrome coronavirus in the lung but fails to inhibit systemic virus spread. J. Gen. Virol. 93, 2601–2605.

Malathi, K., Dong, B., Gale Jr., M., Silverman, R.H., 2007. Small self-RNA generated by RNase L amplifies antiviral innate immunity. Nature 448, 816–819.

Malathi, K., Saito, T., Crochet, N., Barton, D.J., Gale, M., Silverman, R.H., 2010. RNase L releases a small RNA from HCV RNA that refolds into a potent PAMP. RNA 16, 2108–2119.

Masters, P.S., Perlman, S., 2013. Coronaviridae. In: Knipe, D.M., Howley, P.M. (Eds.), Fields Virology, vol. 1. Wolters Kluwer, Philadelphia, pp. 825–858.

Matthews, K.L., Coleman, C.M., van der Meer, Y., Snijder, E.J., Frieman, M.B., 2014. The ORF4b-encoded accessory proteins of Middle East respiratory syndrome coronavirus and two related bat coronaviruses localize to the nucleus and inhibit innate immune signalling. J. Gen. Virol. 95, 874–882.

McBride, R., van Zyl, M., Fielding, B.C., 2014. The coronavirus nucleocapsid is a multifunctional protein. Viruses 6, 2991–3018.

Menachery, V.D., Eisfeld, A.J., Schafer, A., Josset, L., Sims, A.C., Proll, S., et al., 2014a. Pathogenic influenza viruses and coronaviruses utilize similar and contrasting approaches to control interferon-stimulated gene responses. mBio 5, e01174–01114.

Menachery, V.D., Yount Jr., B.L., Josset, L., Gralinski, L.E., Scobey, T., Agnihothram, S., et al., 2014b. Attenuation and restoration of severe acute respiratory syndrome coronavirus mutant lacking 2'-o-methyltransferase activity. J. Virol. 88, 4251–4264.

Mielech, A.M., Kilianski, A., Baez-Santos, Y.M., Mesecar, A.D., Baker, S.C., 2014. MERS-CoV papain-like protease has deISGylating and deubiquitinating activities. Virology 450–451, 64–70.

Minakshi, R., Padhan, K., Rani, M., Khan, N., Ahmad, F., Jameel, S., 2009. The SARS coronavirus 3a protein causes endoplasmic reticulum stress and induces ligand-independent downregulation of the type 1 interferon receptor. PLoS One 4, e8342.

Minskaia, E., Hertzig, T., Gorbalenya, A.E., Campanacci, V., Cambillau, C., Canard, B., et al., 2006. Discovery of an RNA virus 3'→5' exoribonuclease that is critically involved in coronavirus RNA synthesis. Proc. Natl. Acad. Sci. U.S.A. 103, 5108–5113.

Mordstein, M., Kochs, G., Dumoutier, L., Renauld, J.C., Paludan, S.R., Klucher, K., et al., 2008. Interferon-lambda contributes to innate immunity of mice against influenza A virus but not against hepatotropic viruses. PLoS Pathog. 4, e1000151.

Nallagatla, S.R., Toroney, R., Bevilacqua, P.C., 2011. Regulation of innate immunity through RNA structure and the protein kinase PKR. Curr. Opin. Struct. Biol. 21, 119–127.

Narayanan, K., Huang, C., Lokugamage, K., Kamitani, W., Ikegami, T., Tseng, C.T., et al., 2008. Severe acute respiratory syndrome coronavirus nsp1 suppresses host gene expression, including that of type I interferon, in infected cells. J. Virol. 82, 4471–4479.

Neuman, B.W., Kiss, G., Kunding, A.H., Bhella, D., Baksh, M.F., Connelly, S., et al., 2011. A structural analysis of M protein in coronavirus assembly and morphology. J. Struct. Biol. 174, 11–22.

Niemeyer, D., Zillinger, T., Muth, D., Zielecki, F., Horvath, G., Suliman, T., et al., 2013. Middle East respiratory syndrome coronavirus accessory protein 4a is a type I interferon antagonist. J. Virol. 87, 12489–12495.

Omrani, A.S., Saad, M.M., Baig, K., Bahloul, A., Abdul-Matin, M., Alaidaroos, A.Y., et al., 2014. Ribavirin and interferon alfa-2a for severe Middle East respiratory syndrome coronavirus infection: a retrospective cohort study. Lancet Infect. Dis. 14, 1090–1095.

O'Neill, L.A., Golenbock, D., Bowie, A.G., 2013. The history of Toll-like receptors—redefining innate immunity. Nat. Rev. Immunol. 13, 453–460.

Onomoto, K., Jogi, M., Yoo, J.S., Narita, R., Morimoto, S., Takemura, A., et al., 2012. Critical role of an antiviral stress granule containing RIG-I and PKR in viral detection and innate immunity. PLoS One 7, e43031.

Perlman, S., Netland, J., 2009. Coronaviruses post-SARS: update on replication and pathogenesis. Nat. Rev. Microbiol. 7, 439–450.

Pfaller, C.K., Li, Z., George, C.X., Samuel, C.E., 2011. Protein kinase PKR and RNA adenosine deaminase ADAR1: new roles for old players as modulators of the interferon response. Curr. Opin. Immunol. 23, 573–582.

Pham, A.M., Santa Maria, F.G., Lahiri, T., Friedman, E., Marie, I.J., Levy, D.E., 2016. PKR transduces MDA5-dependent signals for type I IFN induction. PLoS Pathog. 12, e1005489.

Pichlmair, A., Schulz, O., Tan, C.P., Rehwinkel, J., Kato, H., Takeuchi, O., et al., 2009. Activation of MDA5 requires higher-order RNA structures generated during virus infection. J. Virol. 83, 10761–10769.

Raj, V.S., Mou, H., Smits, S.L., Dekkers, D.H., Muller, M.A., Dijkman, R., et al., 2013. Dipeptidyl peptidase 4 is a functional receptor for the emerging human coronavirus-EMC. Nature 495, 251–254.

Rasmussen, S.B., Reinert, L.S., Paludan, S.R., 2009. Innate recognition of intracellular pathogens: detection and activation of the first line of defense. APMIS 117, 323–337.

Roth-Cross, J.K., Bender, S.J., Weiss, S.R., 2008. Murine coronavirus mouse hepatitis virus is recognized by MDA5 and induces type I interferon in brain macrophages/microglia. J. Virol. 82, 9829–9838.

Ruch, T.R., Machamer, C.E., 2012. The coronavirus E protein: assembly and beyond. Viruses 4, 363–382.

Runge, S., Sparrer, K.M.J., Lassig, C., Hembach, K., Baum, A., Garcia-Sastre, A., et al., 2014. In vivo ligands of MDA5 and RIG-I in measles virus-infected cells. PLoS Pathog. 10, e1004081.

Rusinova, I., Forster, S., Yu, S., Kannan, A., Masse, M., Cumming, H., et al., 2013. Interferome v2.0: an updated database of annotated interferon-regulated genes. Nucleic Acids Res. 41, D1040–D1046.

Saito, T., Owen, D.M., Jiang, F.G., Marcotrigiano, J., Gale, M., 2008. Innate immunity induced by composition-dependent RIG-I recognition of hepatitis C virus RNA. Nature 454, 523–527.

Samuel, C.E., 2001. Antiviral actions of interferons. Clin. Microbiol. Rev. 14, 778–809.

Sawicki, S.G., Sawicki, D.L., Siddell, S.G., 2007. A contemporary view of coronavirus transcription. J. Virol. 81, 20–29.

Scheuplein, V.A., Seifried, J., Malczyk, A.H., Miller, L., Hocker, L., Vergara-Alert, J., et al., 2015. High secretion of interferons by human plasmacytoid dendritic cells upon recognition of Middle East respiratory syndrome coronavirus. J. Virol. 89, 3859–3869.

Schiller, J.J., Kanjanahaluethai, A., Baker, S.C., 1998. Processing of the coronavirus MHV-JHM polymerase polyprotein: identification of precursors and proteolytic products spanning 400 kilodaltons of ORF1a. Virology 242, 288–302.

Schlee, M., 2013. Master sensors of pathogenic RNA—RIG-I like receptors. Immunobiology 218, 1322–1335.

Schmid, S., Mordstein, M., Kochs, G., Garcia-Sastre, A., Tenoever, B.R., 2010. Transcription factor redundancy ensures induction of the antiviral state. J. Biol. Chem. 285, 42013–42022.

Schneider, W.M., Chevillotte, M.D., Rice, C.M., 2014. Interferon-stimulated genes: a complex web of host defenses. Annu. Rev. Immunol. 32, 513–545.

Schreibelt, G., Tel, J., Sliepen, K.H.E.W.J., Benitez-Ribas, D., Figdor, C.G., Adema, G.J., et al., 2010. Toll-like receptor expression and function in human dendritic cell subsets: implications for dendritic cell-based anti-cancer immunotherapy. Cancer Immunol. Immunother. 59, 1573–1582.

Schulz, O., Pichlmair, A., Rehwinkel, J., Rogers, N.C., Scheuner, D., Kato, H., et al., 2010. Protein kinase R contributes to immunity against specific viruses by regulating interferon mRNA integrity. Cell Host Microbe 7, 354–361.

Shi, C.S., Qi, H.Y., Boularan, C., Huang, N.N., Abu-Asab, M., Shelhamer, J.H., et al., 2014. SARS-coronavirus open reading frame-9b suppresses innate immunity by targeting mitochondria and the MAVS/TRAF3/TRAF6 signalosome. J. Immunol. 193, 3080–3089.

Silverman, R.H., Weiss, S.R., 2014. Viral phosphodiesterases that antagonize double-stranded RNA signaling to RNase L by degrading 2-5A. J. Interferon Cytokine Res. 34, 455–463.

Siu, K.L., Kok, K.H., Ng, M.H., Poon, V.K., Yuen, K.Y., Zheng, B.J., et al., 2009. Severe acute respiratory syndrome coronavirus M protein inhibits type I interferon production by impeding the formation of TRAF3.TANK.TBK1/IKKepsilon complex. J. Biol. Chem. 284, 16202–16209.

Siu, K.L., Yeung, M.L., Kok, K.H., Yuen, K.S., Kew, C., Lui, P.Y., et al., 2014. Middle East respiratory syndrome coronavirus 4a protein is a double-stranded RNA-binding protein that suppresses PACT-induced activation of RIG-I and MDA5 in the innate antiviral response. J. Virol. 88, 4866–4876.

Snijder, E.J., Bredenbeek, P.J., Dobbe, J.C., Thiel, V., Ziebuhr, J., Poon, L.L., et al., 2003. Unique and conserved features of genome and proteome of SARS-coronavirus, an early split-off from the coronavirus group 2 lineage. J. Mol. Biol. 331, 991–1004.

Sommereyns, C., Paul, S., Staeheli, P., Michiels, T., 2008. IFN-lambda (IFN-lambda) is expressed in a tissue-dependent fashion and primarily acts on epithelial cells in vivo. PLoS Pathog. 4, e1000017.

Spiegel, M., Pichlmair, A., Muhlberger, E., Haller, O., Weber, F., 2004. The antiviral effect of interferon-beta against SARS-coronavirus is not mediated by MxA protein. J. Clin. Virol. 30, 211–213.

Spiegel, M., Pichlmair, A., Martinez-Sobrido, L., Cros, J., Garcia-Sastre, A., Haller, O., et al., 2005. Inhibition of Beta interferon induction by severe acute respiratory syndrome coronavirus suggests a two-step model for activation of interferon regulatory factor 3. J. Virol. 79, 2079–2086.

Stark, G.R., Darnell, J.E., 2012. The JAK-STAT pathway at twenty. Immunity 36, 503–514.

Strayer, D.R., Dickey, R., Carter, W.A., 2014. Sensitivity of SARS/MERS CoV to interferons and other drugs based on achievable serum concentrations in humans. Infect. Disord. Drug Targets 14, 37–43.

Stroher, U., DiCaro, A., Li, Y., Strong, J.E., Aoki, F., Plummer, F., et al., 2004. Severe acute respiratory syndrome related coronavirus is inhibited by interferon-alpha. J. Infect. Dis. 189, 1164–1167.

Sun, L., Xing, Y., Chen, X., Zheng, Y., Yang, Y., Nichols, D.B., et al., 2012. Coronavirus papain-like proteases negatively regulate antiviral innate immune response through disruption of STING-mediated signaling. PLoS One 7, e30802.

Tanaka, T., Kamitani, W., DeDiego, M.L., Enjuanes, L., Matsuura, Y., 2012. Severe acute respiratory syndrome coronavirus nsp1 facilitates efficient propagation in cells through a specific translational shutoff of host mRNA. J. Virol. 86, 11128–11137.

tenOever, B.R., 2016. The evolution of antiviral defense systems. Cell Host Microbe 19, 142–149.

Thiel, V., Weber, F., 2008. Interferon and cytokine responses to SARS-coronavirus infection. Cytokine Growth Factor Rev. 19, 121–132.

Thiel, V., Ivanov, K.A., Putics, A., Hertzig, T., Schelle, B., Bayer, S., et al., 2003. Mechanisms and enzymes involved in SARS coronavirus genome expression. J. Gen. Virol. 84, 2305–2315.

Thornbrough, J.M., Jha, B.K., Yount, B., Goldstein, S.A., Li, Y., Elliott, R., et al., 2016. Middle East respiratory syndrome coronavirus NS4b protein inhibits host RNase L activation. mBio 7, e00258.

Totura, A.L., Baric, R.S., 2012. SARS coronavirus pathogenesis: host innate immune responses and viral antagonism of interferon. Curr. Opin. Virol. 2, 264–275.

Ujike, M., Taguchi, F., 2015. Incorporation of spike and membrane glycoproteins into coronavirus virions. Viruses 7, 1700–1725.

van Hemert, M.J., van den Worm, S.H., Knoops, K., Mommaas, A.M., Gorbalenya, A.E., Snijder, E.J., 2008. SARS-coronavirus replication/transcription complexes are membrane-protected and need a host factor for activity in vitro. PLoS Pathog. 4, e1000054.

Versteeg, G.A., Bredenbeek, P.J., van den Worm, S.H., Spaan, W.J., 2007. Group 2 coronaviruses prevent immediate early interferon induction by protection of viral RNA from host cell recognition. Virology 361, 18–26.

Vijay, R., Perlman, S., 2016. Middle East respiratory syndrome and severe acute respiratory syndrome. Curr. Opin. Virol. 16, 70–76.

Wang, Y., Liu, L., 2016. The membrane protein of severe acute respiratory syndrome coronavirus functions as a novel cytosolic pathogen-associated molecular pattern to promote beta interferon induction via a toll-like-receptor-related TRAF3-independent mechanism. mBio 7, e01872-15.

Wang, X.X., Liao, Y., Yap, P.L., Png, K.J., Tam, J.P., Liu, D.X., 2009. Inhibition of protein kinase R activation and upregulation of GADD34 expression play a synergistic role in facilitating coronavirus replication by maintaining de novo protein synthesis in virus-infected cells. J. Virol. 83, 12462–12472.

Wang, X.Q., Zhang, H.M., Abel, A., Nelson, E., 2016. Protein kinase R (PKR) plays a pro-viral role in porcine reproductive and respiratory syndrome virus (PRRSV) replication by modulating viral gene transcription. Arch. Virol. 161, 327–333.

Wathelet, M.G., Orr, M., Frieman, M.B., Baric, R.S., 2007. Severe acute respiratory syndrome coronavirus evades antiviral signaling: role of nsp1 and rational design of an attenuated strain. J. Virol. 81, 11620–11633.

Weber, M., Weber, F., 2014. Segmented negative-strand RNA viruses and RIG-I: divide (your genome) and rule. Curr. Opin. Microbiol. 20, 96–102.

Weber, F., Wagner, V., Rasmussen, S.B., Hartmann, R., Paludan, S.R., 2006. Double-stranded RNA is produced by positive-strand RNA viruses and DNA viruses but not in detectable amounts by negative-strand RNA viruses. J. Virol. 80, 5059–5064.

Wong, L.-Y.R., Lui, P.-Y., Jin, D.-Y., 2016. A molecular arms race between host innate antiviral response and emerging human coronaviruses. Virol. Sin. 31, 12–23.

Wreschner, D.H., McCauley, J.W., Skehel, J.J., Kerr, I.M., 1981. Interferon action—sequence specificity of the ppp(A2′p)nA-dependent ribonuclease. Nature 289, 414–417.

Yang, Y., Zhang, L., Geng, H., Deng, Y., Huang, B., Guo, Y., et al., 2013. The structural and accessory proteins M, ORF 4a, ORF 4b, and ORF 5 of Middle East respiratory syndrome coronavirus (MERS-CoV) are potent interferon antagonists. Protein Cell 4, 951–961.

Yang, Y., Ye, F., Zhu, N., Wang, W., Deng, Y., Zhao, Z., et al., 2015. Middle East respiratory syndrome coronavirus ORF4b protein inhibits type I interferon production through both cytoplasmic and nuclear targets. Sci. Rep. 5, 17554.

Yim, H.C.H., Williams, B.R.G., 2014. Protein kinase R and the inflammasome. J. Interferon Cytokine Res. 34, 447–454.

Yoneyama, M., Jogi, M., Onomoto, K., 2016. Regulation of antiviral innate immune signaling by stress-induced RNA granules. J. Biochem. 159, 279–286.

Zamanian-Daryoush, M., Mogensen, T.H., DiDonato, J.A., Williams, B.R., 2000. NF-kappaB activation by double-stranded-RNA-activated protein kinase (PKR) is mediated through NF-kappaB-inducing kinase and IkappaB kinase. Mol. Cell. Biol. 20, 1278–1290.

Zhao, L., Jha, B.K., Wu, A., Elliott, R., Ziebuhr, J., Gorbalenya, A.E., et al., 2012. Antagonism of the interferon-induced OAS-RNase L pathway by murine coronavirus ns2 protein is required for virus replication and liver pathology. Cell Host Microbe 11, 607–616.

Zhao, X.S., Guo, F., Liu, F., Cuconati, A., Chang, J.H., Block, T.M., et al., 2014. Interferon induction of IFITM proteins promotes infection by human coronavirus OC43. Proc. Natl. Acad. Sci. U.S.A. 111, 6756–6761.

Zhou, J., Chu, H., Li, C., Wong, B.H.Y., Cheng, Z.S., Poon, V.K.M., et al., 2014. Active replication of Middle East respiratory syndrome coronavirus and aberrant induction of inflammatory cytokines and chemokines in human macrophages: implications for pathogenesis. J. Infect. Dis. 209, 1331–1342.

Ziebuhr, J., 2005. The coronavirus replicase. Curr. Top. Microbiol. Immunol. 287, 57–94.

Ziebuhr, J., Schelle, B., Karl, N., Minskaia, E., Bayer, S., Siddell, S.G., et al., 2007. Human coronavirus 229E papain-like proteases have overlapping specificities but distinct functions in viral replication. J. Virol. 81, 3922–3932.

Ziegler, T., Matikainen, S., Ronkko, E., Osterlund, P., Sillanpaa, M., Siren, J., et al., 2005. Severe acute respiratory syndrome coronavirus fails to activate cytokine-mediated innate immune responses in cultured human monocyte-derived dendritic cells. J. Virol. 79, 13800–13805.

Zielecki, F., Weber, M., Eickmann, M., Spiegelberg, L., Zaki, A.M., Matrosovich, M., et al., 2013. Human cell tropism and innate immune system interactions of human respiratory coronavirus EMC compared to those of severe acute respiratory syndrome coronavirus. J. Virol. 87, 5300–5304.

Zinzula, L., Tramontano, E., 2013. Strategies of highly pathogenic RNA viruses to block dsRNA detection by RIG-I-like receptors: hide, mask, hit. Antivir. Res. 100, 615–635.

Zust, R., Cervantes-Barragan, L., Kuri, T., Blakqori, G., Weber, F., Ludewig, B., et al., 2007. Coronavirus non-structural protein 1 is a major pathogenicity factor: implications for the rational design of coronavirus vaccines. PLoS Pathog. 3, e109.

Zust, R., Cervantes-Barragan, L., Habjan, M., Maier, R., Neuman, B.W., Ziebuhr, J., et al., 2011. Ribose 2′-O-methylation provides a molecular signature for the distinction of self and non-self mRNA dependent on the RNA sensor Mda5. Nat. Immunol. 12, 137–143.

CHAPTER EIGHT

Molecular Basis of Coronavirus Virulence and Vaccine Development

L. Enjuanes[1], S. Zuñiga, C. Castaño-Rodriguez, J. Gutierrez-Alvarez, J. Canton, I. Sola[1]

National Center of Biotechnology (CNB-CSIC), Campus Universidad Autónoma de Madrid, Madrid, Spain
[1]Corresponding authors: e-mail address: l.enjuanes@cnb.csic.es; isola@cnb.csic.es

Contents

Abstract

Virus vaccines have to be immunogenic, sufficiently stable, safe, and suitable to induce long-lasting immunity. To meet these requirements, vaccine studies need to provide a comprehensive understanding of (i) the protective roles of antiviral B and T-cell-mediated immune responses, (ii) the complexity and plasticity of major viral antigens, and (iii) virus molecular biology and pathogenesis. There are many types of vaccines including subunit vaccines, whole-inactivated virus, vectored, and live-attenuated virus vaccines, each of which featuring specific advantages and limitations. While nonliving virus vaccines have clear advantages in being safe and stable, they may cause side effects and be less efficacious compared to live-attenuated virus vaccines. In most cases, the latter induce long-lasting immunity but they may require special safety measures to prevent reversion to highly virulent viruses following vaccination. The chapter summarizes the recent progress in the development of coronavirus (CoV) vaccines, focusing on two zoonotic CoVs, the severe acute respiratory syndrome CoV (SARS-CoV), and the Middle East respiratory syndrome CoV, both of which cause deadly disease and epidemics in humans. The development of attenuated virus vaccines to combat infections caused by highly pathogenic CoVs was largely based on the identification and characterization of viral virulence proteins that, for example, interfere with the innate and adaptive immune response or are involved in interactions with specific cell types, such as macrophages, dendritic and epithelial cells, and T lymphocytes, thereby modulating antiviral host responses and viral pathogenesis and potentially resulting in deleterious side effects following vaccination.

1. INTRODUCTION

1.1 Focus of the Review

There are four "common" human coronaviruses (CoVs) that are endemic in the human population: HCoV-229E, HCoV-OC43, HCoV-NL63, and HCoV-HKU1. The first two CoVs have been known since the 1960s, while the emergence of severe acute respiratory syndrome CoV (SARS-CoV) in 2002 led to an active search for novel CoVs and the identification of HCoV-NL63 and HCoV-HKU1 in 2004 and 2005, respectively (van der Hoek et al., 2004; Woo et al., 2005a). The common human CoVs are generally associated with relatively mild clinical symptoms and cause a self-limiting upper respiratory tract disease (common cold) (Walsh et al., 2013). In some cases, common CoVs may also be associated with more severe pathogenesis in the lower respiratory tract, such as bronchiolitis or pneumonia (Pene et al., 2003; van der Hoek et al., 2005; Woo et al., 2005b). Human CoVs cause more serious disease in young, elderly, or immunocompromised individuals, and they may lead to exacerbation of preexisting conditions, such as asthma or chronic obstructive pulmonary disease, frequently requiring hospitalization (Mayer et al., 2016; Varkey and Varkey, 2008). Considering that the prevalence of these viruses ranges between 3% and 16% and that around 70% of the population is infected during childhood, with recurrent infections occurring throughout life (van der Hoek, 2007; Zhou et al., 2013b), common human CoVs represent a significant burden to public health.

More recently, two previously unknown animal CoVs emerged that were shown to cause deadly disease in humans. The first, SARS-CoV, emerged in Southern China and spread around the globe in late 2002 and early 2003, infecting at least 8000 people and killing nearly 10% of the infected individuals (Lee et al., 2003). The second, Middle East respiratory syndrome CoV (MERS-CoV), was first reported in 2012 in the Middle East and there are still ongoing reports of sporadic cases, particularly in Saudi Arabia and the United Arab Emirates. This virus has caused close to 1782 laboratory-confirmed cases, resulting in 634 deaths, and is the cause of an important outbreak in Korea that started in May 2015, leading to more than 186 confirmed cases with a death toll of 36, according to the World Health Organization (WHO) (http://www.who.int/emergencies/mers-cov/en/).

Current treatment strategies for SARS and MERS, and discussion of the discovery and development of new virus-based and host-based therapeutic options for CoV infection have been reviewed recently (Zumla et al., 2016). In this chapter, we will mainly focus on the development of vaccines suitable to prevent infections caused by highly pathogenic CoVs, particularly SARS-CoV and MERS-CoV, in humans. To date, a wide range of vaccine candidates have been developed for these viruses, including subunit, whole-inactivated virus, DNA, and vectored vaccines (see reviews by Du and Jiang, 2015; Enjuanes et al., 2008; Zhang et al., 2014a). However, in many cases, these vaccines were found to induce antibody-dependent enhancement of infectivity (ADEI) and eosinophilia. In contrast, live-attenuated vaccines have a long history of success and are the most frequently used vaccines in humans. This chapter will be focusing on a recently developed next generation of live-attenuated vaccines based on recombinant viruses. Attenuation of viruses generally relies on the previous identification of genes involved in viral virulence in specific hosts. Often, these genes encode proteins that antagonize the innate immune response, and their deletion leads to recombinant viruses that are attenuated in their virulence and, therefore, may be developed into candidate vaccines. The timeline from bench research to approved vaccine use is generally 10 years or more (Papaneri et al., 2015). The use of genetically engineered viruses may significantly reduce both the time and costs required for vaccine development and production. In this chapter, specific features of promising vaccine candidates in meeting the earlier criteria will be reviewed.

1.2 CoV Genome Structure and Protein Composition

CoVs contain the largest genome known among RNA viruses, consisting of a single-stranded positive-sense RNA molecule of around 30 kb in length (Fig. 1) (de Groot et al., 2012). It is similar to cellular mRNAs, as it contains $5'$-capped and $3'$ polyadenylated ends. The $5'$-terminal two-thirds of the genome contain two overlapping open reading frames (ORFs): ORF1a and ORF1b (Fig. 1). Translation of ORF1a yields polyprotein 1a (pp1a), and –1 ribosomal frameshifting allows translation of ORF1b to yield pp1ab (Ziebuhr, 2005). Together, these polyproteins are co- and post-translationally processed into 16 nonstructural proteins (nsps), most of them being involved in viral genome replication and subgenomic mRNA synthesis. The $3'$-third of the genome encodes a series of structural proteins in the

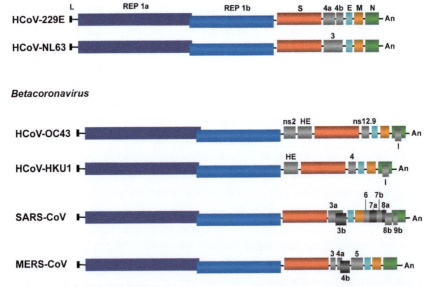

Fig. 1 Genome structure of human CoVs. Each *bar* represents the genomic organization of a human CoV. The tags above the bars indicate the name of each gene. Genus-specific genes are represented in *light* and *dark gray* colors. *An*, poly(A) tail; *I*, internal ORF; *L*, leader sequence; *REP* 1a and *REP* 1b, replicase gene (comprised of ORFs 1a and 1b).

order 5′-S-E-M-N-3′ and genus-specific proteins that vary in number among the different CoV members (Fig. 1) (de Groot et al., 2012; Sola et al., 2015).

Human HCoV-229E and HCoV-NL63 belong to the genus *Alphacoronavirus*, while HCoV-OC43, HCoV-HKU1, SARS-CoV, and MERS-CoV belong to genus *Betacoronavirus*. Although the overall genomic organization is conserved among all coronavirus genera, the members of this virus family encode a unique set of genus-specific proteins (Fig. 1) that, in general, are involved in the modulation of pathogenesis.

1.3 Requirement of B- and T-Cell Responses for Protection

Neutralizing monoclonal antibodies (mAbs) represent a promising therapeutic strategy against emerging CoV infections. Fully human neutralizing antibodies may be developed using different technologies and applied as therapeutic or prophylactic agents (Jiang et al., 2014; Tang et al., 2014; ter Meulen et al., 2004; Ying et al., 2014; Zhang et al., 2005; Zhu et al., 2007). Of special interest are the potent human cross-reactive neutralizing antibodies specific for SARS-CoV (Pascal et al., 2015). Since most of these antibodies target the receptor-binding domain of the spike (S) glycoprotein, antigenic variability within the S gene among human and animal strains must be considered in the design of broad-spectrum neutralizing molecules (Ying et al., 2015b). A combination of mAbs targeting different epitopes may be used to prevent the emergence of escape mutants and, potentially, generate a synergistic neutralizing effect (Ying et al., 2015a). In line with this, passive immunotherapy with mouse or dromedary immune serum was shown to be protective in animal models against SARS-CoV (Subbarao et al., 2004) and MERS-CoV (Zhao et al., 2015b), respectively, suggesting that this approach may be used prophylactically or therapeutically in infected patients.

Neutralizing antibodies induced by the S glycoprotein provide complete protection from lethal CoV infections (Sui et al., 2005). Moreover, an inverse correlation was observed between IgA secretion and MERS-CoV infectivity in patients, suggesting that virus-specific IgA production may be a suitable tool to evaluate the potency of candidate vaccines against MERS-CoV (Muth et al., 2015). However, the IgA response is short lived in patients. In contrast, virus-specific memory CD8$^+$ T cells persisted for up to 6 years after SARS-CoV infection, at which time memory B cells and virus-specific antibodies were undetectable (Yang et al., 2006). It has been shown that memory CD8$^+$ T cells specific for an immunodominant epitope substantially protected aged mice from lethal SARS-CoV infection (Channappanavar et al., 2014). After challenge, memory CD8$^+$ T cells produced effector cytokines (interferon gamma, IFN-γ; tumor necrosis factor alpha, TNF-α; and interleukin 2, IL-2) and cytolytic molecules, reducing viral loads in the lung. However, dysregulation of some of these inflammatory mediators, including type I IFN and inflammatory monocyte–macrophage responses, caused lethal pneumonia in SARS-CoV-infected mice and, therefore, should be considered during vaccine design to minimize immunopathology (Channappanavar et al., 2016). In addition to the protective effect mediated by memory CD8$^+$ T cells, SARS-CoV-specific CD4$^+$ T cell and antibody responses are likely necessary for complete protection. The requirement of T-cell responses in MERS-CoV protection was also suggested by immunization experiments in macaques (Muthumani et al., 2015). Therefore, for effective protection, CoV vaccines should elicit not only antibody responses but also specific memory CD4$^+$ and CD8$^+$ T cells.

1.4 Antigenic Complexity of SARS- and MERS-CoV

Information on the complexity of CoV serotypes is crucial for predictions on whether antibodies against a previous CoV infection or a specific vaccine may protect from reinfection, which has important implications for vaccine design and neutralizing antibody therapy. Antigenic variability in the S protein, the major target of neutralizing antibodies, is extremely low between different MERS-CoV strains (Drosten et al., 2015). A recent serological study using infectious MERS-CoV isolates collected from patients in Saudi Arabia in 2014 showed no significant differences in serum neutralization, indicating that all these isolates belong to the same serotype (Muth et al., 2015). Based on these data, it seems likely that the S genes of all currently circulating MERS-CoVs are interchangeable in candidate vaccine formulations. The potential relevance of neutralizing antibodies directed against other envelope proteins remains to be studied.

MERS-CoV and all the closely related viruses isolated from camels and bats belong to the same viral species, with bat viruses being at the root of the phylogenetic tree. Most likely, the virus circulating in camels was acquired from bats and represents the origin of viruses identified in humans over the past few years. Recombination events within the spike gene of viral ancestors were likely involved in the emergence of MERS-CoV (Corman et al., 2014). Therefore, camels may serve as reservoirs for the maintenance and diversification of the MERS-CoVs responsible for human infection (Sabir et al., 2016). The resultant genetic variability may lead to additional anti-genic diversity, with obvious consequences for vaccine design.

Because SARS-CoV differs immunologically from other betacoro-naviruses with little cross-reactivity of antiviral antibodies (Hou et al., 2010), the development of vaccine candidates suitable to provide broad pro-tection should address the diversity of the main immunogenic determinants of the S protein (Zhou et al., 2013a). For example, SARS-CoV-specific domain in the S protein was found to contain an epitope (80R) that critically determines the sensitivity of a given virus to neutralizing antibodies specific for this epitope. Variants of this epitope have been found in SARS-like-CoVs from civet cats and in human SARS-CoVs that evolved during the epidemics. While the majority of SARS-CoVs from the first outbreak were sensitive to the 80R-specific antibody, the GD03 strain isolated from the index patient of the second outbreak was resistant, confirming the impor-tance of the S protein's natural antigenic variability in eliciting neutralizing antibody responses (Sui et al., 2005).

The SARS-CoV S glycoprotein is a major target of protective immunity in vivo. Two specific human mAbs recognizing the S protein exhibited potent cross-reactivity against isolates from the two SARS outbreaks and palm civets, but not bat strains (Zhu et al., 2007). A combination of two neutralizing mAbs could prevent the emergence of neutralization escape mutants or at least attenuate viral virulence in vivo. However, although neutralizing mAbs directed against epitopes located at the interface bet-ween the viral S protein and its cellular receptor, angiotensin-converting enzyme 2 (ACE2), proved to have great potency and breadth in neutra-lizing multiple viral strains (Sui et al., 2014), both the single and combined use of one or two mAbs, respectively, failed to prevent the emergence of antibody escape variants. Therefore, the use of one or two neutralizing mAbs that target a structurally flexible SARS-CoV epitope may be of limited value for in vivo immunotherapies and should be combined with neutralizing mAbs that bind a second conserved epitope with low structural plasticity.

1.5 Animal Models for CoV Vaccine and Antivirals Studies

Suitable animal models that reproduce the pathology caused by human CoVs are required for studies of pathogenesis and vaccine testing. Unfortu-nately, no appropriate animal models have been developed to date for any of the four common human CoVs. A transgenic (Tg) mouse model expressing human aminopeptidase N was generated for HCoV-229E but was not suitable for pathogenesis studies as it was based on immunodeficient Stat1$^{-/-}$ mice (Lassnig et al., 2005). A mouse model has been extensively used for studies of HCoV-OC43, as this human CoV causes lethal infections in mice (Jacomy and Talbot, 2003). However, following inoculation of mice with respiratory isolates of HCoV-OC43, the virus was found to adapt rapidly to grow in brain tissue, while viral RNA remained nearly undetectable in the lung (Butler et al., 2006; St-Jean et al., 2004), suggesting that the model does not reproduce the respiratory pathology seen in humans, limiting its value for studies of virus-induced pathology.

The use of animal models for SARS-CoV and MERS-CoV for studying pathology in humans has been recently reviewed, including clinical symptoms, viral replication, and pathology in humans, nonhuman primates (NHPs), rabbits, ferrets, marmosets, hamsters, and mice (Gretebeck and Subbarao, 2015; van Doremalen and Munster, 2015). Additional models for MERS-CoV based on dromedary camels and other animal species have also been reported (Falzarano et al., 2014; Haagmans et al., 2016; van Doremalen and Munster, 2015). Their large size and the lack of clear clinical signs of disease make camels a less suitable model for studying MERS-CoV pathology. Marmosets show clinical signs following infection with MERS-CoV virus but, in this case, research animals and appropriate reagents suitable to characterize the immune response are scarce or not available, limiting the use of this model system (van Doremalen and Munster, 2015). Hamsters cannot be naturally infected by MERS-CoV, largely preventing their use as an animal model. MERS-CoV S protein-mediated binding to its receptor, human dipeptidyl peptidase-4 (DPP4), involves interactions with 14 amino acid residues. Appropriate replacements of five residues that differ between hamster and human DPP4 render the hamster DPP4 a functional receptor for MERS-CoV (van Doremalen et al., 2014). Thus far, Tg hamsters have not been used due to the lack of specific gene targeting tools. With the availability of the CRISPR-Cas9 system, the situation may now change and hamsters susceptible to MERS-CoV might be developed but their suitability as animal models of MERS-CoV-induced disease remains unclear at present.

MERS-CoV is able to infect rabbits but does not cause histopathology or clinical symptoms although the virus can be detected in lungs. The virus is shed from the upper respiratory tract, providing a possible route of MERS-CoV transmission in this animal species (Haagmans et al., 2015). Clearly, the large size of rabbits also limits their use in BSL-3 containment laboratories.

Mice are an ideal model for pathogenesis studies of many viruses because of their small size and the availability of suitable genomic and immunological reagents. A key difference between SARS-CoV and MERS-CoV is that SARS-CoV infects several strains of mice, whereas MERS-CoV does not (Coleman et al., 2014b; Gretebeck and Subbarao, 2015). A standard procedure was the adaptation of SARS-CoV and MERS-CoV to grow in mice and reproduce the disease caused in humans. This strategy was directly applied in the case of SARS-CoV using conventional mouse strains without the need for Tg mice expressing the human ACE2 receptor (Day et al., 2009; Frieman et al., 2012; Roberts et al., 2007). Mouse-adapted SARS-CoV obtained by passing the virus 15 times in mice (SARS-CoV-MA15) has been an excellent model as it reproduces very well the pathology caused by SARS-CoV in humans, including mortality (DeDiego et al., 2011, 2014b). In contrast, in the case of MERS-CoV, the virus was grown in Tg or knockin humanized mice susceptible to MERS-CoV (K. Li, P. McCray, and S. Perlman, 2016, personal communication).

The first type of MERS-CoV-susceptible mice includes Tg mice that express hDPP4, the virus receptor, using promoters such as those from surfactant protein C, cytokeratin 18 (Li et al., 2016), or cytomegalovirus (Tao et al., 2015). In these mice, the LD50 has been estimated to be <10 TCID$_{50}$ of MERS-CoV. Although MERS-CoV grows almost equally well in the lungs and the brain in these Tg mice, these animals proved to be very useful for protection studies (Agrawal et al., 2015; Zhao et al., 2015a). In some Tg mouse strains, in which the hDPP4 was expressed under control of the surfactant protein C promoter, the virus was found to grow primarily in the lung, which might extend their use to viral pathogenesis studies (C. Tseng, 2016, personal communication).

The other type of MERS-CoV-susceptible mice available to date includes knockin strains in which 3 or 13 exons of mouse DPP4 have been replaced with the homologous sequences from hDPP4 (Agrawal et al., 2015; Li et al., 2016; Pascal et al., 2015). A major difference to the Tg mice described earlier is that the mouse-adapted MERS-CoV only produced disease in knockin mice after the virus has been passed 30 times in the knockin mouse (K. Li, P. McCray, and S. Perlman, 2016, personal communication), whereas the LD50 of MERS-CoV in these mice was around 10^4 pfu/ mouse. Another mouse lineage was generated using CRISPR-Cas9 technology by altering mDPP4 amino acid residues 288 and 330 that are known to interact with the MERS-CoV S protein, leading to mice that closely reproduced the disease observed in humans, including mortality. In this case, the LD50 was around 10^6 pfu/mouse (A. Crockrell and R. Baric, 2016, personal communication), significantly higher than that observed in the knockin mice (Cockrell et al., 2014; Peck et al., 2015; van Doremalen and Munster, 2015) and (K. Li, P. McCray, and S. Perlman, 2016, personal communication). Interestingly, all knockin mice and those generated with the CRISPR technology supported MERS-CoV replication in the lungs, making them extremely useful models for pathogenesis studies.

An alternative approach for the rapid generation of a mouse model for MERS-CoV has been the transduction of mice with adenoviral vectors expressing the human host-cell receptor DPP4 (Zhao et al., 2014a). These mice developed a pneumonia characterized by extensive inflammatory cell infiltration, with virus clearance occurring 6–8 days after infection. Using these mice the efficacy of a therapeutic intervention (poly I:C) and a potential vaccine based on Venezuelan equine encephalitis (VEE) virus has been demonstrated (Zhao et al., 2014a). An important advantage of this approach is that it may be rapidly adapted to other viruses that may emerge in the future, especially in cases in which a suitable mouse model is not available.

2. SUBUNIT, INACTIVATED, AND VECTORED VACCINES

2.1 Subunit Vaccines

Vaccines based on recombinant MERS-CoV S protein, in particular its RBD, have demonstrated partial efficacy in protecting immunized macaques from MERS-CoV infection, reducing pneumonia, and viral titers (Lan et al., 2015). Several fragments of the MERS-CoV S protein were found to induce MERS-CoV neutralizing antibody responses in mice and rabbits (Du et al., 2013; Jiang et al., 2014; Ma et al., 2014b; Mou et al., 2013), similar to what has been shown previously for SARS-CoV (Du et al., 2008; Wang et al., 2012). The fragment-containing residues 377–588 of MERS-CoV proved to be sufficient to protect Ad5/hDPP4-transduced and hDPP4-Tg mice against MERS-CoV. The immunogenicity of this fragment was further improved, resulting in strong humoral and cellular immune responses, by linking the fragment to human Fc and using an adjuvant (Tang et al., 2015; Zhang et al., 2016). These reports confirm that the MERS-CoV S protein is very well suited for the development of MERS subunit vaccines. The full-length S protein contains several non-neutralizing immunodominant domains that may compromise the immunogenicity of major neutralizing domains or induce harmful immune responses as demonstrated for the S protein of SARS-CoV (Weingartl et al., 2004). Intranasal vaccination with the RBD domain of the MERS-CoV S protein induced much stronger local mucosal immune responses in the lung than subcutaneous immunization (Ma et al., 2014a). Other studies showed that full-length monomeric or trimeric recombinant SARS-CoV and MERS-CoV S proteins are able to induce protective responses in mice (Honda-Okubo et al., 2015; Li et al., 2013) and the presentation of MERS-CoV full-length S protein as nanoparticles in combination with appropriate adjuvants elicits neutralizing antibodies in immunized mice (Coleman et al., 2014a).

Alternatively, DNA vaccines expressing full-length S protein or smaller fragments are effective against MERS-CoV infection. An optimized DNA vaccine encoding the full-length S protein of MERS-CoV elicited antigen-specific neutralizing antibodies in mice, camels, and rhesus macaques, with six of eight vaccinated macaques showing no radiographic evidence of infiltration after MERS-CoV challenge (Muthumani et al., 2015). Potent antigen-specific cellular immune responses were induced in the immunized macaques, suggesting that T-cell responses may also play a role in MERS-CoV protection. Two companies actively working to develop MERS-CoV DNA vaccines (Inovio Pharmaceuticals Inc., Philadelphia, USA and GeneOne Life Science, Seoul, Korea) will soon perform a Phase I clinical trial for this DNA-based vaccine (Inovio News Release, http://ir.inovio.com/news/news-releases/news-releases-details/2015/Inovio-Pharmaceuticals-Partners-with-GeneOne-Life-Science-for-MERS-Immunotherapy-Clinical-Development/default.aspx).

In addition to the DNA-only strategy, DNA priming and protein boosting could be an alternative vaccination approach for MERS-CoV. It was shown that priming of mice and NHPs with DNA encoding the full-length S gene and boosting with the S1 subunit protein-induced robust neutralizing antibody responses against several MERS-CoV strains and protected NHPs from MERS-CoV challenge (Wang et al., 2015).

Chimeric virus-like particles (VLPs) containing SARS-CoV S protein and influenza matrix protein 1 protected mice against challenge with SARS-CoV (Liu et al., 2011) and may induce strong immune responses due to its polymeric nature. To increase their efficacy, subunit vaccines or VLPs have to be administered together with adjuvants (Bolles et al., 2011; Tseng et al., 2012). Subunit vaccines only include subviral components that do not represent the full antigenic complexity of the virus, resulting in limited protective efficacy due to limited Th1-mediated immune responses and, in some cases, unbalanced immune responses that may lead to immunopathology.

2.2 Vaccines Based on Inactivated Whole Virus

Many vaccines based on chemically inactivated SARS-CoV virions have been evaluated in animal models such as hamsters, mice, ferrets, and NHPs (Bolles et al., 2011; Iwata-Yoshikawa et al., 2014; Roberts et al., 2010; Tseng et al., 2012). In all cases, production of neutralizing antibodies with different levels of protection was observed. Challenge of mice with SARS-CoV was reported to cause Th2-type immunopathology, suggesting that SARS-CoV components induced hypersensitivity. Further studies suggested that the immunopathology leading to eosinophilia was linked, at least in part, to the viral nucleoprotein (Bolles et al., 2011). Also, immunization with oligomers of the SARS-CoV S protein were shown to promote eosinophilia following viral challenge in different animal model systems (Tseng et al., 2012), suggesting that potential side effects of vaccines based on inactivated SARS-CoV or MERS-CoV should be carefully evaluated prior to use in humans. Moreover, inactivated whole-virus vaccines raise biosafety concerns due to the risk of vaccine preparations containing infectious virus. To minimize this risk, genetically attenuated viruses may be used as the starting point for the production of killed virus inactivation.

2.3 Vectored Vaccines

Vectored vaccines are generally based on vectors with a proven safety record. These vaccines allow production and release of immunogenic antigens from infected cells for a limited period of time. Vectors based on viruses from different families (poxvirus, adenovirus, measles, and togavirus (VEE)) have been used in the development of vaccines for CoVs. In the case of MERS-CoV, the most advanced and promising candidate is the modified vaccinia virus Ankara (MVA), a viral vector that does not replicate in mammalian cells and, therefore, holds great promise as a vaccine platform (Altenburg et al., 2014; Haagmans et al., 2016; Song et al., 2013; Volz et al., 2015). Using this vector, S protein fragments of different length were expressed: full-length, extracellular S1 domain, or the RBD. In all cases, neutralizing antibodies and T-cell responses for MERS-CoV were induced. One of these vectors induced mucosal immunity and reduced the shedding of MERS-CoV by a factor of one thousand after challenge with the virulent virus in dromedary camels, thus preventing spread from the animal reservoir

(Haagmans et al., 2016). An MVA-vectored vaccine will be entering clinical trials in 2016 supported by the German Center for Infection Research (DZIF) (Paddock, 2015).

A second, more advanced vectored vaccine is based on recombinant adenovirus expressing S protein fragments of different size (Guo et al., 2015; Kim et al., 2014; Shim et al., 2012). Sublingual immunization with a recombinant adenovirus encoding SARS-CoV S protein induced systemic and mucosal immunity in a mouse model system (Shim et al., 2012). Systemic and mucosal immunity were also elicited in mice by single immunization with human adenovirus type 5 or 41 vector-based vaccines carrying the S protein of MERS-CoV (Guo et al., 2015). Whether or not, these immune responses confer protection against viral infection remains to be evaluated.

Immunization with measles virus vectors expressing the SARS-CoV S protein induced neutralizing antibodies and strong Th1-biased responses, a hallmark of live-attenuated viruses and a highly desirable feature for an antiviral vaccine (Escriou et al., 2014), though eradication of measles virus may represent an obstacle for the application of this type of vaccines. VEE replicon particles expressing the SARS-CoV S protein provided protection against lethal homologous and heterologous challenge in an aged mice model (Sheahan et al., 2011).

Therefore, in principle, several well-known vectors offer the possibility of protection against CoV infection. Although these vectors are based on live viruses with a reasonable record of safety, they are limited to presenting one or a reduced number of CoV antigens to the immune system, in contrast to live vaccines based on the whole, attenuated CoV.

3. LIVE-ATTENUATED VACCINES

We consider it likely that the main strategy for producing more efficacious vaccines in the near future will be based on live-attenuated vaccines that lack specific virulence markers. The most rapid laboratory responses involved in vaccine generation is usually achieved by using subunit vaccines, whole-inactivated virus, or by the construction of vectored vaccines. Successful protection is frequently associated with the generation of a strong neutralizing immune response. Nevertheless, control of virus emergence with this type of vaccines frequently fails due to the relatively low titers and short-lived duration of neutralizing antibody responses and, at times, due to induction of immune pathogenesis including eosinophilia reactions. In contrast, live-attenuated vaccines have a long history of success and are generally more immunogenic than nonreplicating vaccines, because they produce a comprehensive spectrum of native viral antigens over a prolonged time span, presenting antigens to the immune system as in natural infections. For virus attenuation, we favor modern approaches that may overcome reversion to virulence, as described later.

3.1 Strategies to Engineer Attenuated CoVs as Vaccine Candidates

An alternative approach to the design of attenuated viruses as vaccine candidates that several laboratories have followed is the identification of virulence-associated viral genes (DeDiego et al., 2014a; Stobart and Moore, 2015; Totura and Baric, 2012) that, in many cases, encode non-essential immunomodulatory proteins. Modification or deletion of these genes leads to the generation of attenuated viruses that may be useful as vaccine candidates. Our laboratory has used this strategy extensively to generate CoV vaccines. Also, other groups have used this approach successfully to produce vaccine candidates for influenza, respiratory syncytial virus, measles, dengue, and mumps (Kirkpatrick et al., 2016; Stobart and Moore, 2015).

Engineering attenuated CoVs as vaccine candidates requires the availability of reverse-genetics systems suitable to introduce deletions of (or mutations in) virulence genes as well as appropriate animal models for the evaluation of efficacy and safety of these candidate vaccines. To date, no vaccines have been developed for common human CoV infections. Nevertheless, infectious cDNA clones have been engineered for the three human CoVs that can be efficiently propagated in tissue culture: HCoV-229E, HCoV-OC43, and HCoV-NL63 (Donaldson et al., 2008; St-Jean et al., 2006; Thiel et al., 2001). Similarly, infectious cDNA clones have been generated for SARS-CoV (Almazan et al., 2000; Yount et al., 2003) and MERS-CoV (Almazan et al., 2013; Scobey et al., 2013), providing an excellent basis for the rational development of live-attenuated vaccines based on recombinant viruses.

3.2 Coronavirus Virulence

Virulence of CoVs is generally associated with specific virulence genes that, in most cases, antagonize cellular innate immune responses but are nor required for efficient virus replication. CoV infection affects many host-cell pathways that modulate pathogenesis. Innate immune response is the first line of host-cell defense against viruses, and any viral factor modulating this pathway may have a strong impact on pathogenesis. CoV proteins nsp3, 4, and 6 are actively involved in the formation of host membrane vesicles and structures that affect CoV replication and evasion of the immune system by hiding the double-stranded RNA (dsRNA) generated during virus replication (Angelini et al., 2013). Pathogen-associated molecular patterns (PAMPs), such as ssRNA, dsRNA, or viral proteins, trigger the activation of transcription factors leading to proinflammatory cytokines and type I IFN induction. PAMPs activate protein kinase RNA-activated, which leads to translation initiation factor eIF2 phosphorylation and inhibition of host translation, and $2'$-$5'$ oligoadenylate synthase (OAS), which triggers RNase L and RNA degradation (Fig. 2). The activation of cytoplasmic sensors RIG-I and MDA-5 triggers the upregulation of IFN regulatory factors (IRF)-3 and IRF-7, and nuclear factor kappa-light-chain-enhancer of activated B cells (NF-kB) through mitochondrial antiviral-signaling protein (Fig. 2). In addition, Toll-like receptors (TLRs) activate the MyD88 response protein and adaptor molecule TRIF-dependent pathways, which also upregulate transcription factors IRF-3, IRF-7, NF-kB, and activator protein 1 (AP-1) (Fig. 2) (DeDiego et al., 2014a). Some CoV proteins act as IFN antagonists by inhibiting its production or signaling (Fig. 2) (DeDiego et al., 2014b; Totura and Baric, 2012; Zhong et al., 2012). Potent cytokines involved in controlling viral infections and priming adaptive immune responses are generated in response to CoV infection (Totura and Baric, 2012; Zhou et al., 2015). However, uncontrolled induction of these proinflammatory cytokines can also increase pathogenesis and disease severity as described for SARS-CoV and MERS-CoV (DeDiego et al., 2011, 2014a,b; Selinger et al., 2014). The cellular pathways mediated by IRF-3, IRF-7, activating transcription factor (ATF)-2/jun, AP-1, NF-kB, and nuclear factor of activated T cells, are the main drivers of the inflammatory response triggered after SARS-CoV infection, with the NF-kB pathway most strongly activated (DeDiego et al., 2011). CoV proteins that elicit this type of innate immune response may be modified to generate attenuated viruses suitable for vaccine development.

3.3 IFN Sensitivity of Human CoVs

Human CoVs are able to suppress IFN induction to different extents and, in general, are sensitive to the addition of exogenous IFN (Kindler et al., 2013). SARS-CoV in particular is highly resistant to the antiviral state induced by

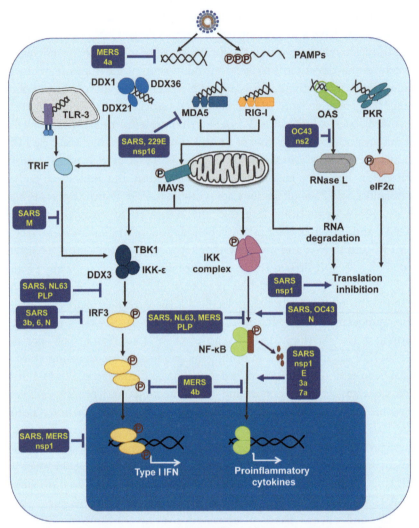

Fig. 2 Innate immunity signaling pathways affected by human CoV proteins. The main pathways leading to IFN and proinflammatory cytokine production are represented in the figure. These signaling routes are activated by dsRNA, which acts as a pathogen-associated molecular pattern (PAMP). The viral proteins affecting these pathways are indicated in the *dark blue boxes*. *SARS*, SARS-CoV; *MERS*, MERS-CoV; *229E*, HCoV-229E; *OC43*, HCoV-OC43; *NL63*, HCoV-NL63.

IFN treatment, suggesting that it possesses many mechanisms to counteract IFN–induced antiviral responses (Zielecki et al., 2013).

The betacoronavirus HCoV–OC43 is an exception for the sensitivity of human CoVs to IFN, as infection of certain cell types by this virus is favored in the presence of exogenous IFN. A possible explanation for this observation may be that HCoV-OC43 uses IFN-inducible transmembrane (IFITM) proteins 2 and 3 for cell entry (Zhao et al., 2014b). Interestingly, IFITM3-mediated antiviral activity was observed for the alphacoronaviruses HCoV-229E and HCoV-NL63, but not for the betacoronaviruses SARS-CoV and MERS-CoV (Wrensch et al., 2014), suggesting that IFITM proteins may have a positive effect on all betacoronavirus infections.

3.4 Innate Immunity Modulators Encoded by Common Human CoVs

A number of nsps have been implicated in human CoV pathogenesis, such as nsp1, nsp3, and nsp16, and, possibly, might be used to produce highly attenuated strains of common human CoVs as vaccine candidates.

Nsp1 protein is only encoded by members of the *Alphacoronavirus* and *Betacoronavirus* genera, with similar functions in all cases despite low-sequence conservation. The nsp1 proteins of HCoV-229E and HCoV-NL63 inhibit protein expression, most likely via their association with 40S ribosomal subunits, and they also act as IFN antagonists (Narayanan et al., 2015; Zust et al., 2007).

The nsp3 transmembrane replicase protein contains several functional domains (Fig. 3) that are conserved among all common human CoVs: two papain-like proteases (PLPs), two ubiquitin-like domains, and an ADP-ribose-1″-phosphatase (ADRP) domain. A role for ADRP in pathogenesis of human CoVs has been proposed, although it is not required for

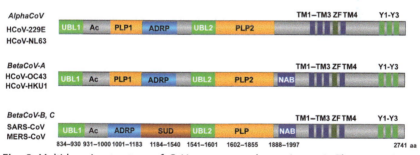

Fig. 3 Multidomain structure of CoV nonstructural protein nsp3. The approximate boundaries of each domain in SARS-CoV protein are indicated underneath by the amino acid numbers in the replicase polyprotein. *Ac*, Glu-rich acidic domain; *ADPR*, ADP-ribose-1″-phosphatase (also called macrodomain or X domain); *NAB*, nucleic acid-binding domain; *SUD*, SARS-unique domain; *TM1–TM4*, transmembrane domains forming an additional domain containing a metal-binding region (*ZF*); *UBL*, ubiquitin-like domain; *Y1–Y3*, Y domains preceding the C-terminal PLP cleavage sequence at nsp3/4.

viral growth in cell culture. HCoV–229E and SARS-CoV mutants lacking ADRP activity grow to similar titers compared to wild-type viruses in MRC-5 and Vero cells, respectively (Kuri et al., 2011; Putics et al., 2005). ADRP-deficient mutants of mouse hepatitis virus (MHV) were attenuated in vivo, most probably due to deficient induction of proinflammatory responses (Eriksson et al., 2008). ADRP mutants of HCoV–229E and SARS-CoV were reported to be more sensitive to IFN compared to wild-type viruses, suggesting that ADRP activity is involved in counteracting IFN activity (Kuri et al., 2011).

The PLP domains of nsp3 have also been linked to the modulation of innate immune response. The HCoV–NL63 PLP2 domain has been extensively characterized and acts as an IFN antagonist (Clementz et al., 2010), and apart from its PLP activity, it has deubiquitinating (DUB) and deISGylating activity (Chen et al., 2007; Clementz et al., 2010). HCoV–NL63 PLP2 DUB activity seems to be involved in modulation of NF-kB activation (Clementz et al., 2010) and p53-mediated modulation of IFN production (Takaoka et al., 2003). PLP2 deubiquitinates the cellular protein MDM2, leading to p53 proteasomal degradation and, as a consequence, decreased IFN production (Yuan et al., 2015).

Nsp16 has ribose $2'-O$-methyltransferase ($2'-O$-MTase) activity that is required to produce the $5'$ cap1 structures on viral RNAs. The methylation proved to be important to avoid viral RNA to be recognized as being "nonself" by the host–cell sensor MDA5 and subsequent activation of cellular innate immune responses. HCoV–229E mutants that lack $2'-O$-MTase activity replicated to 10^2-fold lower titers than the wild-type virus in MRC-5 cells, confirming the relevance of nsp16 activity. Moreover, the $2'-O$-MTase mutant was more sensitive to treatment with type I IFN, and macrophages infected with the mutant virus produced more type I IFN than those infected with the parental virus (Zust et al., 2011). The role of nsp16 in modulation of innate immunity seems to be conserved among all CoVs, as similar results were obtained for MHV and SARS-CoV, and a SARS-CoV mutant lacking $2'-O$-MTase activity was attenuated in vivo (Menachery et al., 2014).

CoVs contain a variable number of genus-specific genes that have been implicated in modulation of pathogenesis. In fact, all common human CoVs contain a genus-specific gene located between the S and envelope (E) genes, named ORF4 (HCoV–229E and HCoV–HKU1), ORF3 (HCoV–NL63), or ns12.9 (HCoV–OC43) (Fig. 1). These genes encode transmembrane proteins that differ significantly in both their sequences and predicted structures

(Muller et al., 2010; Zhang et al., 2014b, 2015), but may have related functions, such as ion channel activity and morphogenesis (Donaldson et al., 2008; Zhang et al., 2014b, 2015). In fact, the SARS-CoV 3a protein, HCoV-NL63 protein 3, and the HCoV-229E 4a protein are each able to complement *in trans* the absence of HCoV-OC43 ns12.9 protein (Zhang et al., 2015). The proteins are not essential for viral growth in cell culture but virus yield is increased between 6- and 25-fold if the protein is expressed (Zhang et al., 2014b, 2015). Also, there is evidence that these protein homologs may play an important role in vivo. For example, all clinical isolates of HCoV-229E contain a full-length ORF4, while in cell culture-adapted strains, this gene is spontaneously divided into two ORFs, 4a and 4b, or replaced by a truncated ORF encoding only the first transmembrane domain (Dijkman et al., 2006). Recombinant HCoV-NL63-Δ3 virus grows to a 10-fold lower titer than the parental virus in primary human airway epithelium (HAE) culture (Donaldson et al., 2008). Finally, a recombinant HCoV-OC43-Δns12.9 virus was attenuated in vivo (Zhang et al., 2015). These results suggest that ORF3 genes can be modified for the generation of attenuated viruses that may be used as vaccine candidates.

The HCoV-OC43 ns2 gene is predicted to encode a $2'$-$5'$-phosphodiesterase (PDE) based on its homology to the MHV ns2 gene (Roth-Cross et al., 2009). The MHV ns2 protein-mediated PDE activity interferes with the OAS-RNase L pathway, affecting pathogenesis by counteracting cellular innate immune responses (Zhao et al., 2012). Moreover, the absence of ns2 PDE activity can be complemented by other viral or cellular PDEs (Zhang et al., 2013).

Common human betacoronaviruses encode a hemagglutinin-esterase (HE) gene (Fig. 1). The HE protein is a transmembrane protein that is incorporated in the viral envelope, acting as a receptor-destroying enzyme with a role in viral entry (Huang et al., 2015). Interestingly, HCoV-OC43 mutants that lack the HE gene or encode an inactive HE protein produced replication-competent propagation-defective viruses (Desforges et al., 2013). These viruses, in principle, may be useful as vaccine candidates.

3.5 SARS-CoV Genes as Modulators of the Innate Immune Response

SARS-CoV encodes many known immunomodulatory proteins, such as nsp1, nsp3, nsp14, membrane (M), E, and nucleocapsid (N) (Fig. 2) (DeDiego et al., 2011, 2014a; Totura and Baric, 2012; Zhong et al., 2012). Modification or deletion of some of them may lead to attenuated viruses that could be used as live vaccine candidates. Many of these proteins also specifically inhibit cellular signaling pathways associated with the innate immune response.

It is generally thought that deletion or genetic modification of SARS-CoV genes encoding proteins that have IFN antagonist activity or other viral defense mechanisms, such as nsp1, nsp3, nsp16, M, 3b, 6, and N, will result in attenuated viruses, as has been shown for SARS-CoV nsp1 (Jimenez-Guardeno et al., 2015) and nsp16 mutants, respectively (Menachery et al., 2014).

As an alternative approach to produce attenuated CoVs to be used as vaccine candidates, the modification of viral replication fidelity has been proposed (Graham et al., 2012; Smith et al., 2013). Live-attenuated RNA virus vaccines are efficacious but subject to reversion to virulence. Among RNA viruses, replication fidelity is recognized as a key determinant of virulence and escape from antiviral therapy, though reduced fidelity is detrimental for some viruses. The replication fidelity of CoVs has been estimated to be approximately 20-fold higher than that of other RNA viruses and is mediated by a $3' \rightarrow 5'$ exonuclease (ExoN) activity (Minskaia et al., 2006) that probably functions in RNA proofreading (Eckerle et al., 2010). It has been demonstrated that engineered abrogation of SARS-CoV ExoN activity results in a stable mutator phenotype with profoundly decreased fidelity both in cell culture and in vivo, and attenuation of pathogenesis in young, aged, and immunocompromised mice (Graham et al., 2012). The ExoN null genotype and mutator phenotype are stable and do not revert to virulence, even after serial passage or long-term persistent infection in vivo. ExoN inactivation thus has potential for broad applications in the stable attenuation of CoVs.

Our laboratory has focused on the role of virulence of the SARS-CoV E and 3a proteins, and of the MERS-CoV E and 5 proteins, and on studying effects of their partial or complete deletion on virus attenuation and possible implications for vaccine development (DeDiego et al., 2007, 2011; Regla-Nava et al., 2015). The safety of these vaccine candidates has been further enhanced by genetic modifications of the nsp1-coding sequence (Jimenez-Guardeno et al., 2015). The CoV E protein is 76–109 aa in length, depending on the virus, and has diverse functions in CoV morphogenesis and the secretory pathway (Westerbeck and Machamer, 2015). The importance of E in different CoV species and genera varies greatly, ranging from being absolutely essential in some members of the genera *Alphacoronavirus* (TGEV) and *Betacoronavirus* (lineage C, MERS-CoV) to having nonessential functions in other betacoronaviruses (MHV (lineage A), SARS-CoV (lineage B)). In SARS-CoV, two strategies have been used to abrogate E function: complete deletion of E protein, and the introduction of small deletions of 8–12 aa at the E protein carboxyl-terminus (Jimenez-Guardeno et al., 2015). A similar strategy is currently being applied to MERS-CoV (J. Gutierrez, I. Sola, and L. Enjuanes, unpublished results). The SARS-CoV E protein is nonessential for virus replication and dissemination, and its deletion has led to attenuated forms of SARS-CoV that are promising vaccine candidates (see later).

Viral infections often induce endoplasmic reticulum (ER) stress, specifically the unfolded protein response (UPR). UPR induces three main signaling pathways to avoid the accumulation of proteins with altered folding in the ER. Stress in the ER is also interconnected with the innate immune response and other host-cell responses (Hetz, 2012). Both HCoV-OC43 and HCoV-HKU1 induce ER stress by a mechanism that is associated with S protein expression leading to virulent CoVs in a mouse model (Favreau et al., 2009; Siu et al., 2014a). Similarly, in the case of SARS-CoV, the E protein has been linked to virulence and the induction of UPR (DeDiego et al., 2011).

E protein has a major role in inflammasome activation and the associated exacerbated inflammation elicited by SARS-CoV in the lung parenchyma (Nieto-Torres et al., 2014). This exacerbated inflammation causes edema leading to acute respiratory distress syndrome and, frequently, to the death of experimentally infected animals or human patients. The E protein is a viroporin that conducts Ca^{2+} ions. Changes in intracellular Ca^{2+} concentration, mediated by E protein ion channel activity, are responsible for inflammasome activation (Nieto-Torres et al., 2015). Elimination of ion channel activity by the introduction of aa substitutions within the transmembrane domain of E protein led to virus attenuation (Nieto-Torres et al., 2014).

At its carboxyl-terminus, the E protein contains another virulence factor, a PDZ domain-binding motif (PBM) that, during SARS-CoV infection, could potentially target more than 400 cellular proteins containing PDZ domains with possible effects on viral pathogenicity. Interestingly, deletion or modification of the E protein PBM resulted in attenuated SARS-CoVs that are good vaccine candidates (Jimenez-Guardeño et al., 2014; Regla-Nava et al., 2015).

Deletion of full-length E protein led to the attenuation of SARS-CoV in several animal models (Lamirande et al., 2008; Netland et al., 2010).

The immunogenicity and protective efficacy of a live-attenuated vaccine based on a recombinant SARS-CoV lacking the E gene (rSARS-CoV-Δ E) was first studied using hamsters. After immunization with rSARS-CoV-ΔE, hamsters developed high serum neutralizing antibody titers and were protected from challenge with homologous (SARS-CoV Urbani) and heterologous (GD03) SARS-CoV in the upper and lower respiratory tract. Deletion of the E protein modestly diminished viral growth in cell culture but abrogated virulence in mice (Netland et al., 2010). We have shown that immunization with rSARS-CoV-ΔE almost completely protected BALB/c mice from fatal respiratory disease caused by mouse-adapted SARS-CoV, and partly protected hACE2 Tg mice from lethal disease, although hACE2 Tg mice, which express the human SARS-CoV receptor, were extremely susceptible to infection. Furthermore, we also showed that rSARS-CoV-ΔE induces antiviral T-cell and antibody responses. To improve vaccine efficacy, we engineered and adapted rSARS-CoV to efficiently grow in mice (rSARS-CoV-MA15). To this end, we incorporated six nucleotide substitutions into the SARS-CoV genome (Frieman et al., 2012; Roberts et al., 2007) using our reverse-genetics system based on bacterial artificial chromosomes. Using the rSARS-CoV-MA as a backbone genome, a second set of E-deletion vaccine candidates was generated (Fett et al., 2013). rSARS-CoV-MA15-ΔE was safe, causing no disease in 6-week-, 12-month-, or 18-month-old BALB/c mice. Immunization with this virus completely protected mice at three ages from lethal disease and induced a more rapid virus clearance. Compared to rSARS-CoV-Δ E, rSARS-CoV-MA-ΔE immunization elicited significantly greater neutralizing antibody titers and virus-specific $CD4^+$ and $CD8^+$ T-cell responses. After challenge, inflammatory cell infiltration, edema, and cell destruction were decreased in the lungs of rSARS-CoV-MA15-ΔE-immunized mice, compared to those from rSARS-CoV-ΔE-immunized 12-month-old mice, suggesting that this mouse-adapted virus is a safe candidate vaccine.

To identify E protein domains that contribute to SARS-CoV-MA15-Δ E attenuation, several rSARS-CoVs with mutations or deletions in the E protein were generated. Substitutions in the amino terminus or deletion of regions in the internal and carboxy-terminal regions of the E protein led to virus attenuation (Regla-Nava et al., 2015). Attenuated viruses induced minimal lung injury, diminished edema, limited neutrophil influx, and increased $CD4^+$ and $CD8^+$ T-cell counts in the lungs of BALB/c mice, compared to animals infected with the wild-type virus. The attenuated viruses completely protected mice against challenge with the lethal parental virus, considerably increasing the survival of infected animals and indicating that these viruses are promising vaccine candidates.

Ideally, a vaccine should provide full protection, and this was the case for the SARS-CoV vaccine candidate that was engineered by deleting the full-length E protein. Nevertheless, a good vaccine must also be genetically stable, and the initial rSARS-CoV-ΔE vaccine was unstable both in cell culture and in vivo. In fact, this mutant virus regained fitness after serial passage in cell culture, resulting in the partial duplication of the membrane gene, and while the chimeric protein increased viral fitness in vitro, the virus remained attenuated in mice. When the full-length E gene was deleted or its PBM coding sequence mutated, revertant viruses either evolved a novel chimeric gene including PBM or restored the sequence of the PBM in the E protein, respectively (Jimenez-Guardeno et al., 2015). During SARS-CoV-ΔE passage in mice, the virus incorporated a mutant variant of the 8a protein including a PBM, resulting in reversion to a virulent phenotype. These data, and additional evidence, led us to conclude that the virus requires a PBM on a transmembrane protein to compensate the removal of this motif from the E protein (Jimenez-Guardeno et al., 2015). Therefore, to increase the genetic stability of the vaccine candidate, we introduced small attenuating deletions in the E gene that did not affect the endogenous PBM, preventing the selection of revertants incorporating novel PBMs into the virus genome and leading to a genetically and functionally stable vaccine candidate. To further increase vaccine safety, we introduced additional attenuating amino acid substitutions in the nsp1 protein. Recombinant viruses including attenuating mutations in the E and nsp1-coding sequences maintained their attenuation after passage in vitro and in vivo. Furthermore, these viruses fully protected mice against challenge with the lethal parental virus, and are therefore safe and stable vaccine candidates for protection against SARS-CoV (Jimenez-Guardeno et al., 2015).

3.6 MERS-CoV Genes as Modulators of the Innate Immune Response

The relevance of MERS-CoV proteins in modulating cellular innate immune responses varies depending on the experimental approach used. Two main strategies have been applied, either the expression of individual virus proteins or the deletion of appropriate protein-coding sequences from the full-length virus genome. Using the first approach, overexpressing MERS-CoV proteins in human 293T transfected cells, the structural, and accessory proteins M, 4a, 4b, and 5 were identified as potential IFN antagonists (Matthews et al., 2014; Yang et al., 2013). In contrast, when the role of these virus proteins was analyzed in cells infected with MERS-CoV deletion mutants in which individual genes (3, 4a, 4b, and 5, respectively) were deleted, it was observed that, among these proteins, 4b was the major antagonist of the innate immune response induction (J. Canton, S. Perlman, L. Enjuanes, and I. Sola, 2016, unpublished results). Deletion of ORF4a resulted in a MERS-CoV mutant with increased INF sensitivity. The effect may be associated with the dsRNA-binding domain of the 4a protein that was shown to interact with PACT in an RNA-dependent manner; thereby, reducing the activation of RIG-I and MDA5 in overexpression assays (Siu et al., 2014b). Also, the combined deletion of the accessory genes 3, 4a, 4b, and 5 in a recombinant MERS-CoV (rMERS-CoV-ΔORF3-5) showed a 1–1.5 log reduction in viral titer compared with the full-length MERS-CoV (Scobey et al., 2013). Therefore, engineering MERS-CoV with deletions in one or more genes could lead to promising vaccine candidates.

The complete deletion of the E gene in MERS-CoV led to the generation of replication-competent propagation-defective viruses that are promising vaccine candidates (Almazan et al., 2013). Because E protein induces cell apoptosis (An et al., 1999; DeDiego et al., 2011), the establishment of stable cell lines constitutively expressing this protein to complement viruses lacking E protein has not been possible. To overcome this limitation, we have generated packaging cell lines that transiently express E protein in an inducible manner and used them to grow replication-competent propagation-defective MERS-CoV-ΔE (J. Gutierrez-Alvarez, I. Sola, and L. Enjuanes, 2016, unpublished results). Additionally, fully stable transformed mammalian cells were generated in which E protein expression was under the control of an optimized Tet-On system for doxycycline-inducible gene expression, drastically reducing leaky expression (Das et al., 2016; Markusic et al., 2005) and (J. Gutierrez, I. Sola, and L. Enjuanes, 2016, unpublished results).

In an alternative strategy, replication- and propagation-efficient rMERS-CoVs were generated that expressed a slightly shortened E protein lacking 8–12 aa at the carboxyl-terminus, as previously reported for SARS-CoV vaccine candidates (J. Gutierrez-Alvarez, S. Perlman, I. Sola, and L. Enjuanes, 2016, unpublished results). Thus, the lessons learned from previous SARS-CoV vaccine biosafety studies, such as the use of attenuated viruses expressing a shortened E protein (see earlier), may also be applicable to engineering MERS-CoV vaccines.

MERS-CoV replicase proteins nsp1, nsp3, and, possibly, nsp14, may also interfere with the signaling pathways associated with the innate immune response through different mechanisms (Lokugamage et al., 2015; Yang et al., 2014). Therefore, modification or deletion of the replicase gene sequences encoding these proteins may also lead to the generation of attenuated viruses. Similar to nsp1 of SARS-CoV, which inhibits host gene expression at the translational level, it has been reported that MERS-CoV nsp1 exhibits a conserved function, inhibiting host mRNA translation and inducing host mRNA degradation. This information could be exploited to produce MERS-CoV vaccine candidates (Lokugamage et al., 2015).

As in the common human CoVs, SARS-CoV and MERS-CoV PLPs have deISGylating and DUB activities and act as interferon antagonists (Barretto et al., 2005; Chen et al., 2007; Clementz et al., 2010; Frieman et al., 2009; Ratia et al., 2006; Zheng et al., 2008). Mutations introduced into the MERS-CoV PLP coding sequence to specifically disrupt ubiquitin binding without affecting viral polyprotein cleavage led to PLP variants without DUB activity that lost their wild-type ability to inhibit IFN-β promoter activation in reporter assays. These findings directly implicate DUB and deISGylating functions of PLP in the inhibition of IFN-β promoter activity, and such modification may lead to attenuated MERS-CoV. In addition, PLP catalytic activity was required by both MERS-CoV and SARS-CoV to reduce induction of endogenous proinflammatory cytokines in infected cells (Mielech et al., 2014), consistent with the important functions of ubiquitination and modification of cellular proteins by interferon-stimulated gene 15 (ISG15) in regulating cellular innate immune pathways. On the other hand, it has been shown that the SARS-CoV PLP interferes with the formation of the signaling complexes including STING (stimulator of interferon genes), TRAF3, TBK1, and IKKε; thus, preventing downstream phosphorylation, dimerization, and nuclear translocation of IRF3 mediated by STING and TBK1 (Fig. 2) (Chen et al., 2014; Yang et al., 2014). This inhibition is not dependent on the PLP catalytic activity (Chen et al., 2014).

These studies provide valuable information on how MERS-CoV PLP-mediated antagonism of the host innate immune response is orchestrated and offers additional attractive options for designing attenuated viruses as vaccine candidates.

4. VACCINE BIOSAFETY

4.1 ADEI and Eosinophilia Induction

In general, subunit vaccines against SARS-CoV and MERS-CoV administered with and without adjuvants provide different degrees of protection (Haagmans et al., 2016; Jaume et al., 2012; Wang et al., 2014). However, as mentioned earlier, these vaccines should be used with caution in humans because of possible ADEI mechanisms, especially when antibody levels are low. Highly concentrated antisera against SARS-CoV were shown to neutralize virus infectivity, whereas diluted mono- and polyclonal anti-S protein antibodies both caused ADEI in human promonocyte cell cultures, leading to cytopathic effects and increased levels of TNF-α, IL-4, and IL-6 (Haagmans et al., 2016; Jaume et al., 2012; Wang et al., 2014).

In addition, it has been documented that immunization of mice with VLPs or inactivated virus, both in the presence or absence of adjuvant, induced eosinophilic immunopathology in young and aged mice (Bolles et al., 2011; Tseng et al., 2012). Using a double-inactivated SARS-CoV vaccine, protection was observed after homologous and heterologous challenge. Protection against a nonlethal heterologous challenge was poor, and enhanced immune pathology was comparable with that seen in SARS-CoV N protein-immunized mice. Importantly, aged mice displayed increased eosinophilic immune pathology in the lungs, and mice were not protected significantly against virus replication.

The induction of immunopathology was also observed during challenge in ferrets and NHPs immunized with candidate vaccines based on VLPs, the whole-inactivated SARS-CoV virus, and rDNA-produced S protein (Tseng et al., 2012). The pulmonary damage after challenge with SARS-CoV was associated with a Th2-type immunopathology with prominent eosinophil infiltration, and upregulation of genes associated with the induction of eosinophilia was observed. Interestingly, this immunopathology in the lungs upon SARS-CoV infection could be avoided by administration of TLR agonist adjuvants (Iwata-Yoshikawa et al., 2014).

To increase the duration of immune responses elicited by vaccines and to prevent the CoV-induced lung immunopathology observed after challenge or natural infection, the effects of adjuvants have been studied. In immunizations either with recombinant CoV S protein or with inactivated whole-virus vaccines, the effects of different adjuvants including alum, CpG, and Adva, a new delta-inulin-based polysaccharide adjuvant, were analyzed

(Honda-Okubo et al., 2015). While all vaccines protected against lethal infection, addition of an adjuvant significantly increased serum neutralizing antibody titers and reduced lung virus titers on day 3 postchallenge. Whereas adjuvant-free or alum-formulated vaccines were associated with a significant increase in lung eosinophilic immunopathology on day 6 postchallenge, this was not seen in mice immunized with vaccines formulated with the delta-inulin adjuvant. The absence of eosinophilic immunopathology in vaccines containing delta-inulin adjuvants was found to correspond to enhanced T-cell IFN-γ-recall responses rather than reduced IL-4 responses, suggesting that immunopathology is primarily caused by an inadequate vaccine-induced Th1 response and illustrating the need to induce durable IFN-γ responses using appropriate adjuvants.

4.2 Interaction of CoV Vaccine Candidates with Cells of the Immune System

The possibility of productive infection in macrophages (MØ) and dendritic cells (DCs) by MERS-CoV is under debate. Whereas some reports provided evidence for the infection of human MØs and DCs (Table 1) (Chu et al., 2014, 2016; Scheuplein et al., 2015; Ying et al., 2016), others have not been able to productively grow virus in MØs from Tg mice-expressing hDPP4 (C. Tseng, personal communication). The different origin of the cells might explain these divergent results but more work is needed to clarify the susceptibility of MØs and DCs to MERS-CoV.

Common human CoVs, such as HCoV-NL63, HCoV-OC43, and HCoV-HKU1, as well as the highly pathogenic SARS-CoV and MERS-CoV infect ciliated epithelial cells in HAE culture, whereas HCoV-229E infects nonciliated cells (Dijkman et al., 2013; Kindler et al., 2013). Human CoVs do not elicit the production of proinflammatory cytokines in human primary respiratory epithelial cells or ex vivo human lung tissue culture, and the production of type I and III IFN was consistently found to be low (Chan et al., 2013; Kindler et al., 2013; Zielecki et al., 2013), suggesting that all human CoVs have evolved mechanisms to antagonize the host innate immune response, mediated by a variety of viral proteins.

Epithelial cells, MØs, and DCs play a role during virus replication in the lung, and may have an impact on pathogenesis. Both HCoV-229E and HCoV-OC43 can infect MØs, although only the former does it efficiently (Collins, 2002; Patterson and Macnaughton, 1982). DCs serve as sentinels in the respiratory tract, and are the connection between innate and adaptive immunity in the lung. HCoV-229E efficiently infects and kills DCs, which

Table 1 Infection of Immune Cells by SARS-CoV and MERS-CoV

			Cells			
Virus	**MØ**		**DC**		**Epithelial**	**T**
MERS-CoV	**PROD**		**PROD**		**PROD**	**ABOR**
	CXCL-10	↑	CXCL-10	↑	CXCL-10	↓	Apoptosis
	CCL-2, -36, -5	↑	CCL-5	↑	CCL-5	↓	Lymphopenia
	IL-8, -12	↑	IL-12	↑	IL-1β	↓	
			IFN-γ	↑	IL-6, -8	↓	
	IFN-α/β	↓	IFN-α/β	↓	IFN-α/β	↓	
					IFN-III	↓	
SARS-CoV	**ABOR**		**ABOR**		**PROD**	**ABOR**
	CXCL-10	↑	CXCL-10	↑	CXCL-10	↓	Apoptosis
	CCL-2	↑	CCL-5	↑	CCL-5	↓	Lymphopenia
	IL-8, -12	↑	CCL-2, -3	↑	IL-1β	↓	
			IL-12		IL-6, -8	↓	
	IFN-α/β	↓	IFN-γ	↑	IFN-α/β	↓	
			IFN-α/β	↓	IFN-III	↓	

has been proposed as a potential mechanism to delay host adaptive immunity, providing time to replicate in the host (Mesel-Lemoine et al., 2012).

The interaction of MERS-CoV with human MØs and DCs (Chu et al., 2014, 2016; Scheuplein et al., 2015; Ying et al., 2016) induces CXCL-10, CCL-2, CCL-3, CCL-5, IL-8, and IL-12 (Zhou et al., 2014). MERS-CoV also infects mouse monocyte-derived dendritic cells, inducing CXCL-10, CCL-5, IL-12, and IFN-γ, although no IFN-β and only marginal IFN-α levels were detected (Chu et al., 2014; Zhou et al., 2015). In contrast, in human plasmacytoid dendritic cells, MERS-CoV induced large amounts of type I and III IFNs, especially IFN-α (Scheuplein et al., 2015), although the infection was abortive.

SARS-CoV also infects human MØs and DCs, but viral replication is abortive and no infectious virus particles are produced (Cheung et al., 2005; Law et al., 2005; Tseng et al., 2005; Yilla et al., 2005; Ziegler et al., 2005). Despite the lack of productive infection in human MØs, SARS-CoV induced the expression of proinflammatory chemokines, including CXCL-10 and CCL-2 but, in contrast, antiviral cytokines such as IFN-α and INF-β were basically absent (Cheung et al., 2005; Law et al., 2005, 2009; Tseng et al., 2005). In unproductive infections of DCs, SARS-CoV induced CXCL-10, CCL-2, CCL-3, CCL-5, and

TNF-α (Law et al., 2005; Tseng et al., 2005). Dysregulated type I IFN and inflammatory monocyte-MØ responses in SARS-CoV-infected mice resulted in high levels of cytokines and chemokines and impaired virus-specific T-cell responses leading to lethal pneumonia (Channappanavar et al., 2016).

Both MERS-CoV and SARS-CoV infect human primary T cells and induce massive apoptosis and lymphopenia, although they do not produce infectious virus in these cells (Chu et al., 2016; Gu et al., 2005; Zhou et al., 2015). Infection of T cells may play a role in controlling the pathogenesis elicited by both MERS-CoV and SARS-CoV (Chen et al., 2010).

Similarly to the field viruses, live SARS-CoV and MERS-CoV vaccines may interact with host MØs and DCs, promoting the synthesis of cytokines and chemokines that could lead to undesired side effects including imbalanced proinflammatory immune responses. The analysis of the type of interleukins and cytokines produced during vaccine administration should be included in safety studies of candidate vaccines and could involve murine leukocytes derived from susceptible Tg mice and human cells collected from healthy donors.

5. CORONAVIRUS ANTIVIRAL SELECTION

Despite extensive efforts commencing with the SARS epidemic in 2003, no antiviral drugs suitable to treat CoV infections have been approved by the FDA (Barnard and Kumaki, 2011; Kilianski and Baker, 2014). Nevertheless, there are several antiviral compounds in preclinical development that may be useful to control human CoV infections (Adedeji and Sarafianos, 2014; Zumla et al., 2016). A number of these compounds target highly conserved replicase proteins, and therefore should be effective against a broad range of CoVs, including common human CoVs. Key enzymatic activities, such as the viral main protease, PLP, and helicase have been employed as targets for CoV-specific antiviral drugs (Kilianski and Baker, 2014). These conserved antiviral targets are generally thought to tolerate fewer mutations compared to structural protein genes due to higher fitness pressure; thus, possibly reducing the risk of resistant variants emerging rapidly during antiviral treatment.

Other compounds inhibiting virus entry or morphogenesis have also been tested though, with few exceptions, these viral life cycle steps are poor antiviral targets, as escape mutants are easily recovered, especially when viral structural proteins are targeted (Kilianski and Baker, 2014). The cell attachment of HCoV-229E and HCoV-NL63 has been inhibited with different antibody combinations (Pyrc et al., 2006, 2007), and subsequent steps of cell entry have been inhibited by S protein-derived heptad-repeat 2 (HR2) peptides in the case of HCoV-NL63 (Pyrc et al., 2006). Similarly, inhibitors of vacuolar acidification were effective against HCoV-229E in cell culture and against HCoV-OC43 both in cells and in vivo (Keyaerts et al., 2009; Pyrc et al., 2007).

The identification of signaling pathways involved in SARS-CoV-mediated pathogenesis has provided selection systems for drugs that significantly increase the survival of infected mice. For example, the identification of NF-kB as the main signaling pathway leading to an exacerbated inflammatory response during SARS-CoV infection enabled the selection of an antiviral suitable to control this infection (DeDiego et al., 2014b). Similarly, the identification of the increased phosphorylation of p38 MAPK by SARS-CoV E protein-activated syntenin also led to a dramatic increase (>80%) in the survival of infected mice treated with p38 MAPK inhibitors (Jimenez-Guardeno et al., 2015) (Fig. 4).

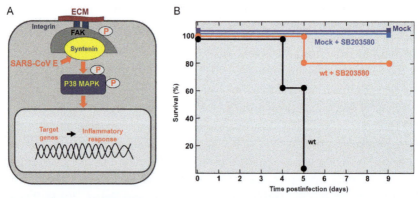

Fig. 4 Inhibitors of p38 MAPK activation protects mice infected with recombinant SARS-CoV. (A) Syntenin initiates a signaling cascade resulting in the phosphorylation (and activation) of p38 MAPK, a protein involved in the expression of proinflammatory cytokines. (B) Inhibition of p38 MAPK phosphorylation led to the survival of 80% of the mice infected with recombinant SARS-CoV, confirming the antiviral potency of this drug (Jimenez-Guardeno et al., 2015). *E*, SARS-CoV envelope protein; *ECM*, extracellular matrix; *FAK*, signaling adhesion kinase protein; *Mock*, noninfected mice; *P*, phosphorylated residue; *P38 MAPK*, p38 MAP kinase; *SB203580*, inhibitor of p38 MAPK; *Wt*, mice infected with virulent SARS-CoV.

6. CONCLUSIONS

New animal and human CoVs are constantly emerging or reemerging, as there are animal reservoirs that maintain them, including bats and birds present in high numbers and with high mobility. As a consequence, the development of technologies suitable to respond swiftly to newly emerging CoVs by producing vaccines is highly desirable. This goal could be achieved with the production of subunit or inactivated vaccines, but optimal combinations of antigen and adjuvant need to be established to minimize the risk of ADEI or eosinophilia that frequently occurred with CoV vaccines developed in the past. We believe that the vaccines of the future will be mainly based on live-attenuated viruses because of their superior potential to induce balanced Th1/Th2 immune responses, the potent and long-lasting immunity, and the comprehensive B and T-cell repertoire induced by this type of vaccines. However, the development of safe live-attenuated virus vaccines requires strong experimental support to confirm sufficient attenuation in the target species, possibly resulting in longer development and production times.

ACKNOWLEDGMENTS

This work was supported by grants from the Government of Spain (BIO2013-42869-R), a U.S. National Institutes of Health (NIH) project (2P01AI060699-06A1), and financial support of IMI and European Commission and in-kind contributions from EFPIA partners (ZAPI project, IMI Grant Agreement n°115760). S.Z., C.C.R., J.G.A., and J.C. received contracts from NIH, Fundacion La Caixa, European Commission and Government of Spain, respectively. We thank Marga González for her technical assistance.

REFERENCES

Adedeji, A.O., Sarafianos, S.G., 2014. Antiviral drugs specific for coronaviruses in preclinical development. Curr. Opin. Virol. 8, 45–53.

Agrawal, A.S., Garron, T., Tao, X., Peng, B.H., Wakamiya, M., Chan, T.S., et al., 2015. Generation of a transgenic mouse model of Middle East respiratory syndrome coronavirus infection and disease. J. Virol. 89, 3659–3670.

Almazan, F., Gonzalez, J.M., Penzes, Z., Izeta, A., Calvo, E., Plana-Duran, J., et al., 2000. Engineering the largest RNA virus genome as an infectious bacterial artificial chromosome. Proc. Natl. Acad. Sci. U.S.A. 97, 5516–5521.

Almazan, F., DeDiego, M.L., Sola, I., Zuñiga, S., Nieto-Torres, J.L., Marquez-Jurado, S., et al., 2013. Engineering a replication-competent, propagation-defective Middle East respiratory syndrome coronavirus as a vaccine candidate. MBio 4. e00650-13.

Altenburg, A.F., Kreijtz, J.H., de Vries, R.D., Song, F., Fux, R., Rimmelzwaan, G.F., et al., 2014. Modified vaccinia virus ankara (MVA) as production platform for vaccines against influenza and other viral respiratory diseases. Viruses 6, 2735–2761.

An, S., Chen, C.J., Yu, X., Leibowitz, J.L., Makino, S., 1999. Induction of apoptosis in murine coronavirus-infected cultured cells and demonstration of E protein as an apoptosis inducer. J. Virol. 73, 7853–7859.

Angelini, M.M., Akhlaghpour, M., Neuman, B.W., Buchmeier, M.J., 2013. Severe acute respiratory syndrome coronavirus nonstructural proteins 3, 4, and 6 induce double-membrane vesicles. MBio 4. e00524-13.

Barnard, D.L., Kumaki, Y., 2011. Recent developments in anti-severe acute respiratory syndrome coronavirus chemotherapy. Futur. Virol. 6, 615–631.

Barretto, N., Jukneliene, D., Ratia, K., Chen, Z., Mesecar, A.D., Baker, S.C., 2005. The papain-like protease of severe acute respiratory syndrome coronavirus has deubiquitinating activity. J. Virol. 79, 15189–15198.

Bolles, M., Deming, D., Long, K., Agnihothram, S., Whitmore, A., Ferris, M., et al., 2011. A double-inactivated severe acute respiratory syndrome coronavirus vaccine provides incomplete protection in mice and induces increased eosinophilic proinflammatory pulmonary response upon challenge. J. Virol. 85, 12201–12215.

Butler, N., Pewe, L., Trandem, K., Perlman, S., 2006. Murine encephalitis caused by HCoV-OC43, a human coronavirus with broad species specificity, is partly immune-mediated. Virology 347, 410–421.

Chan, R.W., Chan, M.C., Agnihothram, S., Chan, L.L., Kuok, D.I., Fong, J.H., et al., 2013. Tropism of and innate immune responses to the novel human betacoronavirus lineage C virus in human ex vivo respiratory organ cultures. J. Virol. 87, 6604–6614.

Channappanavar, R., Fett, C., Zhao, J., Meyerholz, D.K., Perlman, S., 2014. Virus-specific memory CD8 T cells provide substantial protection from lethal severe acute respiratory syndrome coronavirus infection. J. Virol. 88, 11034–11044.

Channappanavar, R., Fehr, A.R., Vijay, R., Mack, M., Zhao, J., Meyerholz, D.K., et al., 2016. Dysregulated type I interferon and inflammatory monocyte-macrophage responses cause lethal pneumonia in SARS-CoV-infected mice. Cell Host Microbe 19, 181–193.

Chen, Z., Wang, Y., Ratia, K., Mesecar, A.D., Wilkinson, K.D., Baker, S.C., 2007. Proteolytic processing and deubiquitinating activity of papain-like proteases of human coronavirus NL63. J. Virol. 81, 6007–6018.

Chen, J., Lau, Y.F., Lamirande, E.W., Paddock, C.D., Bartlett, J.H., Zaki, S.R., et al., 2010. Cellular immune responses to severe acute respiratory syndrome coronavirus (SARS-CoV) infection in senescent BALB/c mice: CD4+ T cells are important in control of SARS-CoV infection. J. Virol. 84, 1289–1301.

Chen, X., Yang, X., Zheng, Y., Yang, Y., Xing, Y., Chen, Z., 2014. SARS coronavirus papain-like protease inhibits the type I interferon signaling pathway through interaction with the STING-TRAF3-TBK1 complex. Protein Cell 5, 369–381.

Cheung, C.Y., Poon, L.L., Ng, I.H., Luk, W., Sia, S.F., Wu, M.H., et al., 2005. Cytokine responses in severe acute respiratory syndrome coronavirus-infected macrophages in vitro: possible relevance to pathogenesis. J. Virol. 79, 7819–7826.

Chu, H., Zhou, J., Wong, B.H., Li, C., Cheng, Z.S., Lin, X., et al., 2014. Productive replication of Middle East respiratory syndrome coronavirus in monocyte-derived dendritic cells modulates innate immune response. Virology 454–455, 197–205.

Chu, H., Zhou, J., Wong, B.H., Li, C., Chan, J.F., Cheng, Z.S., et al., 2016. Middle East respiratory syndrome coronavirus efficiently infects human primary T lymphocytes and activates the extrinsic and intrinsic apoptosis pathways. J. Infect. Dis. 213, 904–914.

Clementz, M.A., Chen, Z., Banach, B.S., Wang, Y., Sun, L., Ratia, K., et al., 2010. Deubiquitinating and interferon antagonism activities of coronavirus papain-like proteases. J. Virol. 84, 4619–4629.

Cockrell, A.S., Peck, K.M., Yount, B.L., Agnihothram, S.S., Scobey, T., Curnes, N.R., et al., 2014. Mouse dipeptidyl peptidase 4 is not a functional receptor for Middle East respiratory syndrome coronavirus infection. J. Virol. 88, 5195–5199.

Coleman, C.M., Liu, Y.V., Mu, H., Taylor, J.K., Massare, M., Flyer, D.C., et al., 2014a. Purified coronavirus spike protein nanoparticles induce coronavirus neutralizing antibodies in mice. Vaccine 32, 3169–3174.

Coleman, C.M., Matthews, K.L., Goicochea, L., Frieman, M.B., 2014b. Wild-type and innate immune-deficient mice are not susceptible to the Middle East respiratory syndrome coronavirus. J. Gen. Virol. 95, 408–412.

Collins, A.R., 2002. In vitro detection of apoptosis in monocytes/macrophages infected with human coronavirus. Clin. Diagn. Lab. Immunol. 9, 1392–1395.

Corman, V.M., Ithete, N.L., Richards, L.R., Schoeman, M.C., Preiser, W., Drosten, C., et al., 2014. Rooting the phylogenetic tree of middle East respiratory syndrome coronavirus by characterization of a conspecific virus from an African bat. J. Virol. 88, 11297–11303.

Das, A.T., Zhou, X., Metz, S.W., Vink, M.A., Berkhout, B., 2016. Selecting the optimal Tet-On system for doxycycline-inducible gene expression in transiently transfected and stably transduced mammalian cells. Biotechnol. J. 11, 71–79.

Day, C.W., Baric, R., Cai, S.X., Frieman, M., Kumaki, Y., Morrey, J.D., et al., 2009. A new mouse-adapted strain of SARS-CoV as a lethal model for evaluating antiviral agents in vitro and in vivo. Virology 395, 210–222.

de Groot, R.J., Baker, S.C., Baric, R., Enjuanes, L., Gorbalenya, A.E., Holmes, K.V., et al., 2012. Family coronaviridae. In: King, A.M.Q., Adams, M.J., Carstens, E.B., Lefkowitz, E.J. (Eds.), Virus Taxonomy. Elsevier, Amsterdam, pp. 774–796.

DeDiego, M.L., Alvarez, E., Almazan, F., Rejas, M.T., Lamirande, E., Roberts, A., et al., 2007. A severe acute respiratory syndrome coronavirus that lacks the E gene is attenuated in vitro and in vivo. J. Virol. 81, 1701–1713.

DeDiego, M.L., Nieto-Torres, J.L., Jimenez-Guardeño, J.M., Regla-Nava, J.A., Alvarez, E., Oliveros, J.C., et al., 2011. Severe acute respiratory syndrome coronavirus envelope protein regulates cell stress response and apoptosis. PLoS Pathog. 7, e1002315.

DeDiego, M.L., Nieto-Torres, J.L., Jimenez-Guardeño, J.M., Regla-Nava, J.A., Castaño-Rodriguez, C., Fernandez-Delgado, R., et al., 2014a. Coronavirus virulence genes with main focus on SARS-CoV envelope gene. Virus Res. 194, 124–137.

DeDiego, M.L., Nieto-Torres, J.L., Regla-Nava, J.A., Jimenez-Guardeño, J.M., Fernandez-Delgado, R., Fett, C., et al., 2014b. Inhibition of NF-kappaB mediated inflammation in severe acute respiratory syndrome coronavirus-infected mice increases survival. J. Virol. 88, 913–924.

Desforges, M., Desjardins, J., Zhang, C., Talbot, P.J., 2013. The acetyl-esterase activity of the hemagglutinin-esterase protein of human coronavirus OC43 strongly enhances the production of infectious virus. J. Virol. 87, 3097–3107.

Dijkman, R., Jebbink, M.F., Wilbrink, B., Pyrc, K., Zaaijer, H.L., Minor, P.D., et al., 2006. Human coronavirus 229E encodes a single ORF4 protein between the spike and the envelope genes. Virol. J. 3, 106.

Dijkman, R., Jebbink, M.F., Koekkoek, S.M., Deijs, M., Jonsdottir, H.R., Molenkamp, R., et al., 2013. Isolation and characterization of current human coronavirus strains in primary human epithelial cell cultures reveal differences in target cell tropism. J. Virol. 87, 6081–6090.

Donaldson, E.F., Yount, B., Sims, A.C., Burkett, S., Pickles, R.J., Baric, R.S., 2008. Systematic assembly of a full-length infectious clone of human coronavirus NL63. J. Virol. 82, 11948–11957.

Drosten, C., Muth, D., Corman, V.M., Hussain, R., Al Masri, M., HajOmar, W., et al., 2015. An observational, laboratory-based study of outbreaks of middle East respiratory syndrome coronavirus in Jeddah and Riyadh, kingdom of Saudi Arabia, 2014. Clin. Infect. Dis. 60, 369–377.

Du, L., Jiang, S., 2015. Middle East respiratory syndrome: current status and future prospects for vaccine development. Expert. Opin. Biol. Ther. 15, 1647–1651.

Du, L., Zhao, G., Lin, Y., Sui, H., Chan, C., Ma, S., et al., 2008. Intranasal vaccination of recombinant adeno-associated virus encoding receptor-binding domain of severe acute respiratory syndrome coronavirus (SARS-CoV) spike protein induces strong mucosal immune responses and provides long-term protection against SARS-CoV infection. J. Immunol. 180, 948–956.

Du, L., Zhao, G., Kou, Z., Ma, C., Sun, S., Poon, V.K., et al., 2013. Identification of receptor-binding domain in S protein of the novel human coronavirus MERS-CoV as an essential target for vaccine development. J. Virol. 17, 9939–9942.

Eckerle, L.D., Becker, M.M., Halpin, R.A., Li, K., Venter, E., Lu, X., et al., 2010. Infidelity of SARS-CoV nsp14-exonuclease mutant virus replication is revealed by complete genome sequencing. PLoS Pathog. 6, e1000896.

Enjuanes, L., DeDiego, M.L., Alvarez, E., Deming, D., Sheahan, T., Baric, R., 2008. Vaccines to prevent severe acute respiratory syndrome coronavirus-induced disease. Virus Res. 133, 45–62.

Eriksson, K.K., Cervantes-Barragan, L., Ludewig, B., Thiel, V., 2008. Mouse hepatitis virus liver pathology is dependent on ADP-ribose-1″-phosphatase, a viral function conserved in the alpha-like supergroup. J. Virol. 82, 12325–12334.

Escriou, N., Callendret, B., Lorin, V., Combredet, C., Marianneau, P., Fevrier, M., et al., 2014. Protection from SARS coronavirus conferred by live measles vaccine expressing the spike glycoprotein. Virology 452–453, 32–41.

Falzarano, D., de Wit, E., Feldmann, F., Rasmussen, A.L., Okumura, A., Peng, X., et al., 2014. Infection with MERS-CoV causes lethal pneumonia in the common marmoset. PLoS Pathog. 10, e1004250.

Favreau, D.J., Desforges, M., St-Jean, J.R., Talbot, P.J., 2009. A human coronavirus OC43 variant harboring persistence-associated mutations in the S glycoprotein differentially induces the unfolded protein response in human neurons as compared to wild-type virus. Virology 395, 255–267.

Fett, C., DeDiego, M.L., Regla-Nava, J.A., Enjuanes, L., Perlman, S., 2013. Complete protection against severe acute respiratory syndrome coronavirus-mediated lethal respiratory disease in aged mice by immunization with a mouse-adapted virus lacking E protein. J. Virol. 87, 6551–6559.

Frieman, M., Ratia, K., Johnston, R.E., Mesecar, A.D., Baric, R.S., 2009. Severe acute respiratory syndrome coronavirus papain-like protease ubiquitin-like domain and catalytic domain regulate antagonism of IRF3 and NF-kappaB signaling. J. Virol. 83, 6689–6705.

Frieman, M., Yount, B., Agnihothram, S., Page, C., Donaldson, E., Roberts, A., et al., 2012. Molecular determinants of severe acute respiratory syndrome coronavirus pathogenesis and virulence in young and aged mouse models of human disease. J. Virol. 86, 884–897.

Graham, R.L., Becker, M.M., Eckerle, L.D., Bolles, M., Denison, M.R., Baric, R.S., 2012. A live, impaired-fidelity coronavirus vaccine protects in an aged, immunocompromised mouse model of lethal disease. Nat. Med. 18, 1820–1826.

Gretebeck, L.M., Subbarao, K., 2015. Animal models for SARS and MERS coronaviruses. Curr. Opin. Virol. 13, 123–129.

Gu, J., Gong, E., Zhang, B., Zheng, J., Gao, Z., Zhong, Y., et al., 2005. Multiple organ infection and the pathogenesis of SARS. J. Exp. Med. 202, 415–424.

Guo, X., Deng, Y., Chen, H., Lan, J., Wang, W., Zou, X., et al., 2015. Systemic and mucosal immunity in mice elicited by a single immunization with human adenovirus type 5 or 41 vector-based vaccines carrying the spike protein of Middle East respiratory syndrome coronavirus. Immunology 145, 476–484.

Haagmans, B.L., van den Brand, J.M., Provacia, L.B., Raj, V.S., Stittelaar, K.J., Getu, S., et al., 2015. Asymptomatic Middle East respiratory syndrome coronavirus infection in rabbits. J. Virol. 89, 6131–6135.

Haagmans, B.L., van den Brand, J.M., Raj, V.S., Volz, A., Wohlsein, P., Smits, S.L., et al., 2016. An orthopoxvirus-based vaccine reduces virus excretion after MERS-CoV infection in dromedary camels. Science 351, 77–81.

Hetz, C., 2012. The unfolded protein response: controlling cell fate decisions under ER stress and beyond. Nat. Rev. Mol. Cell Biol. 13, 89–102.

Honda-Okubo, Y., Barnard, D., Ong, C.H., Peng, B.H., Tseng, C.T., Petrovsky, N., 2015. Severe acute respiratory syndrome-associated coronavirus vaccines formulated with delta inulin adjuvants provide enhanced protection while ameliorating lung eosinophilic immunopathology. J. Virol. 89, 2995–3007.

Hou, Y.X., Peng, C., Han, Z.G., Zhou, P., Chen, J.G., Shi, Z.L., 2010. Immunogenicity of the spike glycoprotein of bat SARS-like coronavirus. Virol. Sin. 25, 36–44.

Huang, X., Dong, W., Milewska, A., Golda, A., Qi, Y., Zhu, Q.K., et al., 2015. Human coronavirus HKU1 spike protein uses O-acetylated sialic acid as an attachment receptor determinant and employs hemagglutinin-esterase protein as a receptor-destroying enzyme. J. Virol. 89, 7202–7213.

Iwata-Yoshikawa, N., Uda, A., Suzuki, T., Tsunetsugu-Yokota, Y., Sato, Y., Morikawa, S., et al., 2014. Effects of Toll-like receptor stimulation on eosinophilic infiltration in lungs of BALB/c mice immunized with UV-inactivated severe acute respiratory syndrome-related coronavirus vaccine. J. Virol. 88, 8597–8614.

Jacomy, H., Talbot, P.J., 2003. Vacuolating encephalitis in mice infected by human coronavirus OC43. Virology 315, 20–33.

Jaume, M., Yip, M.S., Kam, Y.W., Cheung, C.Y., Kien, F., Roberts, A., et al., 2012. SARS CoV subunit vaccine: antibody-mediated neutralisation and enhancement. Hong Kong Med. J. 18, 31–36.

Jiang, L., Wang, N., Zuo, T., Shi, X., Poon, K.M., Wu, Y., et al., 2014. Potent neutralization of MERS-CoV by human neutralizing monoclonal antibodies to the viral spike glycoprotein. Sci. Transl. Med. 6, 234ra259.

Jimenez-Guardeño, J.M., Nieto-Torres, J.L., DeDiego, M.L., Regla-Nava, J.A., Fernandez-Delgado, R., Castaño-Rodriguez, C., et al., 2014. The PDZ-binding motif of severe acute respiratory syndrome coronavirus envelope protein Is a determinant of viral pathogenesis. PLoS Pathog. 10, e1004320.

Jimenez-Guardeno, J.M., Regla-Nava, J.A., Nieto-Torres, J.L., DeDiego, M.L., Castano-Rodriguez, C., Fernandez-Delgado, R., et al., 2015. Identification of the mechanisms causing reversion to virulence in an attenuated SARS-CoV for the design of a genetically stable vaccine. PLoS Pathog. 11, e1005215.

Keyaerts, E., Li, S., Vijgen, L., Rysman, E., Verbeeck, J., Van Ranst, M., et al., 2009. Antiviral activity of chloroquine against human coronavirus OC43 infection in newborn mice. Antimicrob. Agents Chemother. 53, 3416–3421.

Kilianski, A., Baker, S.C., 2014. Cell-based antiviral screening against coronaviruses: developing virus-specific and broad-spectrum inhibitors. Antivir. Res. 101, 105–112.

Kim, E., Okada, K., Kenniston, T., Raj, V.S., AlHajri, M.M., Farag, E.A., et al., 2014. Immunogenicity of an adenoviral-based Middle East respiratory syndrome coronavirus vaccine in BALB/c mice. Vaccine 32, 5975–5982.

Kindler, E., Jonsdottir, H.R., Muth, D., Hamming, O.J., Hartmann, R., Rodriguez, R., et al., 2013. Efficient replication of the novel human betacoronavirus EMC on primary human epithelium highlights its zoonotic potential. MBio 4, e00611–e00612.

Kirkpatrick, B.D., Whitehead, S.S., Pierce, K.K., Tibery, C.M., Grier, P.L., Hynes, N.A., et al., 2016. The live attenuated dengue vaccine TV003 elicits complete protection against dengue in a human challenge model. Sci. Transl. Med. 8, 330–336.

Kuri, T., Eriksson, K.K., Putics, A., Zust, R., Snijder, E.J., Davidson, A.D., et al., 2011. The ADP-ribose-1″-monophosphatase domains of severe acute respiratory syndrome coronavirus and human coronavirus 229E mediate resistance to antiviral interferon responses. J. Gen. Virol. 92, 1899–1905.

Lamirande, E.W., DeDiego, M.L., Roberts, A., Jackson, J.P., Alvarez, E., Sheahan, T., et al., 2008. A live attenuated SARS coronavirus is immunogenic and efficacious in golden Syrian hamsters. J. Virol. 82, 7721–7724.

Lan, J., Yao, Y., Deng, Y., Chen, H., Lu, G., Wang, W., et al., 2015. Recombinant receptor binding domain protein induces partial protective immunity in rhesus macaques against Middle East respiratory syndrome coronavirus challenge. EBioMedicine 2, 1438–1446.

Lassnig, C., Sanchez, C.M., Egerbacher, M., Walter, I., Majer, S., Kolbe, T., et al., 2005. Development of a transgenic mouse model susceptible to human coronavirus 229E. Proc. Natl. Acad. Sci. U.S.A. 102, 8275–8280.

Law, H.K., Cheung, C.Y., Ng, H.Y., Sia, S.F., Chan, Y.O., Luk, W., et al., 2005. Chemokine up-regulation in SARS-coronavirus-infected, monocyte-derived human dendritic cells. Blood 106, 2366–2374.

Law, H.K., Cheung, C.Y., Sia, S.F., Chan, Y.O., Peiris, J.S., Lau, Y.L., 2009. Toll-like receptors, chemokine receptors and death receptor ligands responses in SARS coronavirus infected human monocyte derived dendritic cells. BMC Immunol. 10, 35.

Lee, N., Hui, D., Wu, A., Chan, P., Cameron, P., Joynt, G., et al., 2003. A major outbreak of severe acute respiratory syndrome in Hong Kong. N. Engl. J. Med. 348, 1986–1994.

Li, J., Ulitzky, L., Silberstein, E., Taylor, D.R., Viscidi, R., 2013. Immunogenicity and protection efficacy of monomeric and trimeric recombinant SARS coronavirus spike protein subunit vaccine candidates. Viral Immunol. 26, 126–132.

Li, K., Wohlford-Lenane, C., Perlman, S., Zhao, J., Jewell, A.K., Reznikov, L.R., et al., 2016. Middle East respiratory syndrome coronavirus causes multiple organ damage and lethal disease in mice transgenic for human dipeptidyl peptidase 4. J. Infect. Dis. 213, 712–722.

Liu, Y.V., Massare, M.J., Barnard, D.L., Kort, T., Nathan, M., Wang, L., et al., 2011. Chimeric severe acute respiratory syndrome coronavirus (SARS-CoV) S glycoprotein and influenza matrix 1 efficiently form virus-like particles (VLPs) that protect mice against challenge with SARS-CoV. Vaccine 29, 6606–6613.

Lokugamage, K.G., Narayanan, K., Nakagawa, K., Terasaki, K., Ramirez, S.I., Tseng, C.T., et al., 2015. Middle East Respiratory Syndrome coronavirus nsp1 inhibits host gene expression by selectively targeting mRNAs transcribed in the nucleus while sparing mRNAs of cytoplasmic origin. J. Virol. 89, 10970–10981.

Ma, C., Li, Y., Wang, L., Zhao, G., Tao, X., Tseng, C.T., et al., 2014a. Intranasal vaccination with recombinant receptor-binding domain of MERS-CoV spike protein induces much stronger local mucosal immune responses than subcutaneous immunization: implication for designing novel mucosal MERS vaccines. Vaccine 32, 2100–2108.

Ma, C., Wang, L., Tao, X., Zhang, N., Yang, Y., Tseng, C.T., et al., 2014b. Searching for an ideal vaccine candidate among different MERS coronavirus receptor-binding fragments—the importance of immunofocusing in subunit vaccine design. Vaccine 32, 6170–6176.

Markusic, D., Oude-Elferink, R., Das, A.T., Berkhout, B., Seppen, J., 2005. Comparison of single regulated lentiviral vectors with rtTA expression driven by an autoregulatory loop or a constitutive promoter. Nucleic Acids Res. 33, e63.

Matthews, K.L., Coleman, C.M., van der Meer, Y., Snijder, E.J., Frieman, M.B., 2014. The ORF4b-encoded accessory proteins of Middle East respiratory syndrome coronavirus and two related bat coronaviruses localize to the nucleus and inhibit innate immune signalling. J. Gen. Virol. 95, 874–882.

Mayer, K., Nellessen, C., Hahn-Ast, C., Schumacher, M., Pietzonka, S., Eis-Hubinger, A.M., et al., 2016. Fatal outcome of human coronavirus NL63 infection despite successful viral elimination by IFN-alpha in a patient with newly diagnosed ALL. Eur. J. Haematol. 97, 208–210.

Menachery, V.D., Yount Jr., B.L., Josset, L., Gralinski, L.E., Scobey, T., Agnihothram, S., et al., 2014. Attenuation and restoration of severe acute respiratory syndrome coronavirus mutant lacking 2'-o-methyltransferase activity. Eur. J. Haematol. 88, 4251–4264.

Mesel-Lemoine, M., Millet, J., Vidalain, P.O., Law, H., Vabret, A., Lorin, V., et al., 2012. A human coronavirus responsible for the common cold massively kills dendritic cells but not monocytes. J. Virol. 86, 7577–7587.

Mielech, A.M., Kilianski, A., Baez-Santos, Y.M., Mesecar, A.D., Baker, S.C., 2014. MERS-CoV papain-like protease has deISGylating and deubiquitinating activities. Virology 450–451, 64–70.

Minskaia, E., Hertzig, T., Gorbalenya, A.E., Campanacci, V., Cambillau, C., Canard, B., et al., 2006. Discovery of an RNA virus 3'→5' exoribonuclease that is critically involved in coronavirus RNA synthesis. Proc. Natl. Acad. Sci. U.S.A. 103, 5108–5113.

Mou, H., Raj, V.S., van Kuppeveld, F.J., Rottier, P.J., Haagmans, B.L., Bosch, B.J., 2013. The receptor binding domain of the new Middle East respiratory syndrome coronavirus maps to a 231-residue region in the spike protein that efficiently elicits neutralizing antibodies. J. Virol. 87, 9379–9383.

Muller, M.A., van der Hoek, L., Voss, D., Bader, O., Lehmann, D., Schulz, A.R., et al., 2010. Human coronavirus NL63 open reading frame 3 encodes a virion-incorporated N-glycosylated membrane protein. Virol. J. 7, 6.

Muth, D., Corman, V.M., Meyer, B., Assiri, A., Al-Masri, M., Farah, M., et al., 2015. Infectious Middle East respiratory syndrome coronavirus excretion and serotype variability based on live virus isolates from patients in Saudi Arabia. J. Clin. Microbiol. 53, 2951–2955.

Muthumani, K., Falzarano, D., Reuschel, E.L., Tingey, C., Flingai, S., Villarreal, D.O., et al., 2015. A synthetic consensus anti-spike protein DNA vaccine induces protective immunity against Middle East respiratory syndrome coronavirus in nonhuman primates. Sci. Transl. Med. 7, 301ra132.

Narayanan, K., Ramirez, S.I., Lokugamage, K.G., Makino, S., 2015. Coronavirus nonstructural protein 1: common and distinct functions in the regulation of host and viral gene expression. Virus Res. 202, 89–100.

Netland, J., DeDiego, M.L., Zhao, J., Fett, C., Alvarez, E., Nieto-Torres, J.L., et al., 2010. Immunization with an attenuated severe acute respiratory syndrome coronavirus deleted in E protein protects against lethal respiratory disease. Virology 399, 120–128.

Nieto-Torres, J.L., Dediego, M.L., Verdia-Baguena, C., Jimenez-Guardeño, J.M., Regla-Nava, J.A., Fernandez-Delgado, R., et al., 2014. Severe acute respiratory syndrome coronavirus envelope protein ion channel activity promotes virus fitness and pathogenesis. PLoS Pathog. 10, e1004077.

Nieto-Torres, J.L., Verdia-Baguena, C., Jimenez-Guardeno, J.M., Regla-Nava, J.A., Castano-Rodriguez, C., Fernandez-Delgado, R., et al., 2015. Severe acute respiratory syndrome coronavirus E protein transports calcium ions and activates the NLRP3 inflammasome. Virology 485, 330–339.

Paddock, C., 2015. MERS vaccine 'ready for human trials'. Retrieved from, http://www.medicalnewstoday.com/articles/296023.php.

Papaneri, A.B., Johnson, R.F., Wada, J., Bollinger, L., Jahrling, P.B., Kuhn, J.H., 2015. Middle East respiratory syndrome: obstacles and prospects for vaccine development. Expert Rev. Vaccines 14, 949–962.

Pascal, K.E., Coleman, C.M., Mujica, A.O., Kamat, V., Badithe, A., Fairhurst, J., et al., 2015. Pre- and postexposure efficacy of fully human antibodies against Spike protein in a novel humanized mouse model of MERS-CoV infection. Proc. Natl. Acad. Sci. U.S.A. 112, 8738–8743.

Patterson, S., Macnaughton, M.R., 1982. Replication of human respiratory coronavirus strain 229E in human macrophages. J. Gen. Virol. 60, 307–314.

Peck, K.M., Cockrell, A.S., Yount, B.L., Scobey, T., Baric, R.S., Heise, M.T., 2015. Glycosylation of mouse DPP4 plays a role in inhibiting Middle East respiratory syndrome coronavirus infection. J. Virol. 89, 4696–4699.

Pene, F., Merlat, A., Vabret, A., Rozenberg, F., Buzyn, A., Dreyfus, F., et al., 2003. Coronavirus 229E-related pneumonia in immunocompromised patients. Clin. Infect. Dis. 37, 929–932.

Putics, A., Filipowicz, W., Hall, J., Gorbalenya, A.E., Ziebuhr, J., 2005. ADP-ribose-1″-monophosphatase: a conserved coronavirus enzyme that is dispensable for viral replication in tissue culture. J. Virol. 79, 12721–12731.

Pyrc, K., Bosch, B.J., Berkhout, B., Jebbink, M.F., Dijkman, R., Rottier, P., et al., 2006. Inhibition of human coronavirus NL63 infection at early stages of the replication cycle. Antimicrob. Agents Chemother. 50, 2000–2008.

Pyrc, K., Berkhout, B., van der Hoek, L., 2007. Antiviral strategies against human coronaviruses. Infect. Disord. Drug Targets 7, 59–66.

Ratia, K., Saikatendu, K.S., Santarsiero, B.D., Barretto, N., Baker, S.C., Stevens, R.C., et al., 2006. Severe acute respiratory syndrome coronavirus papain-like protease: structure of a viral deubiquitinating enzyme. Proc. Natl. Acad. Sci. U.S.A. 103, 5717–5722.

Regla-Nava, J.A., Nieto-Torres, J.L., Jimenez-Guardeno, J.M., Fernandez-Delgado, R., Fett, C., Castano-Rodriguez, C., et al., 2015. Severe acute respiratory syndrome coronaviruses with mutations in the E protein are attenuated and promising vaccine candidates. J. Virol. 89, 3870–3887.

Roberts, A., Deming, D., Paddock, C.D., Cheng, A., Yount, B., Vogel, L., et al., 2007. A mouse-adapted SARS-coronavirus causes disease and mortality in BALB/c mice. PLoS Pathog. 3, 23–37.

Roberts, A., Lamirande, E.W., Vogel, L., Baras, B., Goossens, G., Knott, I., et al., 2010. Immunogenicity and protective efficacy in mice and hamsters of a beta-propiolactone inactivated whole virus SARS-CoV vaccine. Viral Immunol. 23, 509–519.

Roth-Cross, J.K., Stokes, H., Chang, G., Chua, M.M., Thiel, V., Weiss, S.R., et al., 2009. Organ-specific attenuation of murine hepatitis virus strain A59 by replacement of catalytic residues in the putative viral cyclic phosphodiesterase ns2. J. Virol. 83, 3743–3753.

Sabir, J.S., Lam, T.T., Ahmed, M.M., Li, L., Shen, Y., Abo-Aba, S.E., et al., 2016. Co-circulation of three camel coronavirus species and recombination of MERS-CoVs in Saudi Arabia. Science 351, 81–84.

Scheuplein, V.A., Seifried, J., Malczyk, A.H., Miller, L., Hocker, L., Vergara-Alert, J., et al., 2015. High secretion of interferons by human plasmacytoid dendritic cells upon recognition of Middle East respiratory syndrome coronavirus. J. Virol. 89, 3859–3869.

Scobey, T., Yount, B.L., Sims, A.C., Donaldson, E.F., Agnihothram, S.S., Menachery, V.D., et al., 2013. Reverse genetics with a full-length infectious cDNA of the Middle East respiratory syndrome coronavirus. Proc. Natl. Acad. Sci. U.S.A. 110, 16157–16162.

Selinger, C., Tisoncik-Go, J., Menachery, V.D., Agnihothram, S., Law, G.L., Chang, J., et al., 2014. Cytokine systems approach demonstrates differences in innate and pro-inflammatory host responses between genetically distinct MERS-CoV isolates. BMC Genomics 15, 1161.

Sheahan, T., Whitmore, A., Long, K., Ferris, M., Rockx, B., Funkhouser, W., et al., 2011. Successful vaccination strategies that protect aged mice from lethal challenge from influenza virus and heterologous severe acute respiratory syndrome coronavirus. J. Virol. 85, 217–230.

Shim, B.S., Stadler, K., Nguyen, H.H., Yun, C.H., Kim, D.W., Chang, J., et al., 2012. Sublingual immunization with recombinant adenovirus encoding SARS-CoV spike protein induces systemic and mucosal immunity without redirection of the virus to the brain. Virol. J. 9, 215.

Siu, K.L., Chan, C.P., Kok, K.H., Woo, P.C., Jin, D.Y., 2014a. Comparative analysis of the activation of unfolded protein response by spike proteins of severe acute respiratory syndrome coronavirus and human coronavirus HKU1. Cell Biosci. 4, 3.

Siu, K.L., Yeung, M.L., Kok, K.H., Yuen, K.S., Kew, C., Lui, P.Y., et al., 2014b. Middle East respiratory syndrome coronavirus 4a protein is a double-stranded RNA-binding protein that suppresses PACT-induced activation of RIG-I and MDA5 in innate antiviral response. J. Virol. 88, 4866–4876.

Smith, E.C., Blanc, H., Surdel, M.C., Vignuzzi, M., Denison, M.R., 2013. Coronaviruses lacking exoribonuclease activity are susceptible to lethal mutagenesis: evidence for proofreading and potential therapeutics. PLoS Pathog. 9, e1003565.

Sola, I., Almazán, F., Zúñiga, S., Enjuanes, L., 2015. Continuous and discontinuous RNA synthesis in coronaviruses. Annu. Rev. Virol. 2, 265–288.

Song, F., Fux, R., Provacia, L.B., Volz, A., Eickmann, M., Becker, S., et al., 2013. Middle East respiratory syndrome coronavirus spike protein delivered by modified vaccinia virus Ankara efficiently induces virus-neutralizing antibodies. J. Virol. 87, 11950–11954.

St-Jean, J.R., Jacomy, H., Desforges, M., Vabret, A., Freymuth, F., Talbot, P.J., 2004. Human respiratory coronavirus OC43: genetic stability and neuroinvasion. J. Virol. 78, 8824–8834.

St-Jean, J.R., Desforges, M., Almazan, F., Jacomy, H., Enjuanes, L., Talbot, P.J., 2006. Recovery of a neurovirulent human coronavirus OC43 from an infectious cDNA clone. J. Virol. 80, 3670–3674.

Stobart, C.C., Moore, M.L., 2015. Development of next-generation respiratory virus vaccines through targeted modifications to viral immunomodulatory genes. Expert Rev. Vaccines 14, 1563–1572.

Subbarao, K., McAuliffe, J., Vogel, L., Fahle, G., Fischer, S., Tatti, K., et al., 2004. Prior infection and passive transfer of neutralizing antibody prevent replication of severe acute respiratory syndrome coronavirus in the respiratory tract of mice. J. Virol. 78, 3572–3577.

Sui, J., Li, W., Roberts, A., Matthews, L.J., Murakami, A., Vogel, L., et al., 2005. Evaluation of human monoclonal antibody 80R for immunoprophylaxis of severe acute respiratory syndrome by an animal study, epitope mapping, and analysis of spike variants. J. Virol. 79, 5900–5906.

Sui, J., Deming, M., Rockx, B., Liddington, R.C., Zhu, Q.K., Baric, R.S., et al., 2014. Effects of human anti-spike protein receptor binding domain antibodies on severe acute respiratory syndrome coronavirus neutralization escape and fitness. J. Virol. 88, 13769–13780.

Takaoka, A., Hayakawa, S., Yanai, H., Stoiber, D., Negishi, H., Kikuchi, H., et al., 2003. Integration of interferon-alpha/beta signalling to p53 responses in tumour suppression and antiviral defence. Nature 424, 516–523.

Tang, X.C., Agnihothram, S.S., Jiao, Y., Stanhope, J., Graham, R.L., Peterson, E.C., et al., 2014. Identification of human neutralizing antibodies against MERS-CoV and their role in virus adaptive evolution. Proc. Natl. Acad. Sci. U.S.A. 111, E2018–E2026.

Tang, J., Zhang, N., Tao, X., Zhao, G., Guo, Y., Tseng, C.T., et al., 2015. Optimization of antigen dose for a receptor-binding domain-based subunit vaccine against MERS coronavirus. Hum. Vaccin. Immunother. 11, 1244–1250.

Tao, X., Garron, T., Agrawal, A.S., Algaissi, A., Peng, B.H., Wakamiya, M., et al., 2015. Characterization and demonstration of the value of a lethal mouse model of Middle East Respiratory Syndrome coronavirus infection and disease. J. Virol. 90, 57–67.

ter Meulen, J., Bakker, A.B., van den Brink, E.N., Weverling, G.J., Martina, B.E., Haagmans, B.L., et al., 2004. Human monoclonal antibody as prophylaxis for SARS coronavirus infection in ferrets. Lancet 363, 2139–2141.

Thiel, V., Herold, J., Schelle, B., Siddell, S., 2001. Infectious RNA transcribed in vitro from a cDNA copy of the human coronavirus genome cloned in vaccinia virus. J. Gen. Virol. 82, 1273–1281.

Totura, A.L., Baric, R.S., 2012. SARS coronavirus pathogenesis: host innate immune responses and viral antagonism of interferon. Curr. Opin. Virol. 2, 264–275.

Tseng, C.T., Perrone, L.A., Zhu, H., Makino, S., Peters, C.J., 2005. Severe acute respiratory syndrome and the innate immune responses: modulation of effector cell function without productive infection. J. Immunol. 174, 7977–7985.

Tseng, C.T., Sbrana, E., Iwata-Yoshikawa, N., Newman, P.C., Garron, T., Atmar, R.L., et al., 2012. Immunization with SARS coronavirus vaccines leads to pulmonary immunopathology on challenge with the SARS virus. PLoS One 7, e35421.

van der Hoek, L., 2007. Human coronaviruses: what do they cause? Antivir. Ther. 12, 651–658.

van der Hoek, L., Pyrc, K., Jebbink, M.F., Vermeulen-Oost, W., Berkhout, R.J., Wolthers, K.C., et al., 2004. Identification of a new human coronavirus. Nat. Med. 10, 368–373.

van der Hoek, L., Sure, K., Ihorst, G., Stang, A., Pyrc, K., Jebbink, M.F., et al., 2005. Croup is associated with the novel coronavirus NL63. PLoS Med. 2, e240.

van Doremalen, N., Munster, V.J., 2015. Animal models of Middle East respiratory syndrome coronavirus infection. Antivir. Res. 122, 28–38.

van Doremalen, N., Miazgowicz, K.L., Milne-Price, S., Bushmaker, T., Robertson, S., Scott, D., et al., 2014. Host species restriction of Middle East respiratory syndrome coronavirus through its receptor, dipeptidyl peptidase 4. J. Virol. 88, 9220–9232.

Varkey, J.B., Varkey, B., 2008. Viral infections in patients with chronic obstructive pulmonary disease. Curr. Opin. Pulm. Med. 14, 89–94.

Volz, A., Kupke, A., Song, F., Jany, S., Fux, R., Shams-Eldin, H., et al., 2015. Protective efficacy of recombinant modified vaccinia virus Ankara delivering Middle East Respiratory Syndrome coronavirus spike glycoprotein. J. Virol. 89, 8651–8656.

Walsh, E.E., Shin, J.H., Falsey, A.R., 2013. Clinical impact of human coronaviruses 229E and OC43 infection in diverse adult populations. J. Infect. Dis. 208, 1634–1642.

Wang, J., Tricoche, N., Du, L., Hunter, M., Zhan, B., Goud, G., et al., 2012. The adjuvanticity of an O. volvulus-derived rOv-ASP-1 protein in mice using sequential vaccinations and in non-human primates. PLoS One 7, e37019.

Wang, S.F., Tseng, S.P., Yen, C.H., Yang, J.Y., Tsao, C.H., Shen, C.W., et al., 2014. Antibody-dependent SARS coronavirus infection is mediated by antibodies against spike proteins. Biochem. Biophys. Res. Commun. 451, 208–214.

Wang, L., Shi, W., Joyce, M.G., Modjarrad, K., Zhang, Y., Leung, K., et al., 2015. Evaluation of candidate vaccine approaches for MERS-CoV. Nat. Commun. 6, 7712.

Weingartl, H., Czub, M., Czub, S., Neufeld, J., Marszal, P., Gren, J., et al., 2004. Immunization with modified vaccinia virus Ankara-based recombinant vaccine against severe acute respiratory syndrome is associated with enhanced hepatitis in ferrets. J. Virol. 78, 12672–12676.

Westerbeck, J.W., Machamer, C.E., 2015. A coronavirus E protein is present in two distinct pools with different effects on assembly and the secretory pathway. J. Virol. 89, 9313–9323.

Woo, P.C., Lau, S.K., Chu, C.M., Chan, K.H., Tsoi, H.W., Huang, Y., et al., 2005a. Characterization and complete genome sequence of a novel coronavirus, coronavirus HKU1, from patients with pneumonia. J. Virol. 79, 884–895.

Woo, P.C., Lau, S.K., Tsoi, H.W., Huang, Y., Poon, R.W., Chu, C.M., et al., 2005b. Clinical and molecular epidemiological features of coronavirus HKU1-associated community-acquired pneumonia. J. Infect. Dis. 192, 1898–1907.

Wrensch, F., Winkler, M., Pohlmann, S., 2014. IFITM proteins inhibit entry driven by the MERS-coronavirus spike protein: evidence for cholesterol-independent mechanisms. Viruses 6, 3683–3698.

Yang, L.T., Peng, H., Zhu, Z.L., Li, G., Huang, Z.T., Zhao, Z.X., et al., 2006. Long-lived effector/central memory T-cell responses to severe acute respiratory syndrome coronavirus (SARS-CoV) S antigen in recovered SARS patients. Clin. Immunol. 120, 171–178.

Yang, Y., Zhang, L., Geng, H., Deng, Y., Huang, B., Guo, Y., et al., 2013. The structural and accessory proteins M, ORF 4a, ORF 4b, and ORF 5 of Middle East respiratory syndrome coronavirus (MERS-CoV) are potent interferon antagonists. Protein Cell 4, 951–961.

Yang, X., Chen, X., Bian, G., Tu, J., Xing, Y., Wang, Y., et al., 2014. Proteolytic processing, deubiquitinase and interferon antagonist activities of Middle East respiratory syndrome coronavirus papain-like protease. J. Gen. Virol. 95, 614–626.

Yilla, M., Harcourt, B.H., Hickman, C.J., McGrew, M., Tamin, A., Goldsmith, C.S., et al., 2005. SARS-coronavirus replication in human peripheral monocytes/macrophages. Virus Res. 107, 93–101.

Ying, T., Du, L., Ju, T.W., Prabakaran, P., Lau, C.C., Lu, L., et al., 2014. Exceptionally potent neutralization of Middle East respiratory syndrome coronavirus by human mono-clonal antibodies. J. Virol. 88, 7796–7805.

Ying, T., Li, H., Lu, L., Dimitrov, D.S., Jiang, S., 2015a. Development of human neutral-izing monoclonal antibodies for prevention and therapy of MERS-CoV infections. Microbes Infect. 17, 142–148.

Ying, T., Prabakaran, P., Du, L., Shi, W., Feng, Y., Wang, Y., et al., 2015b. Junctional and allele-specific residues are critical for MERS-CoV neutralization by an exceptionally potent germline-like antibody. Nat. Commun. 6, 8223.

Ying, T., Li, W., Dimitrov, D.S., 2016. Discovery of T-Cell infection and apoptosis by Mid-dle East Respiratory Syndrome Coronavirus. J. Infect. Dis. 213, 877–879.

Yount, B., Curtis, K.M., Fritz, E.A., Hensley, L.E., Jahrling, P.B., Prentice, E., et al., 2003. Reverse genetics with a full-length infectious cDNA of severe acute respiratory syn-drome coronavirus. Proc. Natl. Acad. Sci. U.S.A. 100, 12995–13000.

Yuan, L., Chen, Z., Song, S., Wang, S., Tian, C., Xing, G., et al., 2015. p53 degradation by a coronavirus papain-like protease suppresses type I interferon signaling. J. Biol. Chem. 290, 3172–3182.

Zhang, M.Y., Choudhry, V., Xiao, X., Dimitrov, D.S., 2005. Human monoclonal anti-bodies to the S glycoprotein and related proteins as potential therapeutics for SARS. Curr. Opin. Mol. Ther. 7, 151–156.

Zhang, R., Jha, B.K., Ogden, K.M., Dong, B., Zhao, L., Elliott, R., et al., 2013. Homol-ogous 2',5'-phosphodiesterases from disparate RNA viruses antagonize antiviral innate immunity. Proc. Natl. Acad. Sci. U.S.A. 110, 13114–13119.

Zhang, N., Jiang, S., Du, L., 2014a. Current advancements and potential strategies in the development of MERS-CoV vaccines. Expert Rev. Vaccines 13, 761–774.

Zhang, R., Wang, K., Lv, W., Yu, W., Xie, S., Xu, K., et al., 2014b. The ORF4a protein of human coronavirus 229E functions as a viroporin that regulates viral production. Bio-chim. Biophys. Acta Biomembr. 1838, 1088–1095.

Zhang, R., Wang, K., Ping, X., Yu, W., Qian, Z., Xiong, S., et al., 2015. The ns12.9 acces-sory protein of human coronavirus OC43 Is a viroporin involved in virion morphogen-esis and pathogenesis. J. Virol. 89, 11383–11395.

Zhang, N., Channappanavar, R., Ma, C., Wang, L., Tang, J., Garron, T., et al., 2016. Iden-tification of an ideal adjuvant for receptor-binding domain-based subunit vaccines against Middle East respiratory syndrome coronavirus. Cell. Mol. Immunol. 13, 180–190.

Zhao, L., Jha, B.K., Wu, A., Elliott, R., Ziebuhr, J., Gorbalenya, A.E., et al., 2012. Antag-onism of the interferon-induced OAS-RNase L pathway by murine coronavirus ns2 protein is required for virus replication and liver pathology. Cell Host Microbe 11, 607–616.

Zhao, J., Li, K., Wohlford-Lenane, C., Agnihothram, S.S., Fett, C., Zhao, J., et al., 2014a. Rapid generation of a mouse model for Middle East respiratory syndrome. Proc. Natl. Acad. Sci. U.S.A. 111, 4970–4975.

Zhao, X., Guo, F., Liu, F., Cuconati, A., Chang, J., Block, T.M., et al., 2014b. Interferon induction of IFITM proteins promotes infection by human coronavirus OC43. Proc. Natl. Acad. Sci. U.S.A. 111, 6756–6761.

Zhao, G., Jiang, Y., Qiu, H., Gao, T., Zeng, Y., Guo, Y., et al., 2015a. Multi-organ damage in human dipeptidyl peptidase 4 transgenic mice infected with Middle East respiratory syndrome-coronavirus. PLoS One 10, e0145561.

Zhao, J., Perera, R.A., Kayali, G., Meyerholz, D., Perlman, S., Peiris, M., 2015b. Passive immunotherapy with dromedary immune serum in an experimental animal model for Middle East respiratory syndrome coronavirus infection. J. Virol. 89, 6117–6120.

Zheng, D., Chen, G., Guo, B., Cheng, G., Tang, H., 2008. PLP2, a potent deubiquitinase from murine hepatitis virus, strongly inhibits cellular type I interferon production. Cell Res. 18, 1105–1113.

Zhong, Y., Tan, Y.W., Liu, D.X., 2012. Recent progress in studies of arterivirus- and coronavirus-host interactions. Viruses 4, 980–1010.

Zhou, P., Han, Z., Wang, L.F., Shi, Z., 2013a. Identification of immunogenic determinants of the spike protein of SARS-like coronavirus. Virol. Sin. 28, 92–96.

Zhou, W., Wang, W., Wang, H., Lu, R., Tan, W., 2013b. First infection by all four non-severe acute respiratory syndrome human coronaviruses takes place during childhood. BMC Infect. Dis. 13, 433.

Zhou, J., Chu, H., Li, C., Wong, B.H., Cheng, Z.S., Poon, V.K., et al., 2014. Active replication of Middle East respiratory syndrome coronavirus and aberrant induction of inflammatory cytokines and chemokines in human macrophages: implications for pathogenesis. J. Infect. Dis. 209, 1331–1342.

Zhou, J., Chu, H., Chan, J.F., Yuen, K.Y., 2015. Middle East respiratory syndrome coronavirus infection: virus-host cell interactions and implications on pathogenesis. Virol. J. 12, 218.

Zhu, Z., Chakraborti, S., He, Y., Roberts, A., Sheahan, T., Xiao, X., et al., 2007. Potent cross-reactive neutralization of SARS coronavirus isolates by human monoclonal antibodies. Proc. Natl. Acad. Sci. U.S.A. 104, 12123–12128.

Ziebuhr, J., 2005. The coronavirus replicase. In: Enjuanes, L. (Ed.), Coronavirus Replication and Reverse Genetics, vol. 287. Springer-Verlag, Berlin, Heidelberg, pp. 57–94.

Ziegler, T., Matikainen, S., Ronkko, E., Osterlund, P., Sillanpaa, M., Siren, J., et al., 2005. Severe acute respiratory syndrome coronavirus fails to activate cytokine-mediated innate immune responses in cultured human monocyte-derived dendritic cells. J. Virol. 79, 13800–13805.

Zielecki, F., Weber, M., Eickmann, M., Spiegelberg, L., Zaki, A.M., Matrosovich, M., et al., 2013. Human cell tropism and innate immune system interactions of human respiratory coronavirus EMC compared to those of severe acute respiratory syndrome coronavirus. J. Virol. 87, 5300–5304.

Zumla, A., Chan, J.F., Azhar, E.I., Hui, D.S., Yuen, K.Y., 2016. Coronaviruses—drug discovery and therapeutic options. Nat. Rev. Drug Discov. 15, 327–347.

Zust, R., Cervantes-Barragan, L., Kuri, T., Blakqori, G., Weber, F., Ludewig, B., et al., 2007. Coronavirus non-structural protein 1 is a major pathogenicity factor: implications for the rational design of coronavirus vaccines. PLoS Pathog. 3, e109.

Zust, R., Cervantes-Barragan, L., Habjan, M., Maier, R., Neuman, B.W., Ziebuhr, J., et al., 2011. Ribose 2'-O-methylation provides a molecular signature for the distinction of self and non-self mRNA dependent on the RNA sensor Mda5. Nat. Immunol. 12, 137–143.

INDEX

Note: Page numbers followed by "*f*" indicate figures, and "*t*" indicate tables.

CPI Antony Rowe
Chippenham, UK
2016-10-01 21:01